Urban Sound Environment

Also available from Taylor & Francis

Environmental Noise Barriers
C. English and B. Kotzen
Hb: ISBN 0–419–23180–3

Environmental and Architectural Acoustics
P. Lord and Z. Maekawa
Hb: ISBN 0–419–15980–0

Auditorium Acoustics and Architectural Design
M. Barron
Hb: ISBN 0–419–17710–8

Sustainable Urban Design
R. Thomas
Hb: ISBN 0–415–28122–9 Pb: ISBN 0–415–28123–7

Urban Sound Environment

Jian Kang

LONDON AND NEW YORK

First published 2017 by Routledge

2 Park Square, Milton Park, Abingdon, Oxfordshire OX14 4RN
52 Vanderbilt Avenue, New York, NY 10017

Routledge is an imprint of the Taylor & Francis Group, an informa business

First issued in paperback 2019

Typeset in Times New Roman by
Bookcraft Ltd, Stroud, Glos

British Library Cataloguing in Publication Data
A catalogue record for this book is available
from the British Library

Library of Congress Cataloging in Publication Data
Kang, J. (Jian).
  Urban sound environment/Jian Kang.
  p. cm.
  Includes bibliographical references and index.
  ISBN 0–415–35857–4 (hbk: alk. paper)
  1. Noise pollution. 2. Noise control. I. Title.
  TD892.K29 2006
  620.2'3091732–dc22 2006008009

ISBN: 978-0-415-35857-6 (hbk)
ISBN: 978-0-367-86520-7 (pbk)

Contents

List of tables		vi
List of illustrations		viii
Preface		xv
List of abbreviations		xvii
1	Fundamentals	1
2	Urban noise evaluation	21
3	Urban soundscape	43
4	Microscale acoustic modelling	107
5	Macroscale acoustic modelling	147
6	Urban noise mitigation	175
7	Sound environment in urban streets and squares	199
	References	247
	Index	273

Tables

2.1   WHO recommended guideline values for community noise in specific environments, data selected from Berglund *et al.* (1999).   33

2.2   Recommended limits in PPG24 for various noise exposure categories for new dwellings near existing noise sources in L. (ODPM 1994).   34

2.3   Emission limits, immission limits and quality targets in L. in Italy, data adopted from Porter *et al.* (1998) and Kang *et al.* (2001a).   36

2.4   Limits in L. for environmental noise in urban areas in China, data from SAC (1993).   38

2.5   Percentage of highly annoyed in the Dutch population, data adopted from Franssen *et al.* (2004).   41

3.1   Basic information of the case study sites.   50

3.2   Measured sound levels, subjective evaluation of sound levels on the site, and subjective evaluation of the sound environment at users' home.   61

3.3   Factor analysis of the overall physical comfort evaluation. Kaiser–Meyer–Olkin measure of sampling adequacy, 0.613; cumulative, 55.1%; extraction method, principal component analysis; rotation method, varimax with Kaiser normalisation; N=9,200.   70

3.4   Classifications for various sounds in the case study sites (%), with F, favourite; N, neither favourite nor annoying; and A, annoying.   79

3.5   Soundscape evaluation form used in the soundscape walk in Sheffield. Boldfaced indices are those used in the second stage survey in the Peace Gardens and the Barkers Pool.   85

3.6   Number of interviewees in the Peace Gardens and the Barkers Pool at the second stage of study.   85

3.7   Factor analysis of the soundscape evaluation – overall results of the Peace Gardens and the Barkers Pool in the two seasonal periods. Kaiser–Meyer–Olkin measure of sampling adequacy, 0.798; cumulative, 53%; extraction method, principal component analysis; rotation method, varimax with Kaiser normalisation; N=491.   88

3.8 Average evaluation score of typical indices in the four squares, by the architectural student group (stage three) and by the general public (stage two).   90

5.1 Comparison between calculated and measured results (dBA).   166

7.1 Ratio of reverberation time between two typical receiver sets (31–90m, 18m, 18m) and (31–90m, 2m, 1m).   202

7.2 Configurations used in the comparison between UK and HK streets.   218

Illustrations

Colour plates can be found between pages 174 and 175.

1.1   Normal equal-loudness-level contours for pure tones, based on binaural free-field listening tests with frontal incidence. Data adopted from ISO (2003b).   4

1.2   Typical spectra of car and train noise. Data adopted from Jonasson and Storeheier (2001), Van Beek *et al.* (2002) and Jonasson *et al.* (2004).   7

1.3   Absorption coefficients of typical boundaries. Data adopted from UK DfES (2003).   9

1.4   Cross section of a Schroeder diffuser.   11

1.5   Transmission loss of typical boundaries. Data adopted from Möser (2004).   13

1.6   Diagram of calculating diffractions over a barrier.   16

2.1   Relationship between L and percentage of annoyed and highly annoyed people. Data adopted from WG-HSEA (2002).   22

2.2   Relationship between traffic flow Q and traffic noise L, calculated using Equation (2.8), where a and b values are adopted from those collected by Barrigón-Morillas *et al.* (2005).   29

2.3   24-h time history of L in the UK, Greater London (Skinner and Grimwood, 2005), Pamplona (Spain) (Arana and García, 1998); Badajoz (Spain) (Barrigón-Morillas *et al.*, 2005), Messina (Italy) (Piccol *et al.*, 2005), and Greater Cairo (Egypt) (Ali and Tamura, 2002). Note that since the measurement conditions were different, the comparison is mainly for the relative changes with time, rather than the absolute values.   40

3.1   (a) Site plan of the Peace Gardens based on the EDINA Digmap. The grey scale in the plan corresponds to sound levels, and the dashed circle indicates where the interviews were conducted; (b) Perspective view.   54

3.2   (a) Site plan of the Barkers Pool based on the EDINA Digmap. The grey scale in the plan corresponds to sound levels, and the dashed circle indicates where the interviews were conducted; (b) Perspective view.   55

3.3   Number of interviewees in various age groups in the four-season field survey in Sheffield.   57

3.4 Main sounds identified by the interviewees in (a) the Peace Gardens and (b) the Barkers Pool. 59

3.5 Relationships between the measured sound level and the mean subjective evaluation of sound level, with linear regressions and correlation coefficients R. (a) Alimos, Greece (b) Thessaloniki, Greece (c) Sesto San Giovanni, Italy (d) Sheffield, UK (e) Kassel, Germany (f) Fribourg, Switzerland. 63

3.6 Relationships between measured L and the subjective evaluation of sound level, and between the measured L and the acoustic comfort evaluation, with binominal regressions and correlation coefficients squared R at (a) The Peace Gardens; and (b) the Barkers Pool. 65

3.7 Comparison between the subjective evaluation of sound level and the acoustic comfort evaluation at (a) The Peace Gardens; (b) the Barkers Pool. 66

3.8 Relationships between (a) the measured sound level and the mean subjective evaluation of sound level; and (b) between the measured sound level and the mean acoustic comfort evaluation under three source conditions in the Peace Gardens, with binominal regressions and R. 68

3.9 Relationships between (a) the measured sound level and the mean subjective evaluation of sound level, and (b) between the measured sound level and the mean acoustic comfort evaluation, under two source conditions in the Barkers Pool with binominal regressions and R. 69

3.10 Sound preferences in the two Sheffield case study sites. 71

3.11 Differences in sound preference amongst age groups in the two Sheffield sites. (a) Bird songs; (b) music from stores; and (c) surrounding speech. 75

3.12 Differences amongst age groups in the two Sheffield sites regarding preferred relaxing sound environment. 75

3.13 Differences in sound preference between males and females in the two Sheffield sites. (a) Church bell; (b) music played on street; and (c) children's shouting. 77

3.14 (a) XiDan Square and (b) ChangChun Yuan Square in Beijing. 82

3.15 Age group distribution at (a) the Peace Gardens and (b) the Barkers Pool. Average of two seasonal periods. 86

3.16 Education and occupation profiles of the interviewees at (a) the Peace Gardens and (b) the Barkers Pool. Average of two seasonal periods. 87

3.17 Comparison between the four squares in terms of the comfortable–uncomfortable index. 91

3.18 Percentage of the architectural students who selected a given environmental factor as 'important'. 92

3.19 A model for describing the soundscape of urban open spaces. 93

3.20 Framework for using ANN for soundscape evaluation. 97

3.21 Comparison of the spectra of three water features in the Chatsworth Garden, England. 99

3.22 (a) Temporal sound fluctuations and (b) spectra (at typical
locations in Meadowhall). 101
3.23 Spectra and temporal characteristics of the sounds played in
the architectural reading room (AR) at the Sheffield University
Main Library. 103
3.24 Typical SPL distribution with time in the main swimming areas. 105
4.1 Distribution of image sources in an idealised street canyon. 108
4.2 Plan view of an idealised urban square. The grid lines show the
division of patches (see Section 4.4). The source and receiver
positions used in a parametric study are also shown (see
Section 7.3), where for the 50 × 50m square, the source is at
(10m,10m), and the positions of four typical receivers are 24
(12.5m, 17.5m), 56 (27.5m, 27.5m), 89 (42.5m, 42.5m), and
100 (47.5m, 47.5m), corresponding to source–receiver
distances of 8, 25, 46 and 53m, respectively. 110
4.3 Distribution of image sources in an idealised square, plan view. 111
4.4 Three-dimensional projection of an idealised street showing an
example of patch division. 116
4.5 Distribution of the energy from a point source to a patch on the
ground. 118
4.6 Determination of the form factor from emitter A to an
orthogonal patch G. 119
4.7 Cross section of an idealised rectangular street with
geometrically reflecting ground, showing the distribution of
source energy and energy exchange between patches. 122
4.8 Comparison of (a) SPL attenuation and (b) RT along the length
between calculation and measurement. 124
4.9 Comparison of decay curves between geometrically and
diffusely reflecting ground at two typical receivers. 127
4.10 Set-up of the coupled FDTD-PE model, redrawn based on Van
Renterghem *et al.* (2006). 132
4.11 Typical square configuration (50 × 50m) used in the parametric
study, showing the source and receiver positions. 135
4.12 Variation in SPL, RT and EDT with increasing ray number
(1kHz and 50 reflection order). 137
4.13 Variation of SPL, RT and EDT with increasing reflection order
(1kHz and 100k ray number). 139
5.1 (a) The idealised square and (b) street, both 50 × 50m, showing
the source (large dot) and receiver (small dots) positions. The
origin of coordinates is at a corner of the square/street. 155
5.2 SPL of the NMS calculation with reflection orders $R = 1$–20,
with reference to the image source model result with $R = 20$;
(a) square; (b) street. 156
5.3 SPL difference between NMS and the image source model in the
square, with $\alpha = 0.1$, 0.5 and 0.9. 157
5.4 Cross section of the pitched roof and the three simplified
building blocks. 158

5.5   Site plan of the two calculation configurations, showing the
       positions of road, buildings and receivers.   159
5.6   SPL with various flat roof heights, with reference to the SPL of
       pitched roof, based on the configuration illustrated in Figure 5.5a.   160
5.7   Three configurations with different building widths used to
       examine the effects of building gaps.   160
5.8   SPL with various building gaps (a) with reference to the SPL
       with a solid block along the street, and (b) the colour map with
       four building arrangements where the building gap is 5m.   161
5.9   Noise maps in a street (see Figure 5.5a) with various reflection orders.   162
5.10   Comparison between noise maps with reflection order 0 and 1
       in an urban area in Sheffield.   163
5.11   Increase in calculation time with increasing reflection order,
       based on the configuration shown in Figure 5.9.   163
5.12   Site plan of an area of Sheffield city centre showing the
       measurement points for (a) sound source data and (b) validation.   165
5.13   Three-dimensional model of an area of Sheffield city centre for
       noise mapping.   166
5.14   Noise map of an area of Sheffield city centre with a reflection
       order of (a) 1 and (b) 3.   167
5.15   Predicted SPL distribution around a plant.   169
5.16   Comparison of SPL distribution between (a) road only (b) plant
       only, and (c) the combination.   170
6.1   Principles and examples of self-protection buildings, cross-
       sectional view.   176
6.2   Schematics of (a) typical absorptive silencers and (b) reactive
       silencers.   179
6.3   Schematics of an 'acoustic lock', plan view.   180
6.4   Effect of aperture width on acoustic and airflow performance
       of a lined aperture for aperture lengths of 0.3, 0.45 and 0.6m;
       (a) variation of effective free area with aperture width;
       (b) variation of element weighted normalised level difference
       D with aperture width. Data adopted from Oldham *et al.*
       (2005d).   181
6.5   Generic/basic configurations of the window system. HL: hood
       length; HD: distance between hood and glass; SSO: source side
       opening; RSO: receiving side opening; MVG: minimum
       ventilation gap; TWW: typical window width; D: distance
       between glass and MPA.   183
6.6   Typical numerical simulation results using FEMLAB, in terms
       of the SPL difference between source and receiving rooms;
       (a) with various opening size SSO-RSO (mm); (b) with
       various air gap TWW (mm) and the effect of louvers
       (see Figure 6.5 – PB003a: 45° louvers with hard surfaces;
       PB003b: 45° louvers with impedance $0.3\,\rho_0 c$; PB003c: 105°
       louvers with impedance $0.3\,\rho_0 c$); and (c) with various hood
       length HL (mm).   185

6.7 Measured acoustic performance of typical/strategic window configurations (Kang *et al.*, 2005), in terms of the SPL difference between source and receiving rooms. Acoustic performance of single- and double-glazing is also shown for comparison. 186

6.8 Schematics of strategic barrier designs, cross-sectional view: (a) multiple-edged barriers; (b) reactive barriers; (c) phase interference barriers; and (d) phase reversal barriers. Adopted from Ekici and Bougdah (2004). 188

6.9 Schematics of various picket barriers and vertically louvered barriers. Adopted from Ekici and Bougdah (2004). 190

6.10 Schematics of barriers with absorptive treatment, cross-sectional view. Adopted from Ekici (2004). 191

6.11 Schematics of dispersive barriers, plan view. 191

6.12 Schematics of strategical architectural/landscape designs, cross-sectional view. (a) Vertical alignment of road; (b) cantilevered and galleried barriers; and (c) suspended panels. Adopted from Ekici (2004). 193

6.13 Preconceptions of various types of barriers on their potential to attenuate noise, with 1 as most effective and 5 as least effective. Data adopted from Joynt (2005). 197

7.1 Basic configuration of (a) the street canyon and (b) the urban element used in the calculation. 201

7.2 Decay curves with increasing source–receiver distance, where the boundaries are diffusely reflective, the street length, width and height are $L = 120$, $W = 20$ and $H = 18$m, respectively, and a point source is positioned at (30m, 6m, 1m). 202

7.3 SPL distribution with a point source at five positions in streets S-M, (a) (60m, 0, 1m); (b) (60m, 15m, 1m); (c) (60m, 30m, 1m); (d) (60m, 45m, 1m); (e) (60m, 60m, 1m); and (f) with nine sources at a spacing of 15m along $y = 0$–120m. The street height is 20m. 203

7.4 Reverberation time distribution with a point source at three positions in streets S-M. (60m, 0, 1m): (a) RT; (a') EDT. (60m, 30m, 1m): (b) RT; (b') EDT. (60m, 60m, 1m): (c) RT; (c') EDT. The street height is 20m. 205

7.5 Comparison of the sound attenuation along the length between diffusely and geometrically reflecting boundaries in two streets. (a) $W = 20$m and $H = 6$m; (b) $W = 20$m and $H = 18$m. 206

7.6 Comparison of (a) RT; (b) EDT; and (c) decay curves between diffusely and geometrically reflecting boundaries. The street width is 20m. 207

7.7 Sound attenuation along the length with different street heights, with diffusely reflecting boundaries. The street width is 20m. 209

7.8 Comparison of reverberation time between two street widths, where the boundaries are diffusely reflective, and the street height is 18m. 209

7.9   Decay curves with increasing street height, where the boundaries
       are diffusely reflective, the street width is 20m, and the source–
       receiver distance is 20m.   210
7.10   SPL distribution with varied street widths in the urban element;
       (a) street S 30m wide and street N 10m wide; (b) street S 10m
       wide and street N 30m wide. The street height is 20m.   211
7.11   SPL changes caused by staggering street S and N; (a) street N
       shifted to $x = 70–90$m; (b) street S shifted to $x = 30–50$m and
       street N shifted to $x = 70–90$m. The street height is 20m.   212
7.12   SPL attenuation along the length with different boundary
       absorption coefficients; (a) diffusely reflecting boundaries;
       (b) geometrically reflecting boundaries. The street width is 20m
       and the street height is 18m.   214
7.13   Decay curves with different boundary absorption coefficients at
       receiver (50m, 2m, 1m), with diffusely reflecting boundaries.
       The street width is 20m and the street height is 18m.   215
7.14   Three-dimensional representation of typical configurations used
       in the comparison between UK and HK streets.   218
7.15   SPL along the length with geometrically reflecting boundaries;
       (a) with receiver at 1.5m and (b) at street height.   220
7.16   Decay curves with geometrically reflecting boundaries at (a) 40m
       and (b) 120m from the source. The receivers are at 1.5m above
       the ground.   221
7.17   Comparison in (a) SPL; (b) RT; and (c) EDT between HK5 and
       UK5 with diffusely reflecting boundaries.   223
7.18   SPL on a horizontal plane at 1.5m above the ground. All the
       façades have a diffusion coefficient of 0.3 whereas the ground is
       geometrically reflective. Point source. Each colour represents
       5dBA.   223
7.19   SPL distribution on a vertical plane at 1m from a façade. All the
       façades have a diffusion coefficient of 0.3 whereas the ground is
       geometrically reflective. Line source. Each colour represents
       1dBA.   224
7.20   SPL distribution on a receiver plane perpendicular to the street
       length. All the façades have a diffusion coefficient of 0.3 whereas
       the ground is geometrically reflective. Line source. Each colour
       represents 3dBA.   225
7.21   Distribution of the (a) SPL; (b) RT; and (c) EDT in a square of
       50 × 50m, with diffusely (dark mesh) and geometrically (light
       mesh) reflecting boundaries. Square height 20m. Source at
       (10m, 10m, 1.5m). Boundary absorption coefficient 0.1.   226
7.22   Comparison of the decay curves between diffusely and
       geometrically reflecting boundaries at receiver 100, where the
       source–receiver distance is 53m.   227
7.23   Variation in SPL, RT and EDT with increasing diffusion
       coefficients at receiver 24, 56 and 89, where the source–receiver
       distances are 8, 25 and 46m respectively.   228

7.24 Comparison of (a) the SPL (and (b) reverberation times between three square heights: 50, 20 and 6m, with diffusely reflecting boundaries. 230

7.25 Comparison of (a) the SPL and (b) reverberation times between three square sizes: 25 × 25m, 50 × 50m and 100 × 100m, with diffusely reflecting boundaries. 231

7.26 Comparison of (a) the SPL and (b) reverberation times between three square sizes: 25 × 25m, 50 × 50m and 100 × 100m, with geometrically reflecting boundaries. 232

7.27 Comparison of (a) the SPL and (b) reverberation times between two square shapes: 50 × 50m and 100 × 25m, with diffusely reflecting boundaries. 234

7.28 Comparison of (a) the SPL and (b) reverberation times between two square shapes: 50 × 50m and 100 × 25m, with geometrically reflecting boundaries. 235

7.29 Effects of boundary absorption on (a) SPL and (b) reverberation times, with diffusely reflecting boundaries. 237

7.30 Effects of boundary absorption on (a) SPL and (b) reverberation times, with geometrically reflecting boundaries. 238

7.31 (a) Four building arrangements and (b) four absorber arrangements in the 50 × 50m square. 239

7.32 SPL distribution with four building arrangements (see Figure 7.31). 240

7.33 SPL distribution with four absorption arrangements (see Figure 7.31). 241

7.34 RT distribution with four building arrangements (see Figure 7.31). 242

7.35 RT distribution with four absorption arrangements (see Figure 7.31). 243

7.36 EDT distribution with four building arrangements (see Figure 7.31). 244

7.37 EDT distribution with four absorption arrangements (see Figure 7.31). 245

Preface

In the last decade or so there have been many major new developments in the field of urban sound environment in terms of research and practice. Whilst large-scale noise mapping software packages have been developed and applied extensively in practice with the advancement of computing resources, various prediction methods for sound propagation in micro- and mesoscale urban areas have also been explored. Correspondingly, there have been a series of new noise control measures and design methods. In the subjective aspect, a number of evaluation methods have been developed with multidisciplinary approaches. In the meantime, the importance of soundscape and sound environment design has been widely recognised, which is a major step forward from simply reducing urban noise level. In terms of environmental policies and regulations, noise problems have been paid great attention at various levels, especially in Europe, leading to a series of substantial actions in noise abatement.

The main motivation of this book is to present the state-of-the-art development in urban sound environment. It also attempts to systematically cover essential knowledge and basic principles in the field. Combining technical presentation and interdisciplinary approach with suggestions for practical application and design, the book is relevant to researchers, practitioners and students in a number of disciplines and sectors, including urban planning, architecture, landscape, acoustics and noise control, environmental science, civil engineering, transport engineering, and environmental psychology/sociology. The book is reasonably self-contained. Whilst the introductory chapter provides some prior knowledge of elementary acoustics, other chapters are made fairly independent, to help those readers who are not familiar with certain fields.

The book is divided into seven chapters, covering three main facets of urban sound environment: sound evaluation and acoustic comfort (Chapters 2 and 3), urban sound modelling/mapping (Chapters 4 and 5), and noise mitigation and sound environment design (Chapters 6 and 7). Chapter 1 briefly introduces fundamental concepts and theories relevant to urban sound environment, including physical properties of sound waves, auditory perception, sound sources, acoustic materials, outdoor sound propagation and room acoustics. Chapter 2 discusses the description and evaluation of urban noise, including an overview of subjective evaluation of urban noise in terms of acoustic/physical factors and social/psychological/economic factors as well as commonly used evaluation methods, objective descriptors relating to urban sounds, key/typical noise standards/regulations and their principles, and current situation of urban noise climate. Chapter 3 focuses on the urban soundscape and acoustic comfort, with particular attention on urban open public spaces. The chapter starts with a review of general soundscape research and evaluation and then describes a series of soundscape studies in Europe and China. With the semantic differential method, main factors that characterise the soundscape are studied. A framework for soundscape description in urban open public spaces is then explored, followed by an overall soundscape evaluation system using artificial neural networks, and systematic considerations on soundscape design. The acoustic comfort is also examined in a series of indoor spaces, which are natural extensions of urban sound environment. Chapter 4 describes a series of simulation techniques as well as related acoustic theories for accurately calculating the sound field for microscale urban areas such as a street or a square. This includes energy-based image source methods for street canyons and urban squares with geometrically (specularly) reflecting boundaries, image source method considering interference, ray tracing, radiosity model for diffusely reflecting boundaries, transport theory, equivalent source method, and some other models. Techniques for urban acoustic animation are also discussed. Physical scale modelling techniques for urban acoustics are then briefly introduced, as well as actual measurements, which are useful for validating the simulation models. Chapter 5 deals with macroscale urban areas, including main algorithms, accuracy, efficiency and strategic application of noise mapping techniques, with case studies, and some other recently developed models for meso-/macroscale urban areas. Chapter 6 presents the main mitigation measures for urban noise, especially those relating to urban and architectural design, including planning considerations, building envelope design, principles and applications of various environmental noise barriers, and nonacoustic issues in designing barriers. Chapter 7 analyses the basic characteristics of sound fields in urban streets and squares and the effects of architectural changes and urban design options, including boundary reflection pattern, street/square geometry, boundary absorption and building arrangements. Finally, as it is not feasible to cover all facets of urban sound environment in great depth within one book, given the multidisciplinary nature of the field, a considerable number of references are provided.

I wish to express my appreciation to Tony Moore for approaching and encouraging me to write this book, and to the editorial team for their support. I am also thankful for the useful discussion with a number of acoustic colleagues including D.J. Oldham, J. Picaut, D. Botteldooren, P.J. Thorsson, M. Ögren, C. Marquis-Favre, H. Bougdah, I. Ekici, B. Shield, C.J. Skinner, K. Attenborough, K.M. Li, K.V. Horoshenkov, A. Bristow, and W. Probst; for the support from colleagues at the University of Sheffield School of Architecture and contribution from researchers at the Acoustics Group, including M.W. Brocklesby, M. Zhang, W. Yang, J. Joynt, K. Chourmouziadou, Y. Meng, L. Yu, C. Yu, J. Huang, B. Chen, Z. Du, R. Huerta, C. Stepan, C. Christophers, S. Keeling-Roberts, C.H. Lin, M.A. Rahim, and L. Thomas; for the support from various research partners including R.J. Orlowski, P. Grasby, K. Harsham, M. Nikolopoulou, M. Kikira, K. Steemers, N.U. Kofoed, G. Scudo, L. Katzschner, R. Compagnon, N. Chrisomallidou, E. Kovani, K. Avdelidi, N. Xiang, D.X. Mao, and J.Y. Tsou; and for the financial support from the Royal Society, European Commission, UK Engineering and Physical Sciences Research Council, British Academy, Lloyd's Foundation, and British Petroleum. Last but not least, I wish to thank my family for their unceasing support and encouragement.

Abbreviations

AI	Articulation index
AI	Artificial intelligence
ANN	Artificial neural networks
BEM	Boundary element method
CNEL	Community noise equivalent level
CNL	Corrected noise level
CNR	Composite noise rating
dB	Decibel
DNL	Day–night average sound level
EDT	Early decay time
ESM	Equivalent sources method
FDTD	Finite difference – time domain method
FEM	Finite element method
FFT	Fast Fourier transform
GIS	Geographical information systems
IL	Insertion loss
ISO	International standard organisation
LCA	Lifecycle assessment
MEM	Maximum error margin
MLS	Maximum length sequence
MPA	Microperforated absorber
NC	Noise criterion
NCB	Balanced noise criterion curve
NEC	Noise exposure category
NEF	Noise exposure forecast
NI	Noisiness index
NNI	Noise and number index
NPL	Noise pollution level
NR	Noise rating
NSDI	Noise sensitivity depreciation index
OD	Origin-destination matrix
PA	Public address
PE	Parabolic equation method
PNC	Preferred noise criterion curve
PNL	Perceived noise level

PPG24	Planning policy guidance note 24
PSIL	Preferred speech interference level
RASTI	Rapid speech transmission index
RT	Reverberation time
S/N	Signal-to-noise ratio
SEL	Sound exposure level
SIL	Speech interference level
SPDF	Single particle distribution function
SPL	Sound pressure level
STD	Standard deviation
STI	Speech transmission index
TNI	Traffic noise index
WECPNL	Weighted equivalent continuous perceived noise level
WFAE	World Forum for Acoustic Ecology
WHO	World Health Organization

Chapter 1

Fundamentals

This chapter briefly introduces the fundamental concepts and theories relating to urban sound environment. It begins with a description of the physical properties of sound waves, followed by explanations of auditory perception, sound sources, acoustic materials, outdoor sound propagation and room acoustics.

1.1   Basic properties of sound

1.1.1   Sound wave

Sound is the transmission of energy through solid, liquid or gaseous media in the form of vibrations. In a medium, each vibrating particle moves only an infinitesimal amount to either side of its normal position. It is first displaced in the direction of propagation of the wave; it will then move back to its undisturbed position and continue towards a maximum negative displacement. In other words, sound is transmitted in the form of a longitudinal wave. The time for completing a full circuit by a displaced particle is called the period, T, and the offset of the wave from a reference point is called a phase. Usually the oscillations are repeated and the repetition rate, namely the number of oscillation per second, is defined as frequency, f. The unit of frequency is hertz (Hz). The continuous oscillations of the source propagates a series of compressions and rarefactions outwards through the medium. The distance between adjacent regions where identical conditions of particle displacement occur is called the wavelength, λ. In other words, λ is the distance a sound wave travels during one cycle of vibration. The velocity with which sound travels through air varies directly with the equilibrium air pressure and inversely with the equilibrium air density. At standard pressure (760mm Hg) and 20°C the velocity of sound propagation is approximately $c = 340$m/s. The relationships between the frequency, period, sound velocity and wavelength are

$$f = \frac{1}{T} = \frac{c}{\lambda} \qquad (1.1)$$

1.1.2   Sound power, pressure and intensity

The sound power of a source, W, is the rate at which acoustic energy is transferred from a vibrating source to a medium. It is measured in watts (W). In a medium, sound can be sensed by the measurement of some physical quantity that is disturbed from its equilibrium value. The sound energy density is the sound energy in a given infinitesimal part of the medium divided by the volume of that part of the medium. The unit is W.

The sound pressure at a point, p, is the incremental change from the static pressure at a given instant caused by the presence of a sound wave. The effective sound pressure at a point is the root mean square (rms) value of the instantaneous sound pressure over a time interval at that point. The unit of sound pressure is N/m. Sound pressures are extremely small. At a distance of a metre from a talker, the average pressure for normal speech is about 0.1N/m above and below atmospheric pressure, whereas the atmospheric pressure is about 1.013N/m at sea level.

The sound intensity, I, is the average rate at which sound energy is transmitted through a unit area perpendicular to the specified direction. It is noted that unlike sound pressure, sound intensity is direction dependent. The unit of sound intensity is W/m. For a free progressive wave, if the ambient density of the medium is ρ_0, the relationship between sound intensity and sound pressure in the direction of propagation is

$$I = \frac{p_{rms}^2}{\rho_0 c} \qquad (1.2)$$

1.1.3   Sound levels

From the threshold of audibility to the threshold of pain the intensity ratio is about 10. It would thus be necessary to represent the huge range of human sensitivity by a scale of small numbers. More importantly, the human ear does not respond linearly to sound intensity or pressure, whereas perceived changes in intensity or pressure tend to be proportional to the ratios between the intensities or pressures concerned. It is thus more convenient to use a logarithmic unit, decibel (dB), to measure sound intensity and pressure. The sound intensity level L_I, the sound pressure level (SPL) L_p, and similarly, the sound power level of a source L_W can be calculated by

$$L_I = 10\log\left(\frac{I}{I_0}\right) \quad (\text{dB}) \qquad (1.3)$$

$$L_p = 10\log\left(\frac{p^2}{p_0^2}\right) = 20\log\left(\frac{p}{p_0}\right) \quad (\text{dB}) \qquad (1.4)$$

$$L_W = 10\log\left(\frac{W}{W_0}\right) \quad (\text{dB}) \qquad (1.5)$$

where $I_0 = 10\text{W/m}$ is the reference intensity or the minimum sound intensity audible to the average human ear at 1kHz. The reference pressure is $p_0 = 2 \times 10\text{N/m}$, which is chosen so that the numerical values for intensity and pressure are approximately the same at standard atmospheric conditions. Similarly, $W_0 = 10\text{ W}$ is the reference sound power.

Very often contributions from more than one sound source are concerned. When two sound waves are in phase, the crests align perfectly and add together, whereas when two waves are 180° out of phase, the crests align with valleys and the waves cancel each other. If the phases between sources of sound are random, the sounds can be added together on a linear energy (pressure squared) basis. To add SPLs, the mean square sound pressure from each source should first be determined according to Equation (1.4). The total SPL can then be obtained from the summation of the mean square sound pressures. Assume there are n sources and L_{pi} is the SPL of each source. According to the above procedure, the total SPL, L_p, can be calculated by

$$L_p = 10\log\left(\sum_{i=1}^{n} 10^{L_{pi}/10}\right) \tag{1.6}$$

1.1.4 Frequency band

Most sounds are complex and contain many frequencies. A sound can be measured in a series of frequency intervals called frequency bands. Octave and fractional octave bands are often used. In each octave band the upper limiting frequency is exactly twice the lower limiting frequency. The centre frequency is the geometric mean of the upper band limit and lower band limit. The centre frequencies that have been standardised for acoustic measurements are 31.5Hz, 63Hz, 125Hz, 250Hz, 500Hz, 1kHz, 2kHz, 4kHz, 8kHz and 16kHz. One-third octave bands are formed by dividing each octave band in three parts. Successive centre frequencies of one-third octave bands are related by factor $\sqrt[3]{2}$. For example, the one-third octave bands contained within the octave centred on 500Hz will have centre frequencies of 400, 500 and 630Hz. A plotted relationship between frequency and sound level is called a sound spectrum.

White noise is a sound that contains uniform energy at every frequency within the range of human hearing and is analogous in spectrum characteristics to white light. Pink noise is a sound that contains uniform energy per octave bandwidth, namely each octave of increasing frequency contains half the power of the preceding one.

1.2 Auditory perception

1.2.1 Level and frequency perception

Frequencies in the audio range are from about 20Hz to 20kHz. With increasing age the upper frequency hearing limit drops continuously. For pure tones, the smallest, just perceptible frequency changes are ±0.35 per cent of the frequency concerned (Zwicker and Feldkeller 1967). A tonal difference of two pairs of tones is perceived equally if the ratio of the two tones in each pair is the same. This feature of human perception is the same as that for loudness.

Although the minimum SPL audible to the average human ear is 0dB at 1kHz, generally speaking, for a constant sound, a level of 10–15dB is barely audible, 130dB will cause a painful sensation, and above 140dB will increase the risk of irreparable nerve damage. In the median range of SPL, a change of 1dB is just perceptible, changes need to be around 3dB or more to be of any significance at all, and an increase of 10dB produces an approximate doubling of the strength of sensation (Schaudinischky 1976).

1.2.2 Loudness and noisiness

There have been several versions of equal-loudness-level contours, and they are obtained by subjective comparative measurements in a free field involving sinusoidal tones (Fletcher and Munson 1933; Fastl and Zwicker 1987; Robinson and Dadson 1956; Suzuki and Takeshima 2004). Figure 1.1 shows the contours based on the current ISO data (ISO 2003b). The unit of loudness level, P, is phon, and their values are the same as the SPL at 1kHz. In Figure 1.1 the

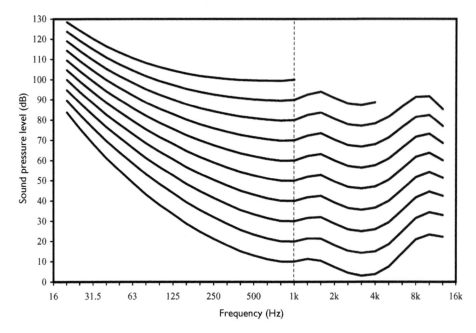

Figure 1.1 Normal equal-loudness-level contours for pure tones, based on binaural free-field listening tests with frontal incidence. Data adopted from ISO (2003b).

lowest curve represents the auditory threshold. Since a change in the loudness level of 10 phon produces approximately double the strength of sensation, a unit for loudness S, sone, is defined as

$$S = 40 + 10\log_2 P \tag{1.7}$$

It has been noted that in many cases loudness and annoyance are two separate and operationally distinct perceived attributes. Based on laboratory subjective tests on noisiness, annoyance or unacceptability, contours of equal noisiness were established (Kryter 1970; Goldstein 1979; US EPA 1971). The unit of subjectively perceived noisiness is the noy, described in terms of perceived noise level (PNL, L_{PN}) in dB. Similar to sone, a sound of 2 noy is subjectively judged to be twice as noisy as a sound of 1 noy. Noisiness and PNL were initially introduced for aircraft noise evaluation, but have since been used for other kinds of noise.

1.2.3   Weighted sound levels

A single value is often desirable when measuring sound. Frequency weighting considers typical human response to sound, such as that shown in Figure 1.1, when the sound level in each frequency is adjusted. The adjusted levels are then added to produce a single number in decibels. Standard weighting networks include A, B, C and D, and the resultant decibel values are called dBA, dBB, dBC and dBD respectively. The A-weighting network, which is commonly used in noise legislation, was originally designed to approximate the response of the human ear at relatively low sound levels.

The 'sound level' at a given receiver is the reading in decibels of a sound level meter. The meter reading corresponds to a value of the sound pressure integrated over the audible frequency range with a specified frequency weighting and integration time.

1.2.4 Masking

The masking effect occurs when a signal is rendered unintelligible or inaudible by a simultaneous sound that exceeds a certain level. In other words, it is the process by which the threshold of audibility for one sound is raised by the presence of another (masking) sound. For pure tones, the masking effect is more significant when the signal frequency is closer to the masking sound. Low-frequency sounds produce a considerable masking effect over higher-frequency sounds, although higher-frequency sounds can also mask lower frequencies to a degree. The frequency range of masking is greater with the increase of the SPL of the masking sound. It is noted that masking produced by narrowband noise is not directly predictable from masking experiments with wideband noise, and *vice versa*.

In addition to spectral masking, temporal masking also exists. Sometimes a signal can be masked by a sound preceding it, called forward masking, or even by a sound following it, called backward masking. Postmasking occurs when a signal is switched off and the ear needs a certain time to recover to its normal sensitivity. For example, a short audible click may become inaudible when it is presented right after a noise burst (Howard and Angus 1996; Rossing 1990).

1.2.5 Sound quality

The term product sound quality was coined in the 1980s. In the very beginning the term just expressed that it was conceived that acoustic emissions had further characteristics than just level (Blauert and Jekosch 1997). Linked to the concept of product quality, sound quality was defined as the 'adequacy of a sound in the context of a specific technical goal and/or task'. Sound quality has three main aspects: (1) stimulus–response compatibility, which is the functional aspect of a sound; (2) pleasantness of sounds, which is based on an instantaneous overall impression emerging from various sound attributes as well as individual preferences and experience; and (3) identifiability of sounds or sound sources, so people know what is going on around them (Guski 1997; Zeitler and Hellbrück 1999). The pleasantness aspect of sound design is commonly evaluated by means of unidimensional rating scales, and the identification aspect by means of decision times in recognition tasks and multidimensional scaling techniques (Susini *et al.* 1999).

Psychoacoustic magnitudes, such as loudness, fluctuation strength or roughness, sharpness, and pitch strength (Zwicker and Fastl 1999), have proved successful for the assessment of sound quality. If the frequencies of two tones only differ slightly amplitude fluctuations or modulations can be perceived. Up to a modulation frequency of about 13Hz one has an impression of regular loudness changes. This perceptual quantity is called fluctuation strength. The sensation of fluctuation strength becomes the impression of roughness when the modulation frequency is ranged from 13Hz to about 300Hz. Roughness is a modulation-based metric that may be described as 'grating'. It is generated by sounds that contain tones spaced within a critical band, amplitude-modulated tones, frequency modulation, or rapidly and repeatedly fluctuating noise. Examples of rough sounds include the humming of an electric razor or a sewing machine. A rough character of a sound usually causes an unpleasant hearing impression. Sharpness is an indication of the spectral balance between low and high frequencies. The more high frequencies a signal contains, the higher its sharpness is. Pitch strength, or tonality, is the distinctness of pure tones in a complex noise. Audible pure tones contained in broadband noise may be annoying, although the contribution to the total loudness may not be significant. There are also a number of other indices, including percentile loudness (Namba *et al.* 1996); intrusiveness, a distortion of the contents of the information (Preis 1996); and impulsiveness (Beidl and Stücklschwaiger 1997).

The psychoacoustic magnitudes allow for an instrumental prediction of attributes of sound perception, although instruments are still far from simulating human sound perception and evaluation in all its facets (Bodden 1997).

1.2.6 Effects of noise

The potential effects of community noise include hearing impairment, startle and defence reactions, aural pain, ear discomfort, speech interference, sleep disturbance, cardiovascular effects, performance reduction and annoyance responses. These health effects, in turn, can lead to social handicap, reduced productivity, decreased performance in learning, absenteeism in the workplace and school, increased drug use and accidents (Berglund *et al.* 1999). Noise could also have economic impacts such as loss of property value (see Section 2.1.2).

1.3 Sound sources

1.3.1 Basic forms of sound sources

A pulsating sphere, alternatively increasing and decreasing its diameter, radiates sound uniformly in all directions and is called a spherical source. If a spherical source is very small, say the source radius is smaller than one-sixth the wavelength, it can be regarded as a point source. Typical point sources include valves, flare stacks and fans.

A line source is defined by its length relative to the separation distance between source and receiver. A line source can be considered to consist of an infinite number of evenly distributed individual point sources. The sound power level of a line source is measured using sound power level per metre. Typical line sources include pipelines, trains and continuous road traffic.

A plane wave is a special case in which the acoustic variables are functions of only one spatial coordinate. It can be created by a rigid piston moving forwards and backwards from the equilibrium position along a very long tube. While an ideal plane (area) source is an infinitely large flat surface that radiates sound, in practice it is defined by its dimensions relative to the separation distance between source and receiver. For example, a façade or roof of an industrial building, or a school playground, can often be regarded as a plane source. The sound power level of a plane source is measured using sound power level per square metre.

1.3.2 Source directivity

When the dimensions of a sound source are much smaller than a wavelength, the effect of the source shape on its radiation will be negligible, providing all parts of the radiator vibrate substantially in phase. For sources that are not small compared to the wavelength, it is important to consider their directional properties. The directivity factor, Q, is defined as the ratio of the intensity at some distance and angle from the source to the intensity at the same distance if the total power from the source were radiated uniformly in all directions. The directivity index, DI, is defined as

$$DI = 10\log Q \tag{1.8}$$

For practical applications it is more convenient to express DI in terms of the SPL:

$$DI = L_{p\theta} - \overline{L}_p \tag{1.9}$$

where $L_{p\theta}$ is the SPL at the direction of interest, and $\overline{L}_p$ is the average SPL in all directions. Both $L_{p\theta}$ and $\overline{L}_p$ are determined for the same fixed distance from the source.

1.3.3   Urban sound sources

There are numerous urban sound sources, ranging from transportation to leisure activities such as concerts and discotheques. The actual sound sources are more complex than the basic forms. Most sound generators can be grouped into the following categories: vibrating solid bodies such as loudspeaker diaphragms, vibrating air columns such as wind instruments, transient forms of mechanical or electrical power such as lightning discharges, impact phenomena such as hammering, sounds from rapidly expanding gases such as jets, complex sonic disturbances resulting from rapidly moving objects in fluids such as fans (White 1975). Figure 1.2 shows the typical spectra of car and train noise.

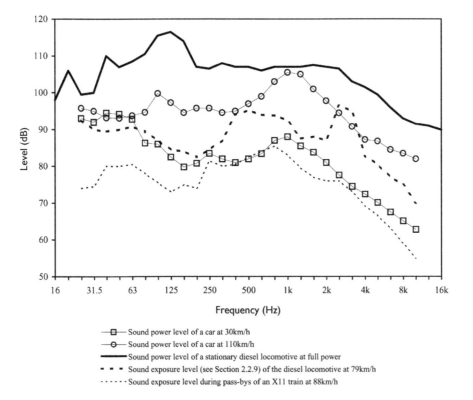

Figure 1.2 Typical spectra of car and train noise. Data adopted from Jonasson and Storeheier (2001), Van Beek et al. (2002) and Jonasson et al. (2004).

The noise of a motor vehicle is caused by the engine, tyres and air turbulence. For urban traffic at low speeds engines are often the main sources of noise. The engine noise is produced by radiation of vibrating surfaces and by individual sources, including primary inlet and exhaust, radiation by inlet and exhaust systems walls, engine vibration emitted noise, gearbox and transmission noise, and cooling fan noise (Lamure 1975). Engine noise is dependent on its rotational speed, the load on the engine and its cubic capacity. Tyres and road interaction contribute to noise at higher speeds and air turbulence is generally unimportant. A wet surface can increase the level by 10dBA (Penn, 1979).

Railway systems comprise long-haul freight and passenger trains and a variety of subway, elevated and surface vehicles. Up to about 50km/h, railway noise is dominated by traction noise, which consists of motor noise and auxiliary noise. At about 50–300km/h noise emission is dominated by rolling noise. Above about 300km/h aerodynamic noise becomes predominant. These transition speeds depend on many parameters such as rail and wheel maintenance conditions for the rolling noise. Noise also occurs when passing bridges, curves and rail joints as well as braking (Van Beek *et al.* 2002).

A series of methods have been developed for the measurement of sound power level, considering free field, semi-free field and diffuse field, under laboratory condition and *in situ* (ISO 1994a, 1994b, 1995, 1998, 1999, 2000, 2003c). Recent developments include sound intensity techniques and microphone array methods (Mast *et al.* 2005).

1.4 Acoustic materials

1.4.1 Reflection, transmission and absorption coefficient

When sound waves fall on a boundary, their energy is partially reflected, partially absorbed by the boundary, and partially transmitted through the boundary to the other side. The reflection coefficient ρ is the ratio of the sound energy that is reflected from the boundary to the sound energy incident on it. The fraction of incident energy that is transmitted through the boundary is called the transmission coefficient, τ. The absorption coefficient α is the ratio of the sound energy that is not reflected from the boundary to the sound energy incident on it, namely the sum of absorbed and transmitted energy. The relationship between the reflection and absorption coefficient is $\rho = 1 - \alpha$. When the sound wave is incident under an angle θ to the normal, the absorption coefficient is called the oblique-incidence absorption coefficient, given as α_θ. When the incident sound is evenly distributed in all directions, the absorption coefficient is called the random-incidence absorption coefficient or statistical absorption coefficient. The absorption characteristics of a material can be measured using an impedance tube or a reverberation room/chamber. The former gives the normal-incidence absorption coefficient and the latter gives the Sabine absorption coefficient, which is usually close to the random-incidence absorption coefficient. Sound reflection, transmission and absorption coefficients are all frequency dependent, and may take on any numerical values between 0 and 1.

1.4.2 Sound absorbers

Basic sound absorbers include porous absorbers, single resonators, perforated panel absorbers and panel and membrane absorbers. The absorption coefficients of typical building boundaries are shown in Figure 1.3.

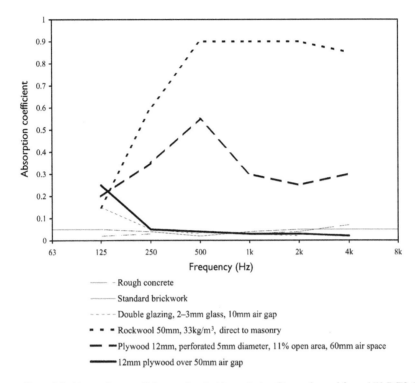

Figure 1.3 Absorption coefficients of typical boundaries. Data adopted from UK DfES (2003).

A porous absorber is characterised by its pores that are accessible from the outside. When sound waves impinge on a porous material, part of the sound energy is converted into thermal energy due to viscous flow losses caused by wave propagation in the material and internal frictional losses caused by motion of the material's fibres. The absorption characteristics of a porous material are dependent upon a number of variables, including its thickness, density, as porosity, flow resistance and fibre orientation. Typically, the absorption of a porous absorber is large at high frequencies and the performance at low frequencies depends mainly on the thickness and the flow resistance. The maximum sound absorption will occur when the particle velocity is at a maximum, namely at a distance of $\lambda / 4$, $3\lambda / 4$, ... from a rigid backing wall. As a result, a greater material thickness is required for a lower frequency, or alternatively, the material may be mounted at some distance from the rigid wall.

A single resonator comprises a cavity within a massive solid connected to the air outside through a restricted neck. The impinging sound energy causes the air in the neck to vibrate in the resonant frequency range and thus, some sound energy is dissipated by means of viscous losses. A single resonator is normally only effective at low frequencies within a narrow frequency range.

Perforated panel absorbers, which comprise a perforated panel mounted some distance from a rigid wall, are commonly used in practice. Such a construction can be regarded as a series of single resonators. The perforation may be in the form of holes or slits. The resonant frequency and the frequency range of absorption depend mainly on the aperture diameter, aperture spacing, panel thickness and the depth of the air space. Similar to single resonators, perforated panel absorbers are usually only effective for a narrow range of frequency. When there is a porous material behind the perforated panel to provide extra damping, the frequency range of absorption can be considerably extended. An alternative way of increasing acoustic damping of the apertures is to use microperforated panels, which have low perforation ratio but many apertures of submillimetre size (Maa 1987). The microperforated absorbers can be used in some special situations, and for example, can be made from transparent materials such as plastic glass.

Panel or membrane absorbers are formed by mounting a panel or membrane at some distance from a rigid wall. This behaves analogously to a mass-spring system. The panel or membrane can be set into vibration when it is struck by a sound wave. Due to the friction in the panel or membrane itself, in its supports and in any space behind it, an energy loss occurs and hence some sound absorption takes place. The resonant frequency of such a system depends mainly on the stiffness, surface density, thickness and elastic modulus of the material, and the depth of the air space behind. Usually panel and membrane absorbers resonant at low to middle frequencies, and the frequency range of absorption is rather narrow. This range can be extended by placing a porous material in the air space because this will provide extra damping (friction).

Absorbers can also be made into three-dimensional units and suspended freely in a space with some distance from the boundaries. Since sound energy is free to impinge on all sides of these units, they could provide a powerful absorbing effect.

Combinations of the above-listed fundamental types of absorbers are also widely used. For example, multiple layers of panel or perforated panel may have multiple resonant frequencies and, thus, the frequency range of absorption can be broadened as compared to a single layer. For a microperforated membrane backed by an air space, the absorption performance can be rather good in a wide range of frequencies due to the resonance from both membrane and apertures (Kang and Fuchs 1999).

1.4.3 Sound reflectors and diffusers

When sound waves strike a boundary, if the boundary is acoustically rigid and smooth and has dimensions considerably greater than the wavelength, the angle of incidence of the wave front is equal to the angle of reflection. In other words, a sound wave from a given source, reflected by a plane surface appears to come from the image of the source in that surface. The reflection pattern from curved surfaces can be similarly determined by the application of geometrical laws.

If there are irregularities on a reflecting boundary, the incident sound energy may be scattered to a particular solid angle. A diffuser disperses reflections both temporally and spatially. It is important to note that the width and depth of an effective diffuser should be considerable relative to the wavelengths interested, although there seems to be strong evidence that even untreated boundaries produce diffuse reflections (Hodgson 1991). Typical diffusers include simple curved surfaces, irregular geometric structures, periodic geometric structures, and mixture of absorptive and reflective materials. Diffusers can be constructed from a wide range of materials, including wood, concrete, metal, brick and glass.

If the directional distribution of the reflected/scattered energy does not depend in any way on the direction of the incident sound, the boundary is called a diffusely reflecting boundary. A practical example is the phase grating diffuser, also called Schroeder diffuser (Schroeder 1975; D'Antonio and Konnert 1984). The diffusers are periodic surface structures with rigid construction. The elements of the structure are wells of different depth but the same width separated by thin fins. Within one period, the depths of the elements vary according to a sequence such as the quadratic residue sequence, primitive root sequence and maximum length sequence. Figure 1.4 shows an example of a Schroeder diffuser where the design

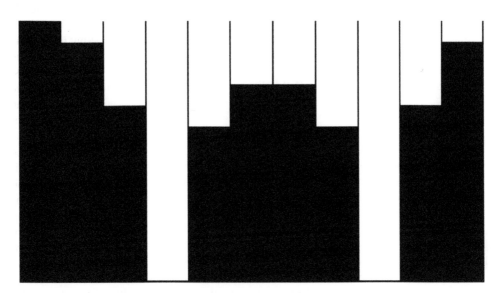

Figure 1.4 Cross section of a Schroeder diffuser.

frequency is 625Hz, the maximum frequency is 2kHz, the well width is 80mm, and within a period of sequences there are 11 wells with depths of 0, 25, 99, 304, 124, 74, 74, 124, 304, 99, 25mm (Kang 2002a). Such diffusers can also be made by varying well depth in both length and width directions and/or combined with absorbers (Cox and D'Antonio 2004).

The terms diffusion and scattering are used in different ways and are interchanged in different subject fields, including room acoustics simulation (Embrechts *et al.* 2001; Vorländer and Mommertz 2000), urban acoustics simulation, diffuser manufacture and building design. Unlike the absorption coefficient, there is no simple physical definition for the diffusion/scattering coefficient that meets the needs of all interest groups (Cox and D'Antonio 2004). Recently two different coefficient definitions and corresponding measurement methods/standards have been developed. A diffusion coefficient measures the quality of reflections produced by a surface. A free-field method, measuring the similarity between the scattered polar response and a uniform distribution, has also been developed (AES 2001). A scattering coefficient is defined as the ratio between the acoustic energy reflected in nonspecular directions and the totally reflected sound energy. To determine the random-incidence scattering coefficient as caused by surface roughness, measurements can be made in a reverberation chamber, either at full scale or in a physical scale model (ISO 2004). A free-field method for measuring the scattering coefficient has also been developed, although it is often more laborious to carry out (Mommertz and Vorländer 1995).

It is noted that the definition of scattering coefficient takes no account of how the scattered energy is distributed, and it is not intended for characterising the spatial uniformity of the scattering from a surface. In room acoustics there is usually a large amount of mixing of different reflections, so that any inaccuracies caused by this simplification would be averaged out. For urban acoustic simulation, however, this may not be applicable since the early reflections often play a dominant role. In this book, for the sake of convenience for computation, the diffusion coefficient, defined as the ratio between the acoustic energy uniformly reflected and the totally reflected energy, is used.

Diffraction is the bending or spreading out of a sound wave after it intersects an aperture, a solid object, a recess or a surface protrusion. Main factors affecting diffraction include the size of the diffraction object relative to the wavelength concerned, edge conditions, and the distances from the object to the source and receiver. Diffraction can increase the sound level behind the diffraction object (see Section 1.5.4), and also increase the diffusion coefficient in front of the diffraction object (Chourmouziadou and Kang 2006).

1.4.4   Airborne sound insulation

Airborne sound insulation is of relevance for urban sound environment, for example, in terms of sound transmission between indoor and outdoor spaces, and sound transmission loss of environmental barriers. In this section two basic boundary forms are discussed.

Single panel

For a homogeneous impervious panel, the sound transmission at low frequencies depends mainly on panel resonance, which is determined by the panel size, elasticity and surface density. At resonance the energy losses are generally small. At medium frequencies, the individual particle vibration is important and the sound transmission depends mainly on impedance and in turn on elasticity and density. At higher frequencies, the panel tends to behave as a series of small masses, and the mass law holds true – the sound transmission loss increases at a rate of about 6dB for each doubling of frequency and by about 6dB for each doubling of surface density. This relationship becomes invalid above the critical frequency, where there is always a certain angle of incidence that will excite coincidence in the panel; thus, the panel is virtually transparent to the exciting sound wave and there is little transmission loss. This so-called coincidence effect is rather significant, particularly for lightweight constructions such as plaster, metal and glass, where the coincidence dip occurs in the 1–4kHz region. Nevertheless, in practice, the sound transmission loss does not become zero since sound energy is randomly incident upon the surface, some loss occurs due to damping factors, and there are internal homogeneities and resonance in the thickness of the wall (Lawrence 1970; Day et al. 1969; Möser 2004). The coincidence dip occurs more obviously if the critical frequency is high. The lowest coincidence frequency is called the limiting frequency or critical frequency. Figure 1.5 shows measured transmission loss of typical boundaries (Möser 2004; ISO 1996), where the above effects can be seen.

Double-leaf

In the mass-controlled frequency range the sound transmission loss depends mainly on impedance mismatch, and most of the energy reduction occurs through reflection at the boundaries between air and the material of the panel. Therefore, an improvement in insulation would be expected if the partition is split into two skins with an air space between so that the number of boundaries at which reflection could take place is doubled. For this reason, ideally the two layers should not be the same. Improvement is more significant for lightweight constructions, especially at relatively high frequencies. It is useful to insert sound absorbent materials in the cavity to increase damping. It is also important to avoid fixed/rigid connections between the two layers, namely structure-borne sound bridges. For

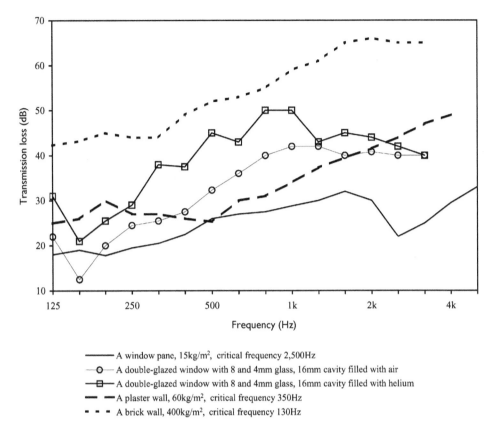

Figure 1.5 Transmission loss of typical boundaries. Data adopted from Möser (2004).

double-glazed units, the gas in the cavity can make significant difference in transmission loss, as shown in Figure 1.5.

1.5   Outdoor sound propagation

The calculation of outdoor noise levels at large distances between source and receiver requires detailed consideration of a number of separate or interactive effects, including source characteristics, source–receiver distance, ground and air attenuation, wind speed and direction, temperature and relative humidity, barrier attenuation and acoustic screening, and surface reflections. The basic principles of outdoor sound propagation are discussed in this section, whereas engineering methods for sound propagation prediction are reviewed in Section 5.1.

1.5.1   Basic equation

In a free field, at a distance d from a point source, the intensity is the sound power of the source divided by the total spherical area of the sound wave at d

$$I = \frac{W}{4\pi d^2} \qquad (1.10)$$

In other words, the intensity at any point is inversely proportional to the square of its distance from the source, which is commonly referred to as the inverse square law. This is equivalent to a reduction of 6dB SPL with each doubling of the distance from the source.

Based on Equation (1.10), the SPL can be calculated in reference to the sound power level, by

$$L_p = L_W - 10\log(4\pi d^2) = L_W - 20\log(d) - 11 \qquad (1.11)$$

For hemispherical sound propagation where the source is located close to hard ground and reflections of the emitted noise occur, Equation (1.11) becomes

$$L_p = L_W - 20\log(d) - 8 \qquad (1.12)$$

For an ideal line source of infinite length in a free field, the SPL can be determined using purely cylindrical sound propagation:

$$L_p = L_W - 10\log(2\pi d) = L_W - 10\log(d) - 8 \qquad (1.13)$$

where L_W is the sound power level per metre. From Equation (1.13) it can be seen that the SPL falls off at 3dB with each doubling of the distance from the source. If a line source is located close to hard ground, Equation (1.13) becomes

$$L_p = L_W - 10\log(d) - 5 \qquad (1.14)$$

The sound radiation from a plane source can be approximately calculated by considering the source as a number of evenly distributed individual point sources.

1.5.2 Atmospheric conditions: air, wind and temperature

There are two mechanisms by which acoustic energy is absorbed by the atmosphere, molecular relaxation and viscosity effects. Air absorption is mainly dependent on temperature and relative humidity. Generally, the effect is only significant at large distance and/or high frequencies. Typical attenuations per 100m caused by air at temperature 20°C and relative humidity of 50 per cent are: 0.032dB at 125Hz; 0.072dB at 250Hz; 0.18dB at 500Hz; 0.42dB at 1kHz; 1.0dB at 2kHz; 2.6dB at 4kHz; and 8.3dB at 8kHz (Lawrence 1970). The air absorption under various temperature and humidity conditions is given by ANSI (1999a).

In the case of wind, there is usually a velocity gradient where the wind speed increases with increasing height above the ground. As a result, sound waves travel upwind at a greater speed near the ground and at progressively slower speeds with increasing height above ground. The sound waves are bent and less sound is received at a point upwind compared to no-wind condition. Conversely, more sound is received downwind (Lawrence 1970).

The effect of temperature gradient on sound propagation is similar to that of wind. Increase in temperature with altitude, which usually occurs at nighttime when the ground air temperature is considerably reduced, results in an increase in sound speed and, consequently, the sound waves will be refracted downward in the absence of any wind. Conversely, the daytime air temperature decreases with increasing altitude and, thus, the sound rays will be continuously bent away from the ground and less sound will be received than when no gradient exists.

In a combined field of temperature and wind gradient their effects will be integrated and it is also possible that they may cancel each other out. Under certain weather conditions skip zones or sound channel conditions may be formed where sound can travel very long distances without much attenuation. Generally speaking, in the case of short-range propagation, say within a mile, the attenuation due to irregularities in wind structure seems to be of major importance in comparison to ordinary temperature and wind refraction, as well as humidity, fog, rain and snow (Ingard 1953).

1.5.3   Ground

Ground attenuation occurs due to the absorption of acoustic energy when a sound wave impinges on the ground, and also due to the ground effect, namely the interference between the direct and reflected sound waves caused by the change in phase of the reflection. Effective ground absorption can be obtained from grass or other vegetation, ploughed fields, snow cover, or other kinds of sound absorbers (see Section 1.4.2). The sound interaction with ground depends on the geometry from source to receiver and the acoustic properties of the ground surface (see Section 5.1.5). For most grounds the normalised characteristic impedance is a sufficient description. However, for grounds where over the first few centimetres the acoustic properties vary significantly with depth or the flow resistivity is relatively low, the propagation constant of sound within the ground layers and the acoustic near-surface structure is also required (Attenborough 1988, 1992). It is also important to consider the combined effects of ground with vector wind and temperature gradients (Parkin and Scholes 1965; Attenborough and Li 1997).

1.5.4   Barrier

A barrier can be considered to be any solid obstacle that impedes the line of sight between source and receiver and thus creates a sound shadow. Considerable theoretical, numerical and experimental research has been carried out on the prediction of sound attenuation of barriers (Maekawa 1968; Rathe 1969; Kurze and Anderson 1971; Koyasu and Yamashita 1973; Hutchins *et al.* 1984a, 1984b; Isei 1980; Isei *et al.* 1980; Seznec 1980; UK DfT 1988; L'Esperance 1989; Hothersall *et al.* 1991a, 1991b; Muradali and Fyfe 1998; ANSI 2003a; Ekici 2004). In this section the basic barrier theory is discussed, whereas more practical considerations and strategic designs are described in Chapter 5.

The basic barrier theory is analogous to optical diffraction theory. The effectiveness of a barrier is primarily dependent on the frequency and path difference, δ, defined as the difference in distance between the direct path through a barrier from source to receiver and the indirect path over the barrier, as shown in Figure 1.6, where R is the distance between source and barrier (m); D is the distance between barrier and receiver (m), and

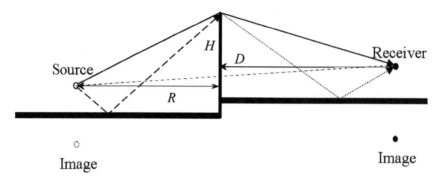

Figure 1.6 Diagram of calculating diffractions over a barrier.

H is the effective height of barrier (m), namely the barrier height above the line between source to receiver. If the source–receiver line is perpendicular to the barrier, δ can be calculated by

$$\delta = R\left[\sqrt{1+\left(\frac{H}{R}\right)^2}-1\right]+D\left[\sqrt{1+\left(\frac{H}{D}\right)^2}-1\right]$$
(1.15)

The extra attenuation of a barrier is closely related to the Fresnel number, N, defined as

$$N = 2\frac{\delta}{\lambda}$$
(1.16)

For a single point source, Kurze and Anderson (1971) gave a simplified equation to calculate the performance of an infinitely long barrier. For $-0.2 < N < 12.5$, the insertion loss (IL), namely the difference in SPL with and without the barrier, can be calculated by

$$IL = 20\log\frac{(2\pi N)^{1/2}}{\tanh(2\pi N)^{1/2}}+5$$
(1.17)

For $N > 12.5$, experimental data show that there is an upper limit of 24dB (Rathe 1969).

In Equation (1.17) the ground effect is not considered. Unless the receiver is much higher than the ground, say more than 2m, the effect of ground reflection should be calculated by applying the above method for the image of the source/receiver (Maekawa 1968; Ekici 2004), if the ground is perfectly reflective.

Diffraction of sound also occurs around the ends of a barrier. The algorithms described above can again be used. The barrier length required for avoiding end effects depends on a number of parameters, but several empirical rules have been recommended for typical configurations: the barrier length should be such that the distance from the source to the ends is at least twice the normal distance of the source to the barrier; or a barrier should cover an angle of 160° subtended from the receiver; or the barrier length should be more than 4–5 times the height (Pirinchieva 1991). It is noted that for relatively low frequencies, the interferences from various sound paths should also be taken into account.

The above discussion is based on a point source. For an incoherent line source parallel to the edge of the barrier, the insertion loss by the barrier is always smaller than the attenuation for the closest point of the line by up to about 5dB (Kurze and Anderson 1971; Kurze and Beranek 1971).

The fundamental theory of barrier diffraction assumes a knife edge at the barrier top. This theory is approximately valid when the barrier thickness is smaller than the wavelength. For thick barriers such as a building or banks, approximation can be made using an equivalent thin barrier, defined by the intersection of two straight lines both just grazing the top edges of the building or banks, one drawn from the reception point and the other drawn from the effective source position (UK DfT 1988). However, this method becomes inaccurate when the thick barrier is high and the angle between the two lines is greater than 90°. In this case, a more accurate method should be used by considering two thin barriers (Zhen 2000).

The effectiveness of a barrier also depends on the sound transmission loss through the barrier. Normally this should be 10dB higher than the diffracted sound. Other factors affecting noise performance include barrier reflections, barrier absorption and meteorological interactions (Van Renterghem and Botteldooren 2003; see also Section 6.3).

1.6   Room acoustics

Since many urban spaces are rather enclosed, the room acoustic theory is of relevance. For small rooms of simple shape it is possible and also necessary to describe the interior sound field in precise mathematical terms by considering the physical wave nature of sound. If a room is large and irregular in shape, however, in the audio-frequency range the number of room modes will be enormous and their pressure distribution will be very complex. Nevertheless, since the wavelengths are short compared to room dimensions, it is reasonable to treat a sound wave as a sound ray, where the physical wave nature of sound such as diffraction and interference can be ignored. A statistically reliable statement can then be made of the average conditions in the room.

1.6.1   Reverberation process

Consider a sound source generating a steady-state sound. After the source starts to supply sound energy in a space, a certain period is needed to build up an equilibrium sound level. Similarly, after the sound source stops, the sound will still be audible for sometime. This decay process is called the reverberation of the space, which can be characterised by the impulse response, the acoustic response at a listening point to a short burst of sound from a source.

After an impulse, the first sound to arrive at the listener will be the sound that travels in a straight line from the source, namely the direct sound. This is followed by a series of sounds that have travelled by paths including one or more reflections from boundaries. Compared to the direct sound, the amplitude of a reflected sound is always less because part of the sound energy is absorbed by the reflective surfaces; also, it travels farther and thus the effects of spherical divergence and air absorption are greater. Reflections that arrive at the listener immediately after the direct sound are called early reflections. After a certain period, the number of reflections becomes so high that individual reflections are no longer distinguishable. These late reflections are called the reverberant sound. The above process can be represented with an echogram, a diagram of sound level against time. The arriving time and the amplitude of a reflection are closely related to the geometry of a space, number of reflections and absorption of the boundaries.

A diffuse sound field is an ideal situation where acoustic energy is uniformly distributed throughout an entire space, and at any point the sound propagation is uniform in all directions.

1.6.2 Reverberation time

The most important index to evaluate a reverberation process is the reverberation time, which is defined as the time taken for a sound to decay 60dB after the source is stopped. It is a function of the volume of the room and the amount of sound absorption within it (Sabine 1993):

$$RT = \frac{0.16V}{A + 4MV} = \frac{0.16V}{\Sigma_{i=1}^{n} \alpha_i S_i + 4MV} \tag{1.18}$$

where RT is the reverberation time (s), V is the volume of the space (m), and A is the total boundary absorption in the space (m), which is calculated by multiplying each individual surface area S_i by its absorption coefficient α_i and adding the whole together. M is the energy attenuation constant due to air absorption (Np/m), which, under typical conditions, namely at air temperature 20°C and relative humidity 50 per cent, is 0.0024 at 2kHz; 0.0061 at 4kHz; 0.0126 at 6.3kHz; and 0.0215 at 8kHz (Harris 1966).

Equation (1.18), the Sabine formula, becomes progressively more inaccurate as the average absorption coefficient increases. The Eyring formula is a modification of the Sabine formula:

$$RT = \frac{0.16V}{-S_0 \ln (1 - \overline{\alpha}) + 4MV} \tag{1.19}$$

where S_0 is the total surface area (m) and $\overline{\alpha} = A / S_0$ is the mean absorption coefficient.

It is important to note that the Sabine and Eyring formulae embody the assumption of a diffuse field. Although there is no real sound field that strictly meets this condition, the above reverberation formulae are accurate enough for many enclosures (Kang and Neubauer 2001).

1.6.3 Measurement of reverberation

A decay curve is used to determine the reverberation time. It is a plot of the decay of SPL in a space as a function of time after the source of sound has ceased and may be measured after the actual cutoff of a continuous sound source in a room. Impulse sources such as a signal gun and balloon popping are often used in practice.

Reverberation time is usually determined using the rate of decay given by the linear regression of the decay curve from a level 5dB below the initial level to 35dB below. It is called the RT30. Since the subjective judgement of reverberation is often well correlated to the early slope of a sound decay, the early decay time (EDT) is also used. It is obtained from the initial 10dB of the decay. For both RT30 and EDT the slope is extrapolated to correspond to a 60dB decay. In a diffuse sound field, a decay curve is perfectly linear and, thus, the RT30 and EDT should have the same value. In this book, RT is based on RT30 except where indicated.

A decay curve can also be derived from the reverse-time integrated squared impulse response of the space, $p(t)$ (ISO 1997). In an ideal situation with no background noise the integration should start at the end of the impulse response and proceed to the beginning. Thus, the energy decay $E(t)$ as a function of time, t, is

$$E(t) = \int_t^\infty p^2(i)\mathrm{d}i = \int_\infty^t p^2(i)\mathrm{d}(-i) \tag{1.20}$$

1.6.4   Sound distribution in a diffuse field

In a diffuse field, the steady-state sound energy distribution can be calculated by (Beranek 1954)

$$L_d = L_W + 10\log\left(\frac{Q}{4\pi d^2} + \frac{4}{R_\mathrm{T}}\right) \tag{1.21}$$

where $R_\mathrm{T} = S_0\alpha_\mathrm{T}/(1-\alpha_\mathrm{T})$ and $\alpha_\mathrm{T} = \bar{\alpha} + 4MV/S_0$, d is the source–receiver distance, Q is the directivity factor of the source, and R_T is the total room constant. From Equation (1.21) it can be seen that the sound field is divided into two distinct parts, $Q/4\pi d^2$ and $4/R_\mathrm{T}$, representing the direct sound field and the reverberant sound field, respectively. The distance with $Q/4\pi d^2 = 4/R_\mathrm{T}$ is known as the reverberation radius. In a diffuse field, the SPL becomes approximately constant beyond the reverberation radius.

1.6.5   Room modes

In the above discussion, the wave effects of the sound are not taken into account. Consider a sound that is supplied to a closed tube and the diameter of the tube is small compared with the wavelength. If the tube length is an integral multiple of a half wavelength, the forward- and backward-travelling waves add in magnitude to produce what is called a standing wave. In other words, the tube resonates at certain frequencies. The frequencies are called resonant frequencies, natural frequencies, normal frequencies or eigenfrequencies.

Similarly, standing waves can also be set up in a room, where they travel not only between two opposite, parallel boundaries, but also around the room involving the boundaries at various angles of incidence. For a rectangular enclosure, the frequencies of these standing waves are given by

$$f_{n_x,n_y,n_z} = \frac{c}{2}\left[\left(\frac{n_x}{L}\right)^2 + \left(\frac{n_y}{W}\right)^2 + \left(\frac{n_z}{H}\right)^2\right]^{\frac{1}{2}} \tag{1.22}$$

where f_{n_x,n_y,n_z} is the resonant frequency; L, W and H are the dimensions of the enclosure; and n_x, n_y and n_z are positive integers (one or two of them may also be zero).

The values found from Equation (1.22) correspond to the room modes. There are three basic room modes: axial modes, two n are zero and the waves travel along one axis; tangential modes, one n is zero and the waves are parallel to one pair of parallel boundaries and are obliquely incident on two other pairs of boundaries; and oblique modes, no n is zero and the waves are obliquely incident on all boundaries.

Chapter 2

Urban noise evaluation

Environmental noise is unwanted or harmful sound, usually generated by human activities including road traffic, railways, air transport, industry, recreation and construction, and is perceived in the domestic environment such as in and near the home, in public parks and schools. This chapter starts with an overview of the subjective evaluation of urban noise in terms of acoustic/physical factors and social/psychological/economic factors, and commonly used evaluation methods (Section 2.1), followed by a series of objective descriptors relating to urban sounds (Section 2.2). It then summarises key/typical noise standards/regulations and their principles (Section 2.3), and the current situation of urban noise climate in some countries (Section 2.4).

2.1 Subjective noise evaluation

The evaluation of sound is a complex system and is related to a number of disciplines including acoustics, physiology, sociology, psychology and statistics. In this section the effects of these factors are discussed, and commonly used evaluation models are outlined (Marquis-Favre *et al.* 2005b).

2.1.1 Acoustic/physical factors

The overall sound level is certainly an important factor for subjective evaluation. Relationships between annoyance and noise exposure, such as those measured by the equivalent continuous sound level, L_{eq} (see Section 2.2.2), have been intensively studied (Schultz 1978; Kryter 1982; Miedema and Vos 1998; Arana and García 1998; Ali and Tamura 2003; Klæboe *et al.* 2004). Lambert *et al.* (1984) divided the daytime traffic noise annoyance into three levels: <55dBA, no annoyance; 55–60dBA, some people annoyed; and >65dBA, definite annoyance. Similarly, Bertoni *et al.* (1993) suggested that for urban road traffic noise, when the day–night average sound level (DNL) (see Section 2.2.3) is greater than 60–62dBA, the correlation between noise level and annoyance becomes stronger, whereas Fields (1993) indicated that even for DNL < 55dBA, there could be a percentage of very annoyed people. Recently, based on a large amount of existing data, a correlation between day–evening–night sound level L_{den} (see Section 2.2.4) and noise annoyance has been derived for various noise types, as shown in Figure 2.1 (WG-HSEA 2002).

Although A-weighted sound level is commonly used in regulations, it is important to consider the characteristics of spectrum. For example, more tonal components may increase

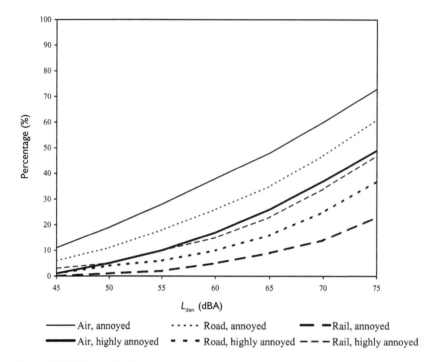

Figure 2.1 Relationship between L_{den} and percentage of annoyed and highly annoyed people. Data adopted from WG-HSEA (2002).

annoyance. Moreover, there is increasing evidence that low-frequency components play an important role in annoyance (Ochiai *et al.* 1999).

With the same energy summation, namely a constant L_{eq}, annoyance may increase with a larger amplitude fluctuation or emergence of occasional events (Namba *et al.* 1996). Fields (1984) suggested that a multiplication of the number of acoustic events by 10 would cause a similar increase in annoyance as an increase of approximately 5dB of the average peak level, although Bjorkman and Rylander (1996) indicated that starting from a certain number of events, the annoyance would not increase. Other factors which affect annoyance include regularity of events, maximum sound level, rise time, duration of occasional events, spectral distribution of energy and number and duration of quiet periods (Guski 1998).

Situational variables (Fields *et al.* 1997) such as relatively long-term changes in noise exposure are also important. Fidell *et al.* (1998) found that a progressive drop of 1.5–3dB near an airport over a long period of time was hardly noticed by the locals. Vallet (1996) showed that a decrease of less than 6dBA did not influence the percentage of very annoyed people. It is also important to consider whether changes to noise exposure create an overreaction phenomenon (Fields 1993). Another situational variable is ambient noise. It seems that the annoyance to a target environmental noise exposure is affected very little by the presence of another sound source qualified as ambient noise. Fields (1998) found that a 20dB increase in ambient noise exposure had no more impact than an approximately 1dB decrease in target noise exposure.

With the same sound level, annoyance may vary with different types of noise such as hown in Figure 2.1. Generally speaking, aircraft noise is more annoying than road traffic noise (Hall *et al.* 1981), whereas road noise is more annoying than railway noise (Fields and Walker 1982; Guski 1998; Miedema and Vos 1998), although Yano *et al.* (1996) suggested that in Japan this might not be true. For road traffic noise, Sattler and Rott (1996) reported that buses were the most annoying, followed by cars, mopeds and trucks, although this might not be applicable in some situations (Recuero *et al.* 1996).

Season and the time of day may also influence annoyance evaluation. Griffiths *et al.* (1980) and Recuero *et al.* (1996) found that noise annoyance was greater in summer than in winter, and Vallet *et al.* (1996) reported that the effects of noise were greater in the evening and at the beginning of night period.

2.1.2   Social/psychological/economic factors

It has been demonstrated by many studies that correlations between noise annoyance and the acoustic/physical factors are often not high. According to Guski (1998), the noise annoyance of inhabitants only depends on approximately 33 per cent of the acoustic parameters such as acoustic energy, number of sound events, and length of moments of calm between intermittent noises. Similarly, Berglund (1998) suggested that a maximum of 30 per cent annoyance variation might be due to the noise exposure expressed in L_{Aeq}, whereas according to Job (1988) and Lercher (1998), this level could be less than 20 per cent. Clearly in addition to the acoustic parameters, other aspects including social/psychological/economic factors play an important role in annoyance evaluation.

In terms of attitude, there are generally six aspects that influence annoyance (Nelson 1987). The first aspect is fear. One may feel that certain noises are annoying because they are linked to dangers such as a plane crash (Kryter 1982). Similarly, people may be more annoyed by a noise source if they believe that their health is affected by this source (Nelson 1987). The second aspect is related to the cause of noise, for example, whether one feels that there are ways to control the noise through various channels including government policies and going to court. According to Miedema and Vos (1999), people may be less annoyed if they are economically dependent on the activities generating the noise. The third aspect is sensitivity to noise, especially for certain source types. It was estimated that the difference in annoyance between groups of people of various sensitivities could be equivalent to about 10dB (Vallet 1996; Miedema and Vos 1999). The fourth aspect is activity; noise may be more disturbing for certain activities such as oral communication, listening to radio and intellectual tasks. The fifth aspect is the perception of the neighbourhood. Annoyance may increase if a neighbourhood is perceived in a negative way, and it is also influenced by the lifestyle chosen by certain people, for whom a certain quantity of noise is part of their life. Moreover, people may get used to certain noises and thus become less annoyed. The final aspect is the global perception of the environment, including the interactions between acoustic and other physical factors such as dust, light, smells, wind and temperature, as well as environmental factors such as view and landscape. Research has been carried out regarding interactions between noise and light (Sanders *et al.* 1974), noise and colour (Puslar-Cudina and Cudina 1999), noise and vibration (Griffin and Howarth 1990), and noise and visual information (Viollon 2003; Abe *et al.* 1999, 2006). For example, it has been shown that the presence of exhaust fumes tends to cause a greater noise annoyance, and so does the vibration of a sound source (Yamashita *et al.* 1991; Sato 1993; Yano *et al.* 1996). In contrast, trees may reduce noise annoyance (Vallet 1996).

The effects of various demographic factors on sound evaluation have been intensively studied. There are varied results regarding the age effect, whereas most studies seem to suggest that the effect of gender is not important (Rylander *et al.* 1972; Sato 1993; Fields 1993; Tonin 1996; Miedema and Vos 1999; Yang and Kang 2005a, 2005b), although according to Verzini *et al.* (1999), low-frequency noises annoy men more than women. Marital status also seems to affect noise annoyance (Fields and Walker 1982). According to Bertoni *et al.* (1993), the house size/type and the family size have no significant influence on annoyance, whereas results from Miedema and Vos (1999) suggest that people living alone are less annoyed compared to those living in a large family. In terms of the education level, some studies show no significant effect on annoyance (Fields 1993; Vallet 1996; Tonin 1996), whereas other studies seem to suggest that people with a higher education level are slightly more annoyed (Miedema and Vos 1999; Verzini *et al.* 1999). Income and economic status appear to be insignificant for annoyance (Maurin and Lambert 1990; Bertoni *et al.* 1993; Fields 1993; Tonin 1996), and so is the general state of health, measured by the frequency of visiting doctors (Bertoni *et al.* 1993).

Noise experience, including exposure to noise at the place of work and over time, could affect residential noise annoyance (Bertoni *et al.* 1993) as well as sleep (Frusthorfer 1983). It seems that the effect of length of residence is not significant for annoyance evaluation (Fields 1993; Tonin 1996), whereas the time spent at home, especially if one lives alone or in a situation of social isolation, is important (Schulte-Fortkamp 1996). Moreover, there is evidence that the type of occupancy, namely owning or renting, might affect noise annoyance (Fields 1993; Tonin 1996; Vallet 1996; Miedema and Vos 1999). Furthermore, the satisfaction about living conditions, such as soundproofing of windows, is important for annoyance evaluation (Maurin and Lambert 1990; Klæboe *et al.* 2005).

Behaviour and habit is another important aspect that could affect annoyance. This includes, for example, opening and closing windows (Bertoni *et al.* 1993; Lercher 1998), using sleeping pills, using balconies or gardens, having a home sound insulated, and frequently leaving for weekends (Lambert *et al.* 1984).

Regional differences, including cultural heritage, construction methods, lifestyle and weather, may also influence noise annoyance (Gjestland 1998; Huang 2004; Xing and Kang 2006). Yano *et al.* (1996) and Kurra *et al.* (1999) have demonstrated the importance of the cultural factor in noise evaluation. A cross-cultural comparison of community responses to road traffic noise in Japan and Sweden suggested that nonacoustic factors, including the various customs of the people living in different countries and in different types of housing were important for annoyance evaluation (Sato *et al.* 1998). Another cross-cultural study on the factors of environmental sound quality, using semantic differential analysis in Japan, Germany, the United States of America and China, also demonstrated notable differences between the four countries (Kuwano *et al.* 1999).

The concept of environmental load is also relevant. At a given noise level inhabitants of small towns seem to be less annoyed than those of large urban communities (Bradley and Jonah 1979; Vallet 1996).

Economic effects of community noise have been studied, especially from the viewpoint of compensation payable on depreciation in property value that can be attributed to noise, among other physical factors (Rosen 1974; Hufschmidt *et al.* 1983; Turner *et al.* 1994; Hawkins 1999; Wilhelmsson 2000; Bateman *et al.* 2001; Navrud 2002; Wardman and Bristow 2004). The hedonic method has often been applied in noise evaluation studies. In order to facilitate comparisons of the results of hedonic price studies, a noise sensitivity depreciation index (NSDI), defined as the marginal percentage depreciation in house prices with respect to dB noise, has been developed (Walters 1975; Nelson 1980, 1982). Based on a series of case studies, some initial relationships have been established between dB increase and house price decrease, with corresponding thresholds/cutoff noise levels, although considerable further work is still needed (Bristow and Wardman 2005).

2.1.3 Methods for sound evaluation

To quantitatively describe sound evaluation such as annoyance, two kinds of methods are commonly used, unidimensional and multidimensional (Marquis-Favre *et al.* 2005b). The former, including category, discrimination and ratio scales, concerns the relationship between an acoustic variable and the perceptual dimension of a stimulus sound, whereas the latter considers various perception dimensions.

The ordered category scale, including absolute judgements, equal-appearing intervals or successive intervals, is most commonly used in annoyance evaluation. The categories can be represented with verbal and/or numeric scales. The formulation of the descriptors, such as 'not at all annoyed' or 'a little annoyed', is very important, more so than the numbers assigned to the descriptors (Yano *et al.* 1996). Analogous scale, namely a line with the two ends clearly defined, is suitable to gather continuous judgement during a time-varying sound sequence.

The discrimination scale, normally with paired comparison, is used to evaluate the relative annoyance of two stimuli. An advantage of the paired comparison method is that for untrained subjects it can give more robust results than the category method (Khan *et al.* 1996), since with the latter there might be confusion between scales.

The ratio scale method includes the magnitude estimation method and the ratio production method. The magnitude estimation method requires a subject to give a real positive number relative to a reference stimulus such as white or pink noise. The technique has been used to calibrate various community noises or their combination into a common unit of subjective evaluation for comparison (Berglund *et al.* 1975, 1976, 1981). When no reference is given, the absolute magnitude estimation can be used. The ratio production method, also called the fractionation method, normally involves an adjustment procedure – the subject is asked to adjust the stimulus so that its value is a fraction or a whole part of the reference stimulus. The equal-sensation matching method, where a sound is adjusted until the sensation (for example, annoyance) is the same as that due to the reference signal, is also used.

Combinations of various methods have also been considered, such as the category partitioning scale method suggested by Heller (1985), which combines category scales and magnitude estimation techniques. In this method there are five verbal categories and each category contains ten levels. Subjects first select a category, and then choose a level within this category.

A commonly used multidimensional evaluation method is the semantic differential method (Osgood 1952), also known as the polarity profile method. A group of antonymous adjectives, each pair defining the two ends of a multiple point scale, represent the multiple dimensions of perception. A related method is selected description, where a series of descriptive adjectives are collected/complied before evaluating a sound, and then respondents only choose some of the most relevant ones for their evaluation (Kuwano and Namba 1995). Since it is often difficult to compile antonymous adjectives, a method to establish spontaneous description of sound perception has been proposed. When listening to a sound, subjects describe their feelings in the form of imaginations, metaphors and comparisons (Schulte-Fortkamp *et al.* 1999), so that a list of representative adjectives can be established. Another technique in multidimensional analysis is the estimation of similarities of pairs of sounds for the description of the auditory space (Axelsson *et al.* 2003). The dimensions of the space are obtained using the multidimensional scaling techniques (Kruskal and Wish 1978).

Some other methods have also been proposed. Preis (1996) proposed a multicomponent approach where annoyance is the result of the linear combination of three terms: (1) annoying loudness – the time-averaged difference between the loudness of the noise and that of background noise; (2) intrusiveness – the time-averaged difference between the sharpness of the noise and that of background noise; and (3) distortion of the informational content – the percentage duration of all sound distortions measured in relation to the total measurement time. Botteldooren and Verkeyn (2002) applied fuzzy mathematics to consider the uncertainty in evaluation. Artificial neural networks have also been explored (see Section 3.9). Whilst various methods have been developed and applied as described above, there is still a recognised need to develop universal methods so that results of different studies can be compared.

It should be noted that although it is often necessary to study the annoyance and the perception of a sound in a totally controlled environment in order to determine the influence of certain parameters, these results cannot always be generalised to real and complex situations, especially in the case of multinuisances, as discussed in Section 2.1.2. In addition, annoyance expressed retrospectively may be different from that felt during the activities.

2.1.4   Multiple sources

There are currently a number of models that describe the combined effects of multiple simultaneous sources (Marquis-Favre *et al.* 2005b):

1   energy summation model – describes the relationship between the overall annoyance and the sound level resulting from the energy summation;
2   independent effects model – expresses annoyance as the linear combination of functions of the L_{eq} of each source;
3   energy difference model – describes annoyance as the function of the total L_{eq} and of the difference between L_{eq} of distinct sources;
4   response summation model (Ollerhead 1978) – a correction factor is added to the total L_{eq} in order to take into account the differences in level between sources;
5   dominant source model – the total annoyance equals that of the most annoying source;
6   subjectively corrected model – uses correction factors to account for differences in the perceived annoyance due to each distinct source;
7   quantitative model (Vos 1992) – similar to model 6 but the correction factors depend on the L_{eq} of individual sources;
8   summation and inhibition model (Powell 1979) – evaluates the total annoyance according to the total L_{eq} with a correction factor; and
9   vector summation model – corresponds to the total annoyance written as the square root of the sum of the squares of each noise source's perceptual variables (Berglund *et al.* 1981).

The above models can be divided into two kinds: (1) mathematical summations of quantities of noise exposure, or corresponding annoyance and loudness, when noises are presented in an isolated way; and (2) models reflecting cognitive and perceptual mechanisms that take into account the sources from which the components of noise come from and the way the information combines to give a global reaction in terms of annoyance or loudness. It has been shown that the latter models better describe real situations (Berglund and Nilsson 1998), although further development of such models is still needed. In addition, the temporal structure of the noise combinations must be taken into account.

2.2   Sound descriptors

In addition to the basic sound descriptors including SPL, weighted sound level, loudness and loudness level, noisiness and PNL, and a series of psychoacoustic indices, as described in Chapter 1, there are also many other indices, considering noise type, such as road traffic noise or aircraft noise, and for sound characteristics (Marquis-Favre *et al.* 2005a). It is often insufficient to characterise the noise environment using descriptors based only on energy summation, because different critical health effects require different descriptions (Berglund *et al.* 1999). For example, the maximum values of noise fluctuations as well as the number of noise events should be considered. Noise exposures in different time periods, including day, evening and night, require separate characterisations. Attention must also be paid if a noise source includes a large proportion of low-frequency components. This section reviews some commonly used descriptors, with a main focus on road traffic noise.

2.2.1   Statistical sound level

L_n is the level of noise exceeded for n per cent of the specified measurement period. In other words, if N measured SPLs are obtained in a time period T with a given time interval and they are sorted in ascending order, then L_n is the $(100n / N)$ th SPL in the order. By convention, L_1, L_{10}, L_{50} and L_{90} are used to give approximate indications of the maximum, intrusive, median and background sound levels, respectively.

2.2.2   Equivalent continuous sound level

The equivalent continuous sound level, $L_{eq,T}$, is a notional sound level. It is widely used to measure noise that varies considerably with time. It is 10 times the logarithm to the base 10 of the ratio of the time-mean-square instantaneous sound pressure, during a stated time interval T, to the square of the standard reference sound pressure (ANSI 1994):

$$L_{eq,T} = 10 \log \left[\frac{1}{T} \int \frac{p^2(t) \mathrm{d}t}{p_0^2} \right] \mathrm{dB} \tag{2.1}$$

If $p(t)$ is A-weighted before $L_{eq,T}$ is calculated then $L_{eq,T}$ will have units of dBA. In many sound level meters $L_{eq,T}$ can be given by implementing Equation (2.1) electronically.

With a series of measured shorter term L_{eq}, an overall $L_{eq,T}$ can be calculated by

$$L_{eq,T} = 10 \log \left[\frac{1}{T} \sum_{i=1}^{N} 10^{0.1 L_{eq,i}} t_i \right] \tag{2.2}$$

where N is the number of shorter-term L_{eq}, and t_i is the time period of the i th L_{eq}. If the time period for all the shorter terms is the same, then Equation (2.2) becomes

$$L_{eq,T} = 10 \log \left[\frac{1}{N} \sum_{i=1}^{N} 10^{0.1 L_{eq,i}} \right] \tag{2.3}$$

If a sound source is only on at a level of L_{eq} for a period of time t, then the $L_{eq,T}$ can be calculated by

$$L_{eq,T} = L_{eq,t} + 10 \log \left(\frac{t}{T} \right) \quad (t \leq T) \tag{2.4}$$

If there are a number of sources, the total $L_{eq,T}$ can be calculated by the decibel addition, as described in Chapter 1.

If the sound level has a normal distribution over a time period, the relationship between $L_{eq,T}$ and L_N is

$$L_{eq,T} = L_{50} + \frac{(L_{10} - L_{90})^2}{60} \tag{2.5}$$

or

$$L_{eq,T} \approx L_{50} + 0.115 \sigma^2 \tag{2.6}$$

where σ is the standard deviation (STD) of the measurement data (Zhen 2000). For fairly noisy traffic

$$L_{eq,T} \approx L_{50} + 3 \tag{2.7}$$

With the increase of σ, $L_{eq,T}$ becomes higher, since in the decibel additions the high levels become dominant. For example, for traffic noise, $L_{Aeq,T}$ is close to L_{30} if σ is 4–5dB.

If L_1 is much higher than the average level, say by 20dB, $L_{eq,T}$ could be close to L_1 (Zhen 2000). However, in some cases, $L_{eq,T}$ still cannot reflect the sound variations sufficiently. For example, a short impulse at night is rather disturbing, but may not be shown if $L_{eq,night}$ is used.

For road traffic noise, the roadside L_{eq} is highly correlated to the traffic flow, Q, in the following form

$$L_{eq} = a \log Q + b \tag{2.8}$$

The values of a and b are affected by a number of factors, such as road and street conditions. Figure 2.2 illustrates a series of curves based on surveys in several cities in Spain (Barrigón-Morillas et al. 2005). It can be seen that the variation is about 10dBA between various curves.

2.2.3   Day–night level

DNL, or day–night equivalent sound level, L_{dn}, is the average over a 24h period but the noise level during the nighttime period, typically 22:00–07:00, is penalised by the addition of 10dBA:

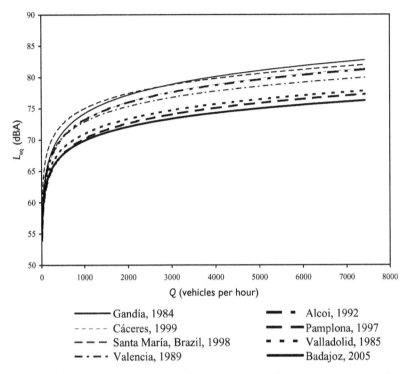

Figure 2.2 Relationship between traffic flow Q and traffic noise L_{eq}, calculated using Equation (2.8), where a and b values are adopted from those collected by Barrigón-Morillas *et al.* (2005).

$$L_{dn} = 10\log\frac{1}{24}\left(15\int_{7}^{22}10^{Lp/10}\mathrm{d}t + 9\int_{22}^{7}10^{(L_p + 10)/10}\mathrm{d}t\right) \quad (\mathrm{dBA})$$

(2.9)

2.2.4   Day–evening–night level

Day–evening–night level, L_{den}, is similar to DNL but an evening period is considered, penalised by an addition of 5dBA. Currently widely used in Europe (EU 2002), L_{den} is defined as

$$L_{den} = 10\log\frac{1}{24}\left(12\times10^{\frac{L_{day}}{10}} + 4\times10^{\frac{(L_{evening}+5)}{10}} + 8\times10^{\frac{(L_{night}+10)}{10}}\right) \quad (\mathrm{dBA})$$

(2.10)

where L_{day}, $L_{evening}$ and L_{night} are the A-weighted long-term average sound levels, determined over all the day periods, evening periods, and night periods of a year, respectively. The time periods can be defined according to the national and regional situations. Typically, the day is 12h long, the evening is 4h and the night is 8h, with default values 07:00–19:00, 19:00–23:00 and 23:00–07:00 local time respectively. A year is the relevant year regarding the emission of sound and an average year should consider the average meteorological conditions over ten or more recent years. The above choice should be identical for all types of noise source.

When applying L_{den} the height of the assessment point depends on the application. For strategic noise mapping of public parks and relatively quiet areas in the open country it is 4.0 ± 0.2m above the ground, whereas for other purposes different heights may be chosen, with a minimum height of 1.5m above the ground.

Similar to L_{den}, in the 1970s the community noise equivalent level (CNEL) was introduced to take evening changes into account (Goldstein 1979).

2.2.5 Traffic noise index

Traffic noise index (TNI) is based on A-weighted sound levels statistically sampled over a 24h day. It depends on fluctuations in noise level over time and the background noise. It is assumed that the former is more important in traffic noise annoyance, and, thus, a heavier weighting factor is given:

$$TNI = 4(L_{10} - L_{90}) + L_{90} - 30 \tag{2.11}$$

where the constant 30 is used to yield a convenient range of numbers (Schultz 1982).

2.2.6 Noise pollution level

Noise pollution level (NPL), L_{NP}, is another noise descriptor that has been found to correlate well with human responses to all types of noise sources (Robinson 1971). It is expressed as the sum of two components:

$$L_{NP} = L_{eq} + 2.56\sigma \tag{2.12}$$

2.2.7 Corrected noise level

Corrected noise level (CNL) is the A-weighted SPL which takes into account any distinguishable characteristics of the noise to be evaluated, such as those specified in British Standard BS4142 (BSI 1997; see Section 2.3.4).

2.2.8 Effective perceived noise level

Effective PNL, L_{EPN}, is the adjusted PNL L_{PN} (see Section 1.2.2), taking into account the presence of pure tones or discrete frequencies and the duration of event. It has mainly been used for aircraft noise evaluation.

2.2.9 Sound exposure level

Sound exposure level, SEL or L_{SE}, is used to quantify short duration noise events such as aircraft flyover, impulsive or impact noise, or single vehicle pass-by, which are often of high intensity, abrupt onset, rapid decay, and cause rapid changes in spectral composition. It is the sound level that if maintained constant for 1s contains the same acoustic energy as a varying noise level:

$$L_{SE} = 10\log \int_0^T \frac{p^2(t)}{p_0^2} dt \tag{2.13}$$

L_{SE} is expressed as L_{AE} when A-weighting is applied. It is noted that although there are different definitions for impulsive sounds, typically, the duration of each impulse is less than 1s. Comparing Equations (2.1) and (2.13), the relationship between L_{SE} and $L_{eq,T}$ can be derived as:

$$L_{eq,T} = L_{SE} - 10\log T \qquad (2.14)$$

2.2.10   Evaluation of aircraft noise

There are a number of indices for evaluating aircraft noise, including composite noise rating (CNR), noise exposure forecast (NEF), noise and number index (NNI), isopsophic index (N), weighted equivalent continuous PNL (WECPNL), mean annoyance level ($\bar{Q}$), noisiness index ($\overline{NI}$) and total noise load (B) (Goldstein 1979; Zhen 2000).

2.2.11   Evaluation of noise disturbance on speech communication

The articulation index (AI) is a method for rating potential communication degradation in the presence of noise (ANSI 1997). To determine the AI, the signal-to-noise (S/N) ratio in each of the 20 one-third octave bands from 200Hz to 5kHz is first measured. The 20 S/N ratios are then individually weighted by an importance function dependent on frequency. The weighted values are combined to give a single overall value – the AI. Further corrections include for reverberation time and for very high background levels. The AI uses a scale of 0 to 1. If the AI is 0 then there will be no understanding, whereas if the AI is 1 there will be complete intelligibility.

The speech transmission index (STI), and its simplified version, rapid speech transmission index (RASTI), are also commonly used. An attractive feature is that the effects of reverberation, ambient noise and the contribution of direct field, which are usually treated individually, are combined in a natural way in the single function (Houtgast and Steeneken 1973, 1985; Steeneken and Houtgast 1980). Similar to the AI, the STI and RASTI also use a scale of 0–1.

The speech interference level (SIL) measures the masking of speech by noise. It is the arithmetic average of the noise levels at several frequency bands relevant to speech. The four octave bands centred on 500Hz, 1kHz, 2kHz and 4kHz are conventionally used (ANSI 1977), although a preferred speech interference level (PSIL) is also used, considering 500Hz, 1kHz and 2kHz only.

2.2.12   Evaluation of noise for interior spaces

Noise criterion (NC) provides a single-number rating of the acceptability of background noise in interior spaces, taking into account perceived loudness or annoyance as well as speech interference (Beranek 1957). An NC value is obtained by plotting the octave band SPL onto the reference curves and determining the lowest curve that is nowhere exceeded by the plotted octave band levels. Revised curves were later developed, including preferred noise criterion (PNC) curves, which extend to a lower frequency and more emphasis is placed on low-frequency noise (Beranek et al. 1971), and balanced noise criterion curves (NCB) (Beranek 1989). Generally speaking, NC and the revised curves should only be used for continuous noises, and the noises should have spectra similar to the curves. Similar to NC, the room-noise criterion (RC) is also a single-number rating, used for assessing heating, ventilation and air-conditioning system (ANSI 1999b). The noise rating (NR) number, or N curves (Kosten and van Os 1962), are also commonly used.

2.3   Standards and regulations

Starting with a discussion regarding general principles and forms of legislation, this section then briefly reviews major and typical standards and regulations at international and national levels, describing the main issues to be considered and typical noise limits. Legislation on environmental noise is divided into two major categories, noise emission by-products such as cars, trucks, aircraft and industrial equipment, and allowable noise levels in the domestic environment. The main focus of this section is on the latter.

2.3.1   Principles and forms of legislation

The main factors in environmental noise legislations include adverse public health effects, annoyance of the residents in the neighbourhood and the risk management strategies of the legislatures. An adverse health effect of noise refers to any temporary or long-term deterioration in physical, psychological or social functioning that is associated with noise exposure. Other general issues to be considered include: population to be protected, type of parameters describing noise and their limits, applicable monitoring methodology and its quality assurance, enforcement procedures to achieve compliance with noise regulatory standards within a defined time frame, emission control measures and regulatory standards, immission standards, authorities responsible for enforcement, costs of compliance, resources commitment, international agreements, and other social, economic and cultural conditions (Berglund *et al.* 1999).

There are two typical approaches to assessing environmental noise impact; one approach is based on absolute sound levels, and the other on the increase of existing ambient sound levels due to a new or expanded development. The advantage of the former is that a noise ceiling is ultimately established, preventing a gradual increase of noise level. With this method, it is assumed that an acceptable minimum noise impact scenario can be achieved. The latter presumes that people are accustomed to the sound environment that presently exists, and if the change does not increase the existing sound level, people would not sense the change and thus would not be significantly impacted.

Regulatory standards have been set at different levels including municipal, regional, national and international. It is common that effect-oriented regulatory standards may be set as a long-term goal, while less stringent standards are adopted for the short term (WHO 1995; Gottlob 1995), balancing the human effects research data and financial and technological constraints.

2.3.2   WHO and ISO

The World Health Organization (WHO) has intensively addressed the problem of community noise. Key issues of noise management include abatement options, models for predicting and assessing source control action, setting noise emission standards for existing and planned sources, noise exposure assessment, and testing the compliance of noise exposure with noise immission standards. In 1992, the WHO Regional Office for Europe convened a task force meeting which set up health-based guidelines for community noise (Berglund *et al.* 1999), aimed at serving as the basis for deriving noise standards within a framework of noise management. Table 2.1 shows the guideline values for selected environments.

Table 2.1 WHO recommended guideline values for community noise in specific environments, data selected from Berglund *et al.* (1999).

Specific environment	Critical health effect(s)	L_{Aeq} (dB)	Time-base (h)	L_{Amax} (dB) (fast)
Outdoor living area	Serious annoyance, daytime and evening	55	16	—
	Moderate annoyance, daytime and evening	50	16	—
Dwelling, indoors	Speech intelligibility and moderate annoyance, daytime and evening	35	16	—
Inside bedrooms	Sleep disturbance, nighttime	30	8	45
Outside bedrooms	Sleep disturbance, window open (outdoor values)	45	8	60
Hospitals, wardrooms, indoors	Sleep disturbance, nighttime	30	8	40
	Sleep disturbance, daytime and evening	30	16	—
Industrial, commercial, shopping and traffic areas, indoors and outdoors	Hearing impairment	70	24	110
Ceremonies, festivals and entertainment events	Hearing impairment (patrons: <5 times/year)	100	4	110

The international standard ISO 1996 (ISO 2003a) is a major international standard concerning the description and measurement of environmental noise. It contains three parts, namely basic quantities and procedures, acquisition of data pertinent to land use, and application to noise limits. It is noted that although the third part of the standard lays down guidelines for the ways in which noise limits should be specified and describes procedures to be used for checking compliance with such limits, no specific noise limits are given. Instead, it is assumed that noise limits are established by local authorities according to these guidelines.

2.3.3 European Commission

The EU Green Paper (EU 1996) aims to stimulate public discussion on the future approach to noise policy. It reviews the overall noise situation and introduces a framework for action including harmonising the methods for assessing noise exposure, encouraging the exchange of information among member states, establishing plans to reduce road traffic noise by applying newer technologies and fiscal instruments, paying more attention to railway noise in view of the future extension of rail networks, introducing more stringent regulation on air transport and using economic instruments to encourage compliance, and simplifying the existing seven regulations on outdoor equipment.

Following the Green Paper, a Directive of the European Parliament and the Council relating to the assessment and management of environmental noise was developed (EU 2002), with the aim of establishing a common EU framework. First, the framework seeks to harmonise noise indicators and assessment methods for environmental noise. Noise from different sources has different dose–effect relations and can thus be defined as different pollutants.

Second, using these common indicators and assessment methods, it seeks to gather noise exposure information in the form of noise maps. Third, it aims to make such information available to the public. The exposure information will form the basis for action plans at the local level. Equally, it will form the basis for goal setting for improvement at the EU level and for the development of an EU strategy including measures. However, it does not seek to set common Europe-wide noise limits.

2.3.4   United Kingdom

Planning Policy Guidance Note 24 (PPG24; ODPM 1994) (which is currently under review and will become PPS24) gives advice to local planning authorities in England on how the planning system can help minimise the impact of noise. It outlines the considerations to be taken into account in determining planning applications for both noise-sensitive developments and those activities which will generate noise, introduces the concept of noise exposure categories (NEC) for residential development, encourages their use and recommends appropriate levels for exposure to different sources of noise, and advises on the use of conditions to minimise the impact of noise. The Planning Advice Note (PAN) 56 is the Scottish equivalent of PPG24 (The Scottish Office 1999).

In PPG24 the NEC ranges from A to D, where Category A represents the circumstances in which noise is unlikely to be a determining factor, Category D relates to the situation in which development should normally be refused, and Categories B and C deal with situations where noise mitigation measures may make development acceptable. In Category B noise should be taken into account when determining planning applications and, where appropriate, conditions are imposed to ensure an adequate level of protection against noise. In Category C planning permission should not normally be granted, but where it is considered that permission should be given, for example, because there are no alternative quieter sites available, conditions should be imposed to ensure a commensurate level of protection against noise.

Table 2.2 shows the noise limits given in PPG24 corresponding to the noise exposure categories for new dwellings. It is noted that in terms of the nighttime noise levels, sites where individual noise events regularly exceed $L_{A,max}$ = 82dB several times in any hour should be treated as being in NEC C, regardless of the $L_{Aeq,8h}$. The levels for mixed sources refer to any combination of road, rail, air and industrial noise sources, and they are based on the lowest

Table 2.2 Recommended limits in PPG24 for various noise exposure categories for new dwellings near existing noise sources in $L_{Aeq,T}$ (ODPM 1994).

Noise source	Time periods	NEC			
		A	B	C	D
Road traffic	07.00–23.00	<55	55–63	63–72	>72
	23.00–07.00	<45	45–57	57–66	>66
Rail traffic	07.00–23.00	<55	55–66	66–74	>74
	23.00–07.00	<45	45–59	59–66	>66
Air traffic	07.00–23.00	<57	57–66	66–72	>72
	23.00–07.00	<48	48–57	57–66	>66
Mixed sources	07.00–23.00	<55	55–63	63–72	>72
	23.00–07.00	<45	45–57	57–66	>66

numerical values of the single source limits in the table. These values should only be used where no individual noise source is dominant, and PPG24 provides a method to check whether any individual noise source is dominant. It is important to note that PPG24 suggests that in some cases it may be appropriate for local planning authorities to determine the range of noise levels which they wish to attribute to any or each of the NEC. For example, where there is a clear need for new residential development in an already noisy area some or all NEC might be increased by up to 3dBA above the recommended levels. In other cases, a reduction of up to 3dBA may be justified.

BS4142 (BSI 1997) describes a method of determining the level of a noise of an industrial nature, together with procedures for assessing whether the noise in question is likely to give rise to complaints from people living in the vicinity. The standard was first published in 1967, and has been revised several times since. The assessment is based on the margin by which it exceeds a background noise level with an appropriate allowance for the acoustic features present in the noise. As this margin increases, so does the likelihood of complaint. The standard is intended to be used for assessing the measured or calculated noise levels, from existing or new/modified premises.

It is stated in BS4142 that a difference of around +10dB or more indicates that complaints are likely. A difference of around +5dB is of marginal significance. If the rating level is more than 10dB below the measured background noise level then this is a positive indication that complaints are unlikely. It is also assumed that certain acoustic features can increase the likelihood of complaint over that expected from a simple comparison between the specific noise level and the background noise level. Where present at the assessment location, such features are taken into account by adding 5dB to the specific noise level to obtain the rating level. This 5dB correction should be applied if one or more of the following features occur, or are expected to be present for new or modified noise sources: the noise contains a distinguishable, discrete, continuous note (whine, hiss, screech, hum, etc.); the noise contains a distinct impulse (bangs, clicks, clatters, or thumps); and the noise is irregular enough to attract attention.

In BS4142 it is indicated that the likelihood that an individual will complain depends on individual attitudes and perceptions in addition to the noise levels and acoustic features present. However, the standard makes no recommendations with respect to the extent to which individual attitudes should be taken into account in any particular case. Moreover, although generally there will be a relationship between the incidence of complaints and the level of general community annoyance, quantitative assessment of the latter is not within the scope of BS4142, nor is the assessment of nuisance.

There is a tendency in the United Kingdom to develop more specific regulations for various types of buildings, from the noise source viewpoint, such as for pubs and clubs (Davies *et al.* 2005), and from the receiver viewpoint, such as for schools (UK DfES 2003).

There are also a number of other UK regulations and legislations relating to noise, including Part III of the Environmental Protection Act 1990, Part III of the Control of Pollution Act 1974, Noise Act 1996, and Licensing Act 2003. Issues such as the investigation of complaints, warning notices, approval of measuring devices, evidence, and penalty notices are included.

2.3.5 Other European countries

Although there is a tendency of convergence, there are still wide differences in standards and regulations between Member States of the EU (Kang *et al.* 2001a). In some countries, such as Belgium and Spain, different regions have their own noise regulations.

The zone categories in terms of land use, for which noise limits vary, are different in different countries. For example, in Belgium there are nine zone categories, more than most other countries. Even for the same number of zone categories, the definitions for various zones vary in different countries.

The definitions of day, night and evening, for specifying noise limits, also vary between countries (Porter *et al.* 1998; Kang *et al.* 2001a). For example, in Italy the day period is 06:00–22:00 and the night period is 22:00–06:00. In Portugal there are three reference periods, day period 07:00–20:00, intermediate period 20:00–24:00, and night period 00:00–07:00. In the Netherlands, day, evening and night are 07:00–19:00, 19:00–23:00 and 23:00–7:00, respectively. In Norway, for Sundays and holidays the evening limits apply for 06:00–22:00.

In terms of noise limit values there are also considerable differences between various countries, although L_{Aeq} is commonly adopted as the evaluation index. In Italy there are three sets of limits, as shown in Table 2.3, including: emission limits – for fixed or moving sound sources not regulated by either standards or legislations, measured at the noisiest receiver position; immission limits – adding the contribution of all the noise sources at the receiver position; and quality targets (Porter *et al.* 1998; Kang *et al.* 2001a). In the Netherlands, an index called $L_{corrected,24h}$ has been used, which is the maximum of L_{day}, $L_{evening}$ + 5dBA, and L_{night} + 10dBA. In Germany, according to TA Lärm (Bundeskabinett 1998), if with the additional acoustic power, the level is less than recommended limit values (RV) -6dBA, then if the combined level is ≤ RV+1dBA, permission is given. Otherwise permission is given only if reduction measures are implemented within three years, but not if the combined level from one emission group is ≥ RV+5dBA.

Particular characteristics of noise are considered differently in various countries (Porter *et al.* 1998). In Portugal, the difference between the background noise represented by L_{95} and the L_{eq} originated by industrial, commercial and service buildings, after undergoing appropriate corrections, may not exceed 10dBA in any given reference period, whereas in Norway, L_{max} must not exceed L_{eq} by more than 10dBA. In Belgium, for noncontinuous noise sources, if the operation time is less than 10 per cent, the limit value can be 15–20dBA higher than that for continuous noise sources. In Switzerland, an adjustment factor, K, is applied, where for industrial noise, $K = K_1 + K_2 + K_3$, with K_1 = 0–10dB considering type of installation, K_2 = 0–6dB for tonality, and K_3 = 0–6dB for impulsiveness. In Norway, with tonal components and impulses the noise limits are corrected by 5dBA.

Table 2.3 Emission limits, immission limits and quality targets in L_{Aeq} in Italy, data adopted from Porter *et al.* (1998) and Kang *et al.* (2001a).

Category of land use	Daytime (06:00–22:00)			Nighttime (22:00–06:00)		
	Emission	Immission	Quality targets	Emission	Immission	Quality targets
I: Noise sensitive premises	45	50	47	35	40	37
II: Residential areas	50	55	52	40	45	42
III: Mixed areas	55	60	57	45	50	47
IV: Intense activity areas	60	65	62	50	55	52
V: Industrial areas and low density of residential buildings	65	70	67	55	60	57
VI: Industrial areas only	65	70	70	65	70	70

Legislation regarding complaints is also an important aspect of environmental noise. In Ireland, for example, given that continual noise from other houses, home workshops and local businesses can be a source of nuisance and distress for people, according to the regulations on neighbourhood noise under the Environmental Protection Agency Act 1992, any individual person, or local authority, may complain to a District Court seeking an Order to deal with the noise nuisance, that is, noise so loud, so continuous, so repeated, of such pitch or duration or occurring at such times that it gives a person reasonable cause for annoyance. First a complainant must give notice to the person making the noise of the intention to make a formal complaint to the District Court and then he/she must serve a notice on the alleged offender that a complaint is being made at least 7 days in advance of the complaint being made to the Court.

2.3.6   Other countries

In the United States of America, the National Environmental Policy Act, which is regarded as a major breakthrough in environmental noise policy, was adopted in 1969. The Noise Control Act was implemented in 1972, and the US Environmental Protection Agency (EPA) subsequently published the Levels Document (US EPA 1974), addressing issues including noise descriptions, human effects resulting from noise exposure, and noise exposure criteria. This has then been supplemented by additional Public Laws, Presidential Executive Orders, and many-tiered noise exposure guidelines, regulations and standards (US FICUN 1980; US EPA 1982; US NRC 1977; Finegold *et al.* 1998; ANSI 2003b). Several major US federal agencies, including the US EPA, the Department of Transportation, the Federal Aviation Administration, the Department of Housing and Urban Development, the National Aeronautics and Space Administration, the Department of Defence, the Federal Highways Agency, and the Federal Interagency Committee on Noise have all published important documents addressing environmental noise and its effects on people.

The DNL was established as the predominant sound descriptor. However, there has been a growing debate about whether or not to continue to rely on the use of DNL, or to supplement it with other noise descriptors. Notably, sound exposure has been introduced. The US EPA Levels Document does not establish regulatory goals, but DNL 55dBA has been selected as that required to totally protect against outdoor activity interference and annoyance.

Noise legislations in the United States of America vary between the states. For example, in New Jersey the peak limit is SPL 80dBA for both day and night, whereas in Connecticut the residential limit measured at property line of source is 80dB for night period and 100dB for day period in terms of unweighted peak, with exemptions including natural phenomena, humans, animals, religious bells/chimes, emergency vehicles, backup alarms, farm equipment, lawn/snow removal equipment, construction and blasting (Kang *et al.* 2001a).

In China, limits for environmental noise in urban areas are given at a national level (SAC 1993), as shown in Table 2.4. It is also specified that single, sudden noises during the night are not allowed to exceed standard values by more than 15dBA. The division of various zones as well as the definition of day and night periods are given by local authorities.

In Japan, noise standards for both general and roadside areas were first set in 1967, through the Basic Law for Environmental Pollution, considering the type of land use and the time of day. In Argentina, the Ordinances consider two types of noise, unnecessary and excessive. The

Table 2.4 Limits in L_{Aeq} for environmental noise in urban areas in China, data from SAC (1993).

Area classification	Day	Night
Special residential areas	50	40
Residential and cultural/education areas	55	45
Mixture of residential, commercial and industrial areas	60	50
Industrial areas	65	55
Arterial roads	70	55

former are forbidden, whereas the latter are classified according to neighbouring activities and limited by maximum levels allowed for daytime (07:00–22:00) and nighttime (22:00–07:00).

Given that existing noise policies and regulations vary considerably across countries and regions, and in many developing countries noise control has not been a first-order priority and only qualitative guidelines are given, it seems that local policies will play an important role for a considerable time period (Dickinson 1993), although noise policies and standards at a global level will ensure that the world population gains the maximum health benefits.

2.4   Urban noise climate

In many developing countries noise pollution is becoming considerably worse, due to various reasons including the lack of regulations and public awareness as well as the difficulty of getting access to noise control. In developed countries, relevant regulations and control measures may diminish the number of people exposed to the very high community noise levels (say, $L_{eq} > 70$dBA), whereas it seems that the number exposed to moderately high levels (say, $L_{eq} = 55$–65dBA) generally continues to increase (Stanners and Bordeau 1995). On the one hand, mitigation efforts such as developing quieter vehicles, introducing noise emission standards for vehicles, moving people to less noise-exposed areas, improving traffic systems, and direct noise abatements including sound insulation and barriers, are helpful for environmental noise reduction. On the other hand, a number of trends are expected to increase environmental noise pollution, including the expanding use of increasingly powerful sources of noise; the wider geographical dispersion of noise sources, together with greater individual mobility and spread of leisure activities; the increasing invasion of noise, particularly into the early morning, evenings and weekends; the increase of roads, traffic, driving speed and distance driven; and the increasing public expectations that are linked to increases in incomes and education levels (OECD 1991, 1995; Skinner and Grimwood 2005). In this section, noise climates in some countries are briefly reviewed.

2.4.1   United Kingdom

During 1990–1991 the UK Building Research Establishment (BRE) undertook the first of a planned series of national noise incidence and attitude surveys, and during 1999–2001 the studies were repeated (Skinner and Grimwood 2005). The latter involved a total of 1,160 24h noise measurements at a sample of dwellings in the United Kingdom, and more than 5,500 in-depth interviews with a sample of the UK adult population. The measurements were made with the microphone positioned at a height of 1.2m above the ground or above floor level of the dwelling, and at a distance of 1m from the front façade of the dwelling. The measurements represent typical weekdays.

According to the 1999–2001 noise measurement, the proportions of the UK population living in dwellings exposed to $L_{den} < 55$, 55–60, 60–65 and 65–70dBA were 33, 38, 16 and 13 per cent respectively. In daytime (07:00–23:00), 54 ± 3 per cent of the UK population were exposed to noise exceeding $L_{eq,16h} = 55$dBA, the WHO recommended level for protecting the majority of people from being seriously annoyed; and this proportion was 67 ± 3 per cent in nighttime (23:00–07:00) considering $L_{eq,8h} = 45$dBA, below which people may sleep with bedroom windows open, according to the WHO guidelines (see Table 2.1).

According to the 1999–2001 social survey, 18 per cent of respondents reported noise as one of the top five from a list of 12 environmental problems that personally affected them. Overall, noise was ranked ninth in this list. It has been reported by 21 per cent of respondents that noise spoilt their home life to some extent, with 8 per cent reporting that their home life was spoilt either 'quite a lot' or 'totally'. The proportions of the respondents who heard noise from road traffic noise; neighbours and/or other people nearby; aircraft; and building, construction, demolition, renovation or road works were 84, 81, 71 and 49 per cent respectively, and the proportions of respondents who were bothered, annoyed or disturbed to some extent by those noise types were 40, 37, 20, and 15 per cent respectively. The evening (19:00–23:00) and nighttime (23:00–07:00) periods were the times when the greatest proportion of respondents reported being particularly bothered, annoyed or disturbed by most types of noise from neighbours and/or other people nearby.

A comparison between the 1990–1991 and 1999–2001 results based on the paired sites in England and Wales indicates that the daytime noise levels measured using L_{Aeq} and L_{A10} have decreased, whilst the nighttime levels measured using L_{A90} have increased. However, the changes have been within 1dBA. These patterns of change are compatible with a model of noise exposure in which the levels of individual events have decreased, but the frequency with which such events occur has increased. A further analysis suggests that the increase in night-time levels is mostly due to the levels experienced at the quietest dwellings in 1999–2000 being higher than those at the equivalent dwellings in 1990–1991.

Social surveys show that from 1990–1991 to 1999–2001 there had been an increase in the proportion of respondents who reported hearing road traffic (from 48 to 54 per cent), neighbours (19 to 25 per cent), and other people nearby (15 to 21 per cent). There had also been an increase in the proportion of people who reported being adversely affected by noise from neighbours and/or other people nearby (21 to 26 per cent).

Compared to the national average, the noise levels in Greater London were generally higher at all times of the day, evening and night, as shown in Figure 2.3. This difference was mainly attributable to higher levels at the quieter sites in Greater London. It is noted that due to the population based sampling used, all the interviews and measurement locations in Greater London were in outer London boroughs, where the population is greater, whereas the noise climate of central London is markedly different from that of outer London in terms of noise levels, noises heard and attitudes to noise (Skinner and Grimwood 2005).

2.4.2   Other countries

The environmental noise levels in Europe are often higher than the legislated limits, partly because noise legislation is not fully enforced, but more importantly, noise pollution is most commonly regulated only for new land use or for the development of transportation systems, whereas enlargements at existing localities may be approved even though noise limits or guideline values are already surpassed (Gottlob 1995).

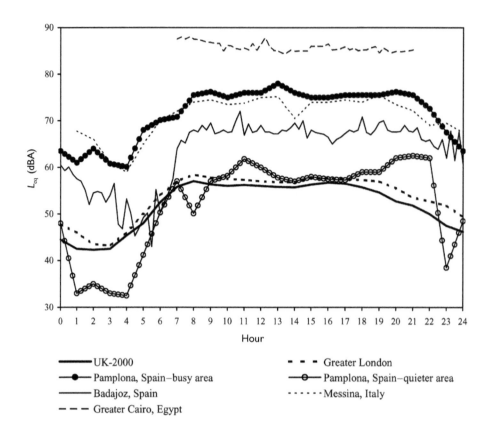

Figure 2.3 24-h time history of L_{Aeq} in the UK, Greater London (Skinner and Grimwood, 2005), Pamplona (Spain) (Arana and García, 1998); Badajoz (Spain) (Barrigón-Morillas *et al.*, 2005), Messina (Italy) (Piccol *et al.*, 2005), and Greater Cairo (Egypt) (Ali and Tamura, 2002). Note that since the measurement conditions were different, the comparison is mainly for the relative changes with time, rather than the absolute values.

Noise surveys were carried out in a number of medium-sized cities in Spain. In Cáceres, with 100,000 inhabitants, the noise level measured in a number of typical streets during working hours was more than 65dBA at 90 per cent of the measurement points (Barrigón *et al.* 2002). In Badajoz city, with 140,000 inhabitants, this value was 72.5 per cent (Barrigón *et al.* 2005). In Pamplona, among the 200,000 inhabitants, 31 per cent were very much annoyed, 23 per cent rather annoyed, and 21 per cent moderately annoyed by road traffic noise, and 91 per cent of the interviewees indicated that environmental noise was a very important factor in the quality of city life (Arana and García 1998).

According to a survey in Messina (Piccolo *et al.* 2005), a medium-sized ancient town in Italy with about 300,000 inhabitants, the daily average sound levels measured at 1m from façades due to road traffic exceeded the environmental standard by about 10dBA, and 25 per cent of the residents were highly disturbed by road traffic noise.

Table 2.5 shows the percentage of highly annoyed people aged 16+ in the Netherlands due to various types of noise, based on a series of surveys in 1987, 1993, 1998 and 2003 amongst 4,000

Table 2.5 Percentage of highly annoyed in the Dutch population, data adopted from Franssen et al. (2004).

Noise sources	1987	1993	1998	2003
Road traffic, urban >50 km/h	12	8	9	11
Road traffic, highways	4	2	2	2
Mopeds	18	15	20	24
Motorcycles	14	10	9	11
Motorcars	8	9	6	6
Trucks	14	11	9	10
Vans	7	6	3	4
Buses	5	5	3	3
Military aircraft	16	9	7	6
Helicopters	6	3	3	3
Stereo/tv/radio neighbours	8	6	10	8
Civil aircraft	5	3	4	4
Leisure activities	5	4	4	5
Rail traffic noise	1.3	1.4	1	1
Construction equipment	3	3	2	3
Industry	1.8	2.2	1	2

inhabitants (Franssen et al. 2004). It can be seen that the most annoying source of noise is road traffic, especially mopeds. Other most annoying noise sources include air traffic and neighbours.

In the United States of America, there have been several major attempts to describe broad environmental noise exposures (US EPA 1974; Galloway et al. 1974; Eldred 1975, 1988; Fidell 1978). A comprehensive plan for measuring community environmental noise and associated human responses was proposed over 30 years ago (Sutherland et al. 1973), although it seems that this has never been implemented in its entirety.

In Brazil, a survey in the city of Curitiba with 1,500,000 inhabitants showed that 93.3 per cent of the 100 evenly distributed measurement locations had $L_{eq} > 65$dBA, and 40.3 per cent had $L_{eq} > 75$dBA (Zannin et al. 2002). Particularly at night, sleep and rest were affected by transient noise signals from electronically amplified sounds, music and propaganda.

In China, according to the survey data in 1995 (China EPA 1995), 71.4 per cent of the kerbside noise level in cities of more than one million population was above 70dBA, and more than two-thirds of the residents in those cities were exposed to noise levels above the standards. The noise level has been increasing with the accelerating growth rate of motor vehicles. A large-scale survey in Beijing showed that the average kerbside L_{eq} was 75.6dBA (Li and Tao 2004). In terms of noise type, the contribution of road traffic, community, construction and industry to urban environmental noise were 61.2, 21.9, 10.1 and 6.8 per cent respectively.

In Japan, a study in Tokyo suggested that noise might be related to the health status of inhabitants living in areas with heavy road traffic (Yoshida et al. 1997). It was shown that a noise level of $L_{eq,24h} = 65$ or 70dBA was the critical point above which respondents indicated increased effects on health and the report of diseases increased. In Thailand, a general survey revealed that 21.4 per cent of the Bangkok population was suffering from sensory neural hearing loss (Berglund et al. 1999).

In Egypt, according to a survey in 2001 in Greater Cairo, 73.8 per cent of the respondents were highly or moderately irritated by road traffic noise (Ali and Tamura 2002). The L_{eq} at 1m from typical noisy façades ranged from 76.4 to 88.9dBA. An interesting study was then carried out regarding the effects of various restrictions (Ali and Tamura 2003). During a ban on horns, the noise level L_{eq} decreased by 9.4–10.8dBA in the downtown area, where there were no buses or commercial vehicles. In an industrial area, a decrease of 7.6dBA was achieved during a ban on horns together with limiting the commercial vehicles to 10 per cent, whereas in two residential areas, during a ban on horns, trucks and noisy buses, the L_{eq} decreased by 6–10.2dBA.

2.4.3   Comparison between surveys

Although a large number of urban noise surveys have been carried out, it is often difficult to make direct comparison between them because of the differences in data-collection methodology. Brown and Lam (1987) reviewed a number of survey procedures and suggested that the surveys can be categorised into four types: random sampling, sampling by land-use category, receptor-oriented sampling and source-oriented sampling. It was suggested that receptor-oriented surveys would offer the best opportunity for gathering noise-level data which can be generalised from site-specific information to the exposure of a population. The random sampling method is strongly dependent on the size of the grid, and variations caused by changes in source conditions are difficult to predict. With land-use-based sampling, comparisons between various cities are often difficult. In surveys using source-oriented sampling, kerbside noise measurements are typically made at locations selected either arbitrarily to represent different road and traffic conditions, or systematically, at equal distances along road ways. It would be inappropriate to compare such results with data obtained using other sampling methods.

In Figure 2.3 the 24h noise variation L_{Aeq} is compared among Greater London, UK average, Pamplona (Spain), Badajoz (Spain), Messina (Italy), and Greater Cairo. It can be seen that in daytime, the SPL is generally rather stable in all cases. The difference between day (07:00–23:00) and nighttime (23:00–07:00) is typically 8–12dBA, except in the quiet area of Pamplona, which is more than 17dBA.

Chapter 3

Urban soundscape

Although considerable efforts have been made in community noise control, recent research has shown that reducing the sound level does not necessarily lead to better acoustic comfort in urban areas (de Ruiter 2000, 2004; Schulte-Fortkamp 2001). For example, when the SPL is below a certain value, people's acoustic comfort evaluation is not related to the sound level, whereas the type of sound sources, the characteristics of users, and other factors play an important role (Ballas 1993; Gaver 1993; Maffiolo *et al.* 1997; Dubois 2000; Yang and Kang 2005a).

The focus of this chapter is soundscape and acoustic comfort, which concentrates on the way people consciously perceive their environment, and involves interdisciplinary efforts including physical, social, cultural, psychological and architectural aspects. Particular attention is paid to urban open spaces. Such spaces are important components in a city. However, almost all cities have some open spaces which are popular whilst others are not. Beside social and visual issues, it is vital to consider the environmental conditions of such places and how they could attract people to them. Recent studies on the soundscape of such spaces have shown that the acoustic environment plays an important role in the overall comfort (Raimbault *et al.* 2003; Kang 2004a).

The chapter starts with a brief review of general soundscape research (Section 3.1) and soundscape evaluation (Section 3.2). It then describes a series of soundscape studies based on field surveys in urban open public spaces in Europe (Sections 3.3–3.5) and in China (Section 3.6) considering acoustic comfort evaluations, sound preferences, as well as the effects of demographic factors, other physical conditions and cultural differences. The main factors that characterise the soundscape are studied using the semantic differential method (Section 3.7). A framework for soundscape description in urban open public spaces is then presented (Section 3.8), followed by an overall soundscape evaluation system using artificial neural networks (Section 3.9). The soundscape design is then systematically examined (Section 3.10). Finally, the acoustic comfort is discussed in a series of indoor spaces, which are natural extensions of the urban sound environment (Section 3.11).

3.1 Soundscape research

The pioneering research in soundscape was carried out by Schafer in the 1960s (Schafer 1977b). A musician and composer, Schafer's early soundscape work had always been about the relationship between the ear, human beings, sound environments and society. In the late 1960s and early 1970s, the World Soundscape Project was carried out. It grew out of Schafer's initial attempt to draw attention to the sonic environment through a course in noise pollution, as well as from his personal distaste for the more raucous aspects of Vancouver's rapidly changing soundscape. The way people perceive their environment consciously and the chance to change the orchestrating of the global soundscape were the focus of the project (Truax 1999). In 1975, Schafer (1977a) led a group on a European tour that included a research project that made detailed investigations of the soundscape of five villages in Sweden, Germany, Italy, France and Scotland. Recently, under the framework of a European research project, the five villages were revisited to undertake comparative studies (Järviluoma 2000).

The World Forum for Acoustic Ecology (WFAE) was founded in 1993, with members who share a common concern about the state of the world soundscape as an ecologically balanced entity, and who represent an interdisciplinary spectrum in the study of the scientific, social and cultural aspects of the natural and human-made sound environment. Its journal, *Soundscape – The Journal of Acoustic Ecology*, was founded in 2000. A number of national societies for acoustic ecology have also been established (Hiramatsu 1999).

Research in soundscape relates to many disciplines (Karlsson 2000), including acoustics, aesthetics, anthropology, architecture, ecology, ethnology, communication, design, human geography, information, landscape, law, linguistics, literature, media arts, medicine, musicology, noise control engineering, philosophy, pedagogics, psychology, political science, religious studies, sociology, technology and urban planning. Recording and listening to natural and human-made sounds is an important aspect of soundscape study (Westerkamp 2000). Linked to this is the environmental conservation of pleasant sounds. For example, in 1996, the Japan Environmental Agency carried out a project called 'one hundred soundscapes of Japan' (Fujimoto *et al.* 1998). Soundscape has also been explored from the sociological viewpoint. The social and cultural environment often shapes common rules of perception of sounds (Corbin 1998; Htouris 2001). In the field of literature, cultural and literary significance of acoustic imagination in intimate relationships between humans and the natural world has been studied (Yuki 2000). There have also been psycholinguistic and cognitive approaches to soundscape research (Dubois 2000; Guastavino and Cheminée 2004; Guastavino *et al.* 2005).

Another important aspect relating to soundscape study is the effect of the acoustic environment on health. In Sweden, a project has been carried out on soundscape support to health (Kihlman *et al.* 2001), where the soundscape refers to the sound variation in space and time caused by the topography of a built-up city and its different sound sources. Two aspects were considered, acoustic and perceived soundscapes, assessed by physical measuring instruments and perceptual scaling methods utilising persons, respectively. Moreover, the detrimental effects of the acoustic environment on human mental health have been investigated (Lercher and Widmann 2001), although it seems that the relationship between them is still undetermined, and it is difficult to give a simple rule. Other works relating to soundscape include sound systems, musical instruments, software for animation and sonic sculptures.

Soundscape study relates closely to the product sound quality for which considerable work has been carried out; examples include cars, construction machines, printers and trains. Psychoacoustic magnitudes, as discussed in Section 1.2.5, have been applied. To a certain degree, an architectural/urban space could be regarded as a product and, consequently, the methodology developed in the field of sound quality is of great importance. This is especially relevant in the field of soundscape reproduction in laboratory conditions (Guastavino and Katz 2004; Guastavino *et al.* 2005).

3.2   Soundscape evaluation

The evaluation of soundscape is rather complicated, involving interactions between various sound sources and between acoustic and other factors. Further to the discussion on urban noise evaluation in Section 2.1, this section outlines some methodologies and results in soundscape evaluation, with particular attention on urban environment.

3.2.1   Sound

An important part of soundscape evaluation is to consider individual sounds. In a research by Southworth (1969), reactions of different population groups to soundscape were studied during a tour around Boston. The study evaluated the identity of the sounds and analysed their pleasantness. It was suggested that the pleasantness of a sound is much more complicated than its physical qualities. Generally speaking, sounds of low to middle frequency and intensity were preferred, but delight increased when sounds were novel, informative, responsive to personal action and culturally approved. It was concluded that the information contained in the sound, the context in which it is perceived and its level, are three aspects that influence people's evaluation of a city's soundscape.

Schafer (1977b) defined sounds as keynotes, foreground sounds and soundmarks. Keynotes are an analogy to music where a keynote identifies the fundamental tonality of a composition around which the music modulates. Foreground sounds, also termed sound signals, are intended to attract attention. Sounds that are particularly regarded by a community and its visitors are called soundmarks, an analogy to landmarks. Natural examples of soundmarks include geysers, waterfalls and wind traps, whereas cultural examples include distinctive bells and the sounds of traditional activities (Smith 2000). Although traffic has become a common feature of many cities, special soundmarks still exist. For example, a soundscape survey with a number of foreign residents in Fukuoka showed that there were considerable differences between the sounds they heard in Japan and in their home countries (Iwamiya and Yanagihara 1998).

A list of sounds in the surroundings was evaluated based on a survey in Japan (Tamura 1998). On the top of the list were the twittering of birds, murmuring of water, insects/frogs, waves, and wind chimes – 45–75 per cent of the subjects found these sounds favourable and 25–65 per cent found them neither favourable nor annoying. The last five sounds were motorbikes, idling engines, constructions, advertising cars and karaoke restaurants – 35–55 per cent of the subjects found these sounds annoying and 45–65 per cent found them neither favourable nor annoying.

It is noted that in addition to the type of sound, the loudness may also influence the categorisation/classification. A study on the relation between loudness and pleasantness shows that the pleasantness of stimuli at intermediate loudness levels is not influenced by its loudness, but for sound at relatively high loudness levels there is a good correlation between the two (Zeitler and Hellbrück 1999; Zwicker and Fastl 1999).

While the basic psychoacoustic magnitudes can be used to evaluate individual sounds (Keiper 1997), the complexity of the sound components in the urban environment should be considered. For example, it has been demonstrated that for sounds with multiple tonal components, the perceptual process is different from that for a single tonal component since the attention of the subjects is not automatically focused (Bodden and Heinrichs 2001). Moreover, the meaning of a sound may considerably influence the evaluation. In order to study this effect, a procedure was proposed to remove the meaning of a sound, but retain the other characteristics such as loudness (Fastl 2001). Furthermore, not only perceptual factors, but also cognitive factors such as memory play an important role in global loudness judgements. It has been demonstrated that the overall loudness is higher than the average of instantaneous loudness judgements (Hellbrück *et al.* 2001).

In an urban environment there are often different sound zones, and in each zone there might be a dominant sound. This is especially important for soundscape evaluation when this sound is related to the users' activities such as group dancing (Kang and Zhang 2005). Also, sounds that are far away, close-up or moving in juxtaposition to the users may provide different information and thus affect the evaluation. In sound quality research, it has been shown that psychoacoustic qualities are different between stable and pass-by sounds (Genuit 2001).

3.2.2 User

It is important to consider the sound sensitivity of individuals (Zimmer and Ellermeier 1999), as well as the meaning of sounds to individuals (Gifford 1996). Ellermeier *et al.* (2001) characterised individual noise sensitivity as a stable personality trait that captures attitudes towards a wide range of environmental sounds. In their research, a sample of 61 unselected listeners was subjected to a battery of psychoacoustic procedures ranging from threshold determinations to loudness scaling tasks. They found small, but systematic differences in participants' verbal loudness estimates, and in the rating of the unpleasantness of natural sounds. The results suggested that what is psychophysically tractable in the concept of noise sensitivity might primarily reflect attitudinal/evaluative rather than sensory components. In other words, there are predictor sound preferences, which affect people's judgement.

Moreira and Bryan (1972) suggested that individuals with high noise susceptibility might be personality types that show a great interest in and have sympathy with others, have a great awareness of their environment, and are likely to be intelligent and creative.

On the other hand, people's attitude could be affected by sounds. For example, it appears that loud noise reduces helping behaviour and induces a lack of sensitivity to others (Gifford 1996; Page 1997). Another closely related field is the psychological effects of sounds on people. For example, research has shown that sounds affect different genders in different ways, and boys and girls perform differently in noisy conditions (Gulian and Thomas 1986; Christie and Glickman 1980).

Demographic factors are important. In Porteous and Mastin's research (1985), on a six-scale rating towards the neighbourhood soundscape elements, more than 70 per cent of the responses had a standard deviation (STD) greater than 1.0 and 23 per cent clustered around 1.5. Whilst it has been shown that there is no correlation between noise sensitivity and demographic characteristics other than age (Weinstein 1978; Taylor 1984), Mehrabian's (1976) research has suggested that, in general, there is a slight tendency for women to be more sensitive than men. It has often been remarked, for example, that women are more emotional than men, meaning that they act with a greater arousal to obviously emotional situations, or that they are emotionally more sensitive to seemingly minute changes in the environment, changes that sometimes are not even perceived by men. There is some evidence to suggest that females generally have a higher arousal level than males, and can hence tolerate sensory deprivation situations better (Croome 1977).

Kariel (1980) examined the effects of sounds on outdoor recreation environments. A group of mountaineers and a group of campers in a developed campground were selected to evaluate some verbally described sounds. Although some difference was found between the two sample groups, the presence of sounds was more significant than the difference between the two groups. Kariel's result suggested that the sounds themselves have an impact that may exceed that of group differences.

Social factors may play an important role in soundscape evaluation (Kang *et al.* 2003b). The assessment of the sound quality of an urban area depends on how long people have been living there, how they define the area, and how much they have been involved in the social life in the area (Schulte-Fortkamp and Nitsch 1999). Expectation is another issue in soundscape evaluation. In fact, noise regulations are based on an assumption that people expect different noise environments depending on the different qualities of their living environment, although it has also been argued that there is no such expectation. Such expectations depend on many social and economic factors and are very difficult to predict, especially for a universal model (Botteldooren *et al.* 2001). Sound experience is also important (Bertoni *et al.* 1993), although a study seems to suggest that recent experience of negative events is not related to reaction (Job *et al.* 1999). Similarly, cultural differences may lead to rather different acoustic comfort evaluation and sound preferences (see Sections 2.1.2 and 3.4.2). In addition, some other special characteristics of the users should be considered. For example, people with stereos may have different sound evaluation from others (Bull 2000).

The idea and experience of an environment is a historically conditioned reflection of cultural life. As Schwartz (1995) argued, nothing quite so dramatic has happened with regard to noise. If there was no traffic noise, the soundscape in cities was filled with church bells, from every direction, day and night. The astonishing success of the nineteenth and early twentieth century campaign to limit the ringing of church bells is most relevant here, for church bells had grown neither louder nor more numerous since, say, the sixteenth century (Girdner 1897). However, church bells were silenced because they belonged to a constellation of sounds whose significance was in the process of being reconfigured. The last 150 years have been witness to a thorough redefinition of the nature of sound and the ambit of noise.

3.2.3 Space

In addition to evaluating sound sources, the acoustic effects of an urban environment should be evaluated. Reverberation is an important index for the acoustic environment in urban streets and squares. It has been demonstrated that with a constant SPL, noise annoyance is greater with a longer reverberation (Kang 1988). On the other hand, a suitable RT, say 1–2s, can make street music more enjoyable (Kang 2001). Depending on the usage of an urban open space, an appropriate reverberation might be determined, although the requirements are much less critical than those in auditoria and also, the concept and perception of reverberation in outdoor spaces may not be the same as that for enclosed spaces.

3.2.4 Interactions between acoustic and other physical conditions

The interaction between acoustic and other physical conditions is an important aspect of the soundscape evaluation in urban environment (Mudri and Lenard 2000). For example, if a place is very hot or very cold, the acoustic comfort could become less critical in the overall comfort evaluation. Of various physical conditions the aural–visual interactions have been intensively studied.

It is essential to understand the differences between aural and visual perception. Although sound and light are both wave phenomena, aural perception differs from visual perception in many ways (Porteous and Mastin 1985; Porteous 1996; Apfel 1998). First, sound is ubiquitous. Unlike visual space, which is sectorial, acoustic space is nonlocational, spherical and all-surrounding. Acoustic space has no obvious boundaries and tends to emphasise a space itself rather than objects in the space. Aural harmonization is temporal, whereas visual harmonization is spatial. Sounds, compared with things seen, are more transitory, more fluid, more unfocused, more lacking in context, less precise in terms of orientation and localization, and less capturable. Therefore, audition is a fairly passive sense. Sound provides dynamism and a sense of reality, helping people to get the sense of the progression of time and the scale of space. Moreover, compared to vision, sound perception is usually information-poor but emotion-rich. People are often moved by a piece of music, or soothed by certain natural sounds such as from water and leaves.

The interaction between aural and visual stimuli is therefore important. In a research by Carles *et al.* (1999), 36 combinations of sound and image of natural/semi-natural settings and urban green spaces were presented to 75 subjects. Affective response was measured in terms of pleasure, resulting in a rank of preferences running from natural to human-made sounds, with the nuance of a potential alert or alarm-raising component of the sound. It was suggested that there are two main functions of sound in the landscape, which provide information in addition to visual data. One function is related to the interpretation of the sound identified, like water, birdsong, voice, cars, and the other function is related to the abstract structure of sound information. In certain places with a distinct environmental identity, any acoustic disturbance can lead to a rapid deterioration in quality. Natural sounds, meanwhile, may improve the quality of built-up environments to a certain extent.

In Southworth's (1969) research, it was found that when aural and visual settings were coupled, attention to the visual form reduced the conscious perception of sound, and *vice versa*. The interactions between aural and visual perception, especially when the sounds are related to the scenes, give people a sense of involvement and lead to a more comfortable feeling.

A study under laboratory conditions with controlled aural and visual stimuli suggested that the visual parameter was a predominant variable as regards aural–visual interactions (Viollon 2003). All the visual information had different ways and different efficiencies in affecting the auditory judgement. The more urban the visual settings were, the more contaminated the auditory judgement was (Viollon *et al.* 2002; Carles *et al.* 1992). This auditory dependence with visual information was many sided: all the human sounds, involving either footsteps or voices, were not influenced, whereas all the nonhuman sounds, involving no human presence, were significantly influenced.

The aural–visual interaction was researched in gardens, and it was shown that a positive evaluation of the landscape reduces annoyance of the soundscapes whereas a negative evaluation of the landscape increases annoyance (Maffiolo *et al.* 1999). For most environmental sounds, including birdsong, cicada vocalisation, music, water flow, wind ring, frogs, barks, vehicles and waves, it was demonstrated experimentally that, good or moderate sights can enhance people's sense of favour (Tsai and Lai 2001).

The aural–visual interaction was also studied in the field of product sound quality. For noise in cars, it was demonstrated that the effect of visual image reduced the negative impression of sound quality and the amount was sometimes equivalent to 10dB reduction in SPL. Also, seat/floor/steering wheel vibrations strengthened the unpleasantness (Hashimoto and Hatano 2001). Similarly, a study on the sound quality evaluation of construction machines showed that the results obtained by presenting sound only were more unpleasant, more powerful and sharper than those obtained by presenting sound with scenery (Hatano *et al*. 2001).

The above results seem to suggest that the evaluation of soundscape in urban open spaces should not be conducted by audio recording and then laboratory listening tests. One alternative could be to evaluate the soundscape with simultaneously recorded video, but this still ignores some other factors, such as humidity and temperature, which could have similar effects to those of the visual factors. From this viewpoint, a more appropriate method would be to carry out field surveys.

3.3   Case studies in urban open public spaces in Europe

Given that the perception of an outdoor environment depends not only on the physical features, but also on the characteristics of the users, it is important to study their interactions. From summer 2001 to spring 2002, an intensive questionnaire survey and objective measurements on soundscapes were carried out for four seasons in 14 urban open public spaces of five European countries, namely Greece, Italy, United Kingdom, Germany and Switzerland. Seven cities – Alimos, Thessaloniki, Sesto San Giovanni, Sheffield, Cambridge, Kassel and Fribourg – were selected and for each city two urban open public spaces were chosen for the field survey. Among the 14 case study sites, there was a wide variation in climatic conditions and urban morphology. The soundscape study was carried out as part of an overall physical comfort investigation, including thermal, lighting and visual aspects (Kang *et al*. 2003b, 2004; Nikolopoulou *et al*. 2003; Steemers *et al*. 2003). This section outlines the methodology of this study, and the results are presented in Sections 3.4 and 3.5 (Yang and Kang 2005a, 2005b).

3.3.1   Sites

Table 3.1 shows the site plan, main functions, major sound sources and the number of interviews for each site. It can be seen that the 14 case study sites exhibited a wide variation. In terms of function, the sites included residential squares (for example, Kritis Square, Petazzi Square and Jardin de Perolles), cultural and tourism squares (for example, the Seashore of Alimos, Peace Gardens, Barkers Pool, All Saint's Garden, Silver Street Bridge and Florentiner Square), railway station squares (for example, Bahnhofsplatz and Place de la Gare), and multifunctional squares (for example, Karaiskaki Square, Makedonomahon Square and IV Novembre Square). In terms of soundscape, traffic noise appeared in all the case study sites, although in some squares it was the main sound source (for example, IV Novembre Square and Place de la Gare), whereas in other squares it could be regarded as the general background noise (for example, Peace Gardens and Jardin de Perolles). On the other hand, a number of sites were featured by their unique sound elements, for example, water sounds in the IV Novembre Square, Peace Gardens, Silver Street Bridge and Bahnhofsplatz, music in the Barkers Pool, church bells in the Petazzi Square, construction/demolition sounds in the Makedonomahon Square, Kritis Square and Peace Gardens. Users' activities were another source of sounds, such as footsteps, surrounding speech and children's shouting.

A more detailed study was carried out in the two Sheffield sites, namely the Peace Gardens and the Barkers Pool. Sheffield is located in the north of England, with a population of 0.65 million. In the hottest month of July, the average daily maximum temperature is about 21°C. In winter, the daily temperature seldom drops below zero. The city centre, where the two case study sites are located, is the commercial and official centre of the city. It is mainly a pedestrian area.

Table 3.1 Basic information of the case study sites.

Sites (survey areas in grey)			Main functions	Main sound sources	Number of interviews
Karaiskaki Square Alimos, Greece			Residential, commercial	Traffic, footsteps, surrounding speech, children	655
Seashore, Alimos, Greece			Tourism, recreation	Traffic, footsteps, surrounding speech, children	848
Makedonomahon Square Thessaloniki, Greece			Residential, office, relaxation	Traffic, construction, children	1037
Kritis Square Thessaloniki, Greece			Residential, relaxation	Traffic, construction, surrounding speech, children	777

IV Novembre Square, Sesto San Giovanni, Italy 	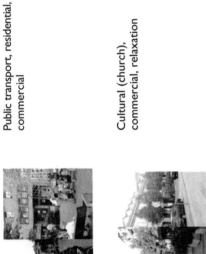	Public transport, residential, commercial	Traffic, surrounding speech, water (fountains)	574
Petazzi Square, Sesto San Giovanni, Italy		Cultural (church), commercial, relaxation	Traffic, church bell, surrounding speech, children	599
The Peace Gardens, Sheffield, UK		Recreation, cultural (historical buildings)	Water (fountains), demolition, children, surrounding speech, traffic	510
The Barkers Pool, Sheffield, UK		Commercial, cultural (music hall)	Traffic, footsteps, music, surrounding speech	499

Table 3.1 contd.

All Saint's Garden Cambridge, UK		Relaxation, commercial, cultural (historical buildings), residential	Traffic, surrounding speech, birds	459
Silver Street Bridge Cambridge, UK		Tourism, relaxation	Traffic, water (river), surrounding speech	489
Florentiner Square Kassel, Germany		Commercial, recreation	Traffic, footsteps, surrounding speech	406

 Bahnhofsplatz Kassel, Germany	Railway station square	Traffic, footsteps, surrounding speech, water (fountains)	418
 Jardin de Perolles Fribourg, Switzerland	Residential, recreation	Footsteps, surrounding speech, children, traffic	888
 Place de la Gare Fribourg, Switzerland	Railway station square	Traffic, footsteps, surrounding speech	1041

The Peace Gardens, as illustrated in Figure 3.1, was opened in 1998. Covering an area of about 3,000m², it is one of the most popular squares in Sheffield. There are large areas of grass for sitting. Benches are also provided around the grassy areas. The water features are the most attractive characteristic of the Peace Gardens, notably the 89 individual jets of the Goodwin Fountain at the centre, together with the Holberry Cascades, which are positioned on the top of the stone staircases that lead visitors from street level down into the garden. The square attracts hundreds of visitors and locals on a fine day to relax among the dramatic water features, intricate stone carvings and colourful flowers. Especially during lunchtime, people from surrounding offices and shops come to the square to have a rest. The Peace Gardens thus acts as a central focal point in the city centre. It

(a)

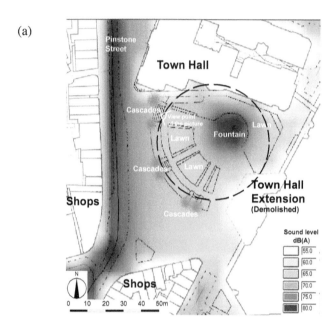

(b)

Figure 3.1 (a) Site plan of the Peace Gardens based on the EDINA Digmap. The grey scale in the plan corresponds to sound levels, and the dashed circle indicates where the interviews were conducted; (b) Perspective view.

affords an opportunity for adults to relax and children to play. Most of the buildings around this sunken square date from the turn of the nineteenth century. The background scene of the square is the old Town Hall, which is a Grade 1 listed building of outstanding architectural interest. On the west side of the Peace Gardens is a busy one-way road. Most of the vehicles passing through are buses. The concrete Town Hall Extension to the rear of the Peace Gardens was demolished in the autumn/winter survey period, causing a considerable change in soundscape, mainly from diggers' rumbling. In most of the survey period, all the fountains were operating, but in the autumn season, due to the preparation of the demolition work, the fountains were often not operating.

The Barkers Pool, as shown in Figure 3.2, is adjacent to the Peace Gardens. This rectangular square is shaped by the Sheffield City Hall and Cole Brothers (now John Lewis). The former is a

(a)

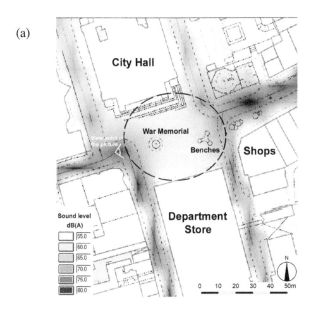

(b)

Figure 3.2 (a) Site plan of the Barkers Pool based on the EDINA Digmap. The grey scale in the plan corresponds to sound levels, and the dashed circle indicates where the interviews were conducted; (b) Perspective view.

1930s neo-classical styled building and the latter is one of the largest and finest department stores in Sheffield. An important design feature of this pedestrian square is the city's War Memorial, which helps to create a solemn and peaceful atmosphere. There are some benches in front of Cole Brothers, which are frequently used, especially by the customers. The large steps in front of the City Hall are also a popular sitting place. Music gives a special atmosphere in this square. During 35 per cent of the survey time, classic music from the City Hall, jazz music from the music store, or street music (that is, saxophone or country music) could be heard in the square. On a fine day, young people like to sit on the steps and enjoy the music played in the City Hall. On two sides of the square there is some low-density traffic. During 65 per cent of the survey period the main sound sources were users' talking, footsteps, skateboarding, wind and traffic.

In Figures 3.1a and 3.2a, the SPL distributions of the two case study sites are also shown. The SPL is calculated using noise mapping (see Chapter 5) software, Cadna/A (DataKustik 2005). The sound sources considered are traffic and fountains at the Peace Gardens, and traffic only at the Barkers Pool. Whilst after the survey periods there have been some new developments at the case study sites, the analyses in this chapter are based on the survey results only.

3.3.2   Questionnaire survey

In total 9,200 interviews were made over the 1 year survey period in the 14 urban open public spaces. For each site around 400–1,000 interviews were carried out with an identical questionnaire. The questionnaire was initially developed in English, and then translated into other languages by native speakers.

The questionnaire was not introduced as a soundscape survey, but as an enquiry relating to general environmental conditions including thermal, lighting, wind, humidity and visual environment. Such an integrative consideration of various factors is useful for avoiding any possibility of bias towards the acoustic aspect.

The interviewees were the users, not passers-by, of the squares, and were selected randomly. Demographic data and temporal status were also obtained through the questionnaire survey, either completed by the interviewees or recorded by the interviewers on a separate sheet. The interviewees were classified into five categories: student, working person, pensioner, housekeeper and others (for example, unemployed).

Interviewees were asked to evaluate the sound environment of the site and their homes. Five scales were used: 1, very quiet; 2, quiet; 3, neither quiet nor noisy; 4, noisy; and 5, very noisy. In order to make sure that the interviewees have a similar understanding of the linear scale, the subjective descriptions in the questionnaire were given together with the linear scale numbers.

The interviewees were also asked to classify at least three sounds as 'favourite' (F), 'neither favourite nor annoying' (N), or 'annoying' (A). The sounds were those frequently heard in a particular site, and were selected from a list of 15 typical sound sources in urban open public spaces.

A subjective evaluation of various physical indices was also carried out, including temperature (very cold, cold, neither cold nor warm, warm, very hot), sunshine (prefer more, ok, too much sunshine), brightness (very dark, dark, neither dark nor bright, bright, very bright), wind (static, little wind, ok, windy, too much wind), view (negative, neither negative nor positive, positive), and humidity (damp, ok, dry).

In the two squares in Sheffield, additional questions relating to soundscape were taken into account, including evaluation of acoustic comfort, identification of sounds and corresponding sound preference classification, as well as semi-structured questions related to the preferred relaxing sound environment. In the acoustic comfort evaluation, five linear scales similar to the one above were used: 1, very comfortable; 2, comfortable; 3, neither comfortable nor uncomfortable; 4, uncomfortable; and 5, very uncomfortable. Fifteen sounds were listed to be classified, which were most likely to be heard in the two squares, although during the interview these sounds were not necessarily heard.

Figure 3.3 shows the age distribution of the interviewees in the two squares in Sheffield. It can be seen that the age range is rather wide. In both squares around 50 per cent of the interviewees were young people aged 18–34. Because of the difficulty in filling in the questionnaire, the percentage of children surveyed was very low in both squares, at around 1 per cent. Therefore, the data for children are treated as missing values in the further analyses. The numbers of male and female interviewees were generally similar, but in the winter season the percentage of male users is about 60 per cent in both squares.

In the Peace Gardens and the Barkers Pool 74.5 and 81.6 per cent of the interviewees were local people, respectively. The Peace Gardens attracts more tourists – 19.9 per cent of the nonlocal users were from overseas. The education level of the users was generally high, with about 40–50 per cent at university level. In terms of professions, the distributions of various categories in both squares are rather similar. Overall 43 and 35 per cent of the interviewees were students and working persons respectively. This is because the two sites are located in the business centre of Sheffield and the two universities in Sheffield are nearby. Of those

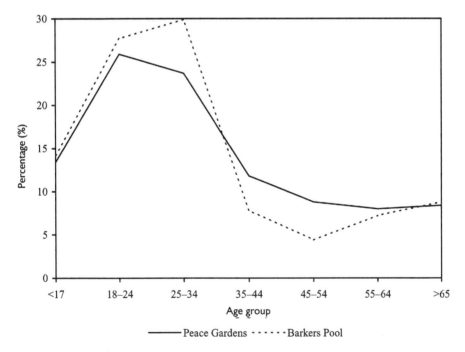

Figure 3.3 Number of interviewees in various age groups in the four-season field survey in Sheffield.

interviewed 12 per cent were pensioners, 3 per cent housekeepers and 7 per cent others. It was also observed that many students came in the afternoon after their classes, whereas most pensioners came in the morning.

3.3.3   Measurements

A one-minute L_{eq} was measured for each interview, either when the interviewee filled in the questionnaire silently, or immediately after the interview. Correspondingly, statistical indices including, $L_{eq,90}$, $L_{eq,50}$ and $L_{eq,10}$ were calculated, based on the measurements for all interviews.

To give an overall picture of the sound distribution at a site, the SPL was measured at five typical positions for a period of 20min each at several periods during the day, and additionally in the vicinity of the sound sensitive receiver(s)/location(s).

Reverberation measurements were made at selected sites. The procedure was to break a balloon or use a piston to generate an impulse, and at the same time record the process of sound decay. A number of source and receiver positions were considered at each site.

Typical sounds were recorded and some psychoacoustic indices were analysed using 01dB software (01dB 2005), including loudness, sharpness and roughness (see Section 1.2.5). It was shown that the sounds in the case study sites represented a reasonably wide range of values in those indices.

As well as acoustic measurement, a weather station was used to record the microclimate data in a 30s interval during the survey. It is noted that although seasonal variation has been a main consideration in the overall comfort evaluation, generally speaking, no significant difference has been found between different seasons in terms of acoustic evaluation. Therefore, in the following sections no separate analysis for different seasons has been presented, except where indicated.

Software SPSS (Field 2000) version 12.0 was used to establish a database with all subjective and objective results.

3.4   Acoustic comfort

3.4.1   Identified sounds

The interviewees were asked to describe up to three sounds they heard in the square during the interview period. The results in the Peace Gardens and the Barkers Pool are illustrated in Figures 3.4a and 3.4b respectively, where those sounds that were mentioned less than ten times are not included. In the Peace Gardens, the soundmark, namely the water sound from the fountains, was heard at the highest frequency and was most often mentioned as the first-noticed sound. The foreground sounds, from the demolition work, digger machines and lawn mowers, also gave a high level of awareness from the interviewees. The keynote sounds of the Peace Gardens showed a wide variation, including traffic, human and natural sounds, which were more likely to be noticed secondly or thirdly.

Compared to the Peace Gardens, the soundscape in the Barkers Pool was more fragmented. The relatively low level of traffic noise gave people more chances to hear other sounds, such as surrounding speech, footsteps, wind, hawkers shouting, skateboarders playing, birds singing and leaves rustling. Music from the buildings and streets was always played quite smoothly and can just be described as noticeable. It is interesting to note that the music sounds were more frequently noticed in the first instance, as a dramatic soundmark, although in terms of sound level, they could not mask the other sounds.

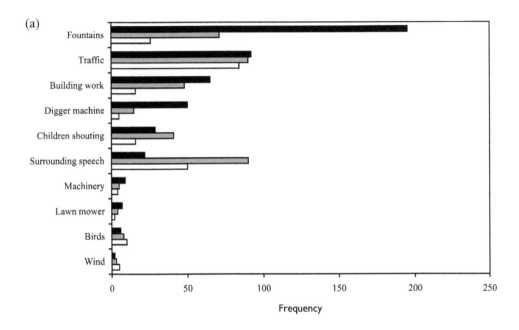

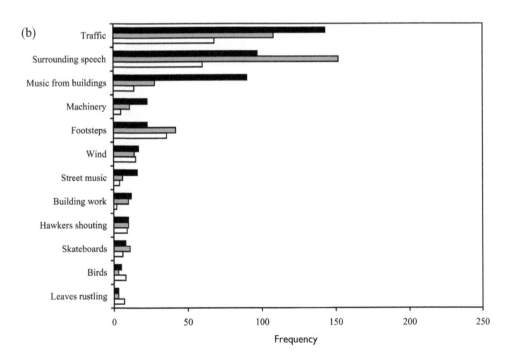

Figure 3.4 ■ First-noticed sound ▦ Second-noticed sound □ Third-noticed sound
Main sounds identified by the interviewees in (a) the Peace Gardens and (b) the Barkers Pool.

The way that people list their heard sounds provides an insight into how they perceive sounds in urban squares. It isn't just researchers who are keen to discover the significant features of the soundscape, but ordinary people also have the instinct to distinguish what researchers define as keynote sounds, foreground sounds and soundmarks.

In order to check the effects of demographic variables on the frequency of noticed sounds, Pearson's chi-square test was carried out, showing that there were no significant differences between males and females, different age groups, local and nonlocal people, as well as different professions. This suggests that people in different demographic groups have a similar ability to notice the sounds in the urban squares. However, whether they have a similar way to evaluate these sounds is a different issue, which will be analysed below (Yang and Kang 2005a; Kang *et al.* 2003b).

3.4.2 Evaluation of sound level

Long-term environmental experience and cultural background

The measured sound levels at each site are shown in Table 3.2, including the mean and STD of the one-minute L_{eq} values, as well as their statistical levels $L_{eq,90}$, $L_{eq,50}$ and $L_{eq,10}$. It can be seen that the Jardin de Perolles in Fribourg is the quietest, with a mean L_{eq} of 55.9dBA. The acoustic environment is also rather stable, with an STD of 4dBA. The noisiest sites are the Makedonomahon Square in Thessaloniki and the IV Novembre Square in Sesto San Giovanni. The mean is L_{eq} over 69.1dBA and the STD is over 5.2dBA.

The mean subjective evaluations of sound level are also shown. It is interesting to note that in terms of the subjective evaluation, rather than the Jardin de Perolles, the Petazzi Square is the quietest site, although the mean L_{eq} of the latter, 66.2dBA, is much higher.

One possible reason for the difference is the influence of the long-term sound environment. In Table 3.2 a comparison is also made amongst different cities in terms of the subjective evaluation of the sound environment at interviewees' homes. Significant differences have been found amongst various cities ($p < 0.001$). Kassel has the quietest home environment, whereas Alimos has the noisiest. Comparing the survey results of Alimos and Kassel, namely between the Karaiskaki Square and the Florentiner Square, and between the Seashore and the Bahnhofsplatz, it is noted that with a similar mean L_{eq}, the mean score of evaluation in Kassel is much higher than that in Alimos. It is possible that people from a noisy home environment adapt more to noisy urban open public spaces. Another possible reason for the difference between the two cities is cultural and lifestyle differences. Probably people in Germany are more aware and/or less tolerant of urban noises, whereas people from warm climates where windows have to be open learn to be tolerant. Due to possible influences of environmental and cultural factors as well as personal preferences (see Section 3.5), it may not be appropriate to directly compare the results of different cities/countries.

Background sound level

Figure 3.5 shows the relationships between the sound level and the subjective evaluation of sound levels in 12 sites, with linear regressions and correlation coefficients R. In the figure each symbol represents the average of the subjective evaluations at a one-dBA scale. To minimise personal bias, the dB scales which have less than ten responses are not included in this figure (and the other figures in this chapter).

Table 3.2 Measured sound levels, subjective evaluation of sound levels on the site, and subjective evaluation of the sound environment at users' home.

		Alimos		Thessaloniki		Sesto San Giovanni		Sheffield		Cambridge		Kassel		Fribourg	
		Kara. Square	Seashore Square	Maked. Square	Kritis Square	Petazzi Square	IV Nov. Square	Peace Gardens	Barkers Pool	All St. Garden	Silver St. Bridge	Flor. Square	Bahns.-platz	Place de la Gare	Jardin de Perolles
L_{eq}, (dBA)	Mean	62.8	64.4	69.3	66	66.2	69.1	67.4	60.2	—	—	61.3	64.7	67.9	55.9
	STD	3.9	3.6	6	7.9	4.8	5.2	6.3	3.4	—	—	4	4.2	2.9	4
	$L_{eq,90}$	57.7	60.5	63.5	57.4	60.7	65	57.9	56.5	—	—	57.1	60.1	63.9	51.2
	$L_{eq,50}$	62.8	64.1	68.2	64.1	66.2	68.4	68.5	59.9	—	—	60.7	63.7	67.9	55.5
	$L_{eq,10}$	67.8	68.5	76.4	78.9	72.1	76.5	74.5	63.6	—	—	65.8	70.1	71.2	61.1
Evaluation: site	Mean	2.7	2.79	3.85	2.79	2.46	3.25	3.4	2.92	2.48	3.2	3.18	3.42	3.59	2.49
	STD	0.86	0.95	0.89	0.89	0.92	0.99	0.92	0.76	0.83	0.88	0.62	0.64	0.83	0.8
Evaluation: home	Mean	2.65	2.65	2.49	2.29	2.50	2.43	2.50	2.55	2.55	2.32	2.18	2.10	2.13	2.30
	STD	1.05	1.04	0.98	0.77	1.12	1.14	0.98	1.10	1.27	1.16	0.84	0.79	1.03	1.02
	Mean	2.65		2.4		2.47		2.52		2.43		2.14		2.21	
	STD	1.04		0.9		1.13		1.04		1.22		0.82		1.03	

(a)

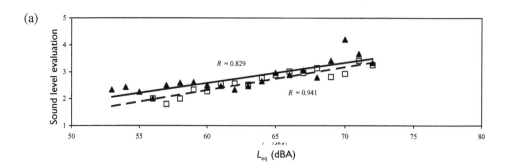

(b)

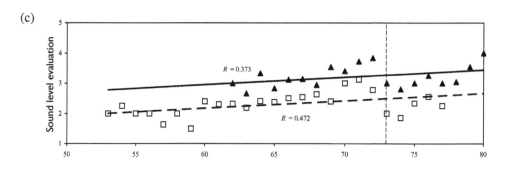

(c)

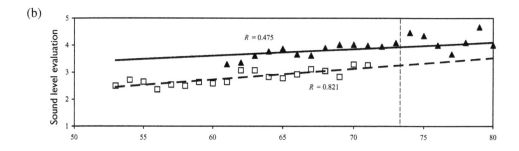

(d)

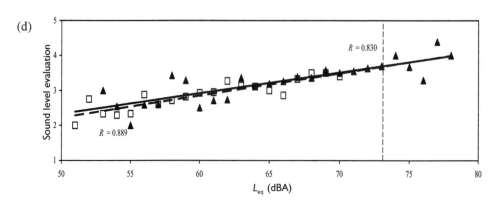

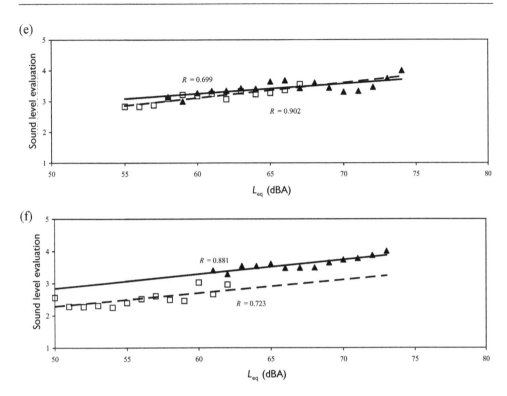

Figure 3.5 Relationships between the measured sound level and the mean subjective evaluation of sound level, with linear regressions and correlation coefficients *R*. (a) Alimos, Greece: ▲ and — Karaiskaki square; □ and - - - Seashore (b) Thessaloniki, Greece: ▲ and — Makedonomahon square; □ and - - - Kritis square   (c) Sesto San Giovanni, Italy: ▲ and — IV Novembre square; □ and - - - Petazzi square (d) Sheffield, UK: ▲ and   — Peace Gardens; □ and - - - Barkers Pool (e) Kassel, Germany: ▲ and — Bahnhofsplatz; □ and - - - Florentiner square (f) Fribourg, Switzerland: ▲ and — Place de la Gare; □ and - - - Jardin de Perolles

Corresponding to conventional understanding, there is generally a strong positive correlation between the sound level and the subjective evaluation ($p < 0.001$). With the increase of L_{eq}, the mean evaluation score also becomes higher. However, from Figure 3.5 it can be seen that although all the linear regressions have a similar tendency, their positions are rather different. This means that with a given sound level, the subjective evaluations are different. It is interesting to note that the differences exist not only between cities, which might be caused by the sound environment at home and by cultural differences, as discussed above, but also between two sites in the same city. For example, in Figure 3.5b, with an L_{eq} of approximately 70dBA, the mean subjective evaluation score is about 3 (neither quiet nor noisy) in the Kritis Square, and about 4 (noisy) in the Makedonomahon Square. The difference between the two squares might be caused by their different sound level ranges. In the Kritis Square, the L_{eq} varies from 53 to 71dBA, whereas in the Makedonomahon Square it ranges from 61 to 80dBA. The difference in $L_{eq,90}$ is 6.1dBA, as shown in Table 3.2. Therefore, the results suggest that with a lower overall sound level range and a lower $L_{eq,90}$, people may feel quieter at a given sound level. Similar situations can also be seen in Figures 3.5c and 3.5f.

The importance of $L_{eq,90}$ is also seen in Figure 3.5d. Although the Peace Gardens and the Barkers Pool have a 7.2dBA difference in the mean L_{eq}, with a similar $L_{eq,90}$, the linear regressions of the two squares are very close. In other words, with a given sound level, the subjects' evaluation scores are about the same in the two squares. Similar results can also be found in Figures 3.5a and 3.5e. Overall, although the L_{eq} over a time period has been widely adopted as a general purpose index for evaluating environmental noise (see Chapter 2), for urban open public spaces, the results in this study suggest that the background sound level is another essential index. A lower background sound level can make people feel quieter, even when the foreground sounds reach a rather high level.

Threshold

Another interesting phenomenon is that below a certain sound level, say 73dBA, there is generally a good correlation between L_{eq} and the subjective evaluation, but the correlation coefficient becomes rather low beyond this sound level. In Figures 3.5b and 3.5c, if only the range of $L_{eq} < 73$dBA is considered, the correlation coefficient increased from 0.475 to 0.879 ($p < 0.01$) in the Makedonomahon Square, from 0.472 to 0.803 ($p < 0.01$) in the Petazzi Square and from 0.373 to 0.802 ($p < 0.01$) in the IV Novembre Square. This suggests that when the sound level reaches a certain value, which is 73dBA on the basis of this study, subjects' evaluation varies significantly and becomes more unpredictable.

3.4.3 Evaluation of acoustic comfort

Figure 3.6 compares two relationships in the Peace Gardens and the Barkers Pool, between measured L_{eq} and the subjective evaluation of sound level, and between the measured L_{eq} and the acoustic comfort evaluation, with binominal regressions and the correlation coefficients squared R. There is again a strong positive correlation between the measured sound level and the subjective evaluation of sound level ($p < 0.01$). The R is 0.772 at the Peace Gardens and 0.795 at the Barkers Pool, indicating that the sound level variation accounts for 77.2 and 79.5 per cent of the variability in the sound level evaluation. However, it is interesting to note that the R between the measured sound level and the acoustic comfort evaluation is much lower, at only 0.541 at the Peace Gardens and 0.404 at the Barkers Pool. From Figure 3.6 it can be seen that the regression of the sound level evaluation is nearly linear, whereas the regression of acoustic comfort evaluation is curved. In particular, when the sound level is lower than a certain value, say 70dBA, there is no significant change in acoustic comfort evaluation with increasing L_{eq}, whereas the sound level evaluation changes continuously.

In addition to the differences between the evaluation of sound level and acoustic comfort, the results in Figure 3.6 also indicate people's tolerance in terms of the acoustic comfort in urban open public spaces. For example, as shown in Figure 3.6a, with a L_{eq} of 61dBA, people felt 'neither quiet nor noisy' in terms of the sound level, whereas they also evaluated the sound environment as 'comfortable'. When the L_{eq} became 76dBA, people felt that it was 'noisy', but they evaluated the sound environment as 'neither comfortable nor uncomfortable'. In the Peace Gardens, the average score is 3.68 for sound level and 2.50 for acoustic comfort, and in the Barkers Pool these values are 2.91 and 2.30 respectively. One important reason for these phenomena is the effect of sound source type. Another reason for the satisfaction in terms of acoustic comfort is that users can choose a location in a square according to their preferences and activities. In the Peace Gardens, for example, teenagers and parents of young children were mostly near the fountains, whereas older people were halfway between the fountains and traffic.

(a)

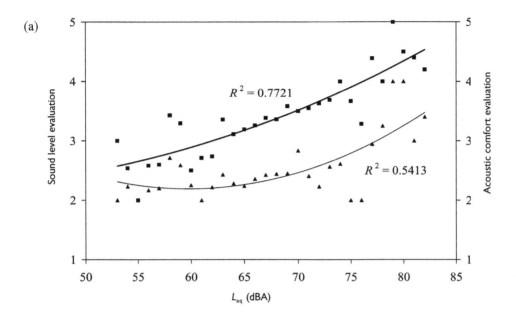

(b)

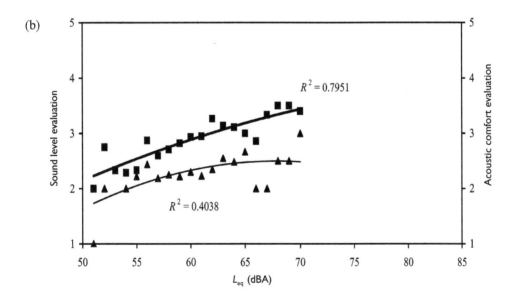

Figure 3.6 ■ and— Subjective evaluation of sound level
▲ and—Acoustic comfort evaluation
Relationships between measured L_{eq} and the subjective evaluation of sound level, and between the measured L_{eq} and the acoustic comfort evaluation, with binominal regressions and correlation coefficients squared R^2 at (a) the Peace Gardens; and (b) the Barkers Pool.

Corresponding to Figure 3.6, in Figure 3.7 a comparison is made between the percentage of people in each category of the sound level evaluation and acoustic comfort evaluation. At the Peace Gardens, as shown in Figure 3.7a, 31 and 41 per cent of the interviewees felt 'neither quiet nor noisy' and 'noisy' respectively, whereas 56.5 per cent of them thought it was 'comfortable'. At the Barkers Pool, similarly, 51.5 per cent of the interviewees rated the sound

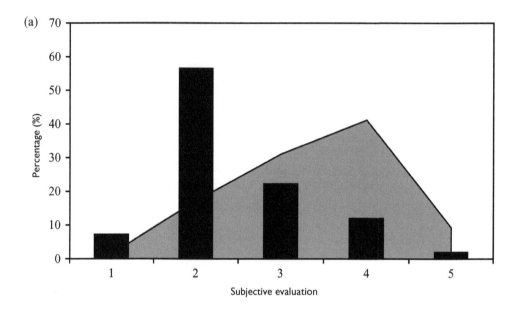

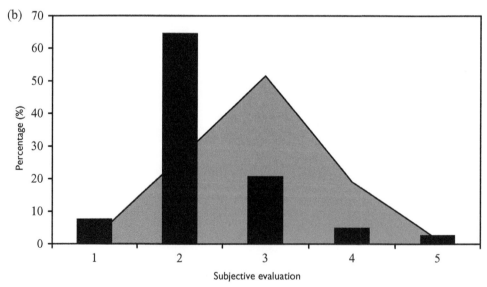

Figure 3.7 ▦ Subjective evaluation of sound level
■ Acoustic comfort evaluation
Comparison between the subjective evaluation of sound level and the acoustic comfort evaluation at (a) the Peace Gardens; (b) the Barkers Pool.

level as 'neither quiet nor noisy', whereas 64.5 per cent of them evaluated the site as 'comfortable'.

Overall, the analysis shows that in urban open public spaces there are considerable differences between the subjective evaluation of sound level and the acoustic comfort evaluation. People's evaluation towards sound level changes corresponds to the changes of measured sound levels, whereas the acoustic comfort evaluation is much more complex. People tend to be more tolerant in this respect, perhaps because the acoustic comfort is determined by more factors than just the sound level. Therefore, in the next section, effects of individual sounds are analysed.

3.4.4   Effects of individual sounds

In the Peace Gardens, fountains and demolition sounds were two main foreground sound elements during the survey period. There were three typical soundscape situations: fountains only, with a mean L_{eq} of 67.8dBA and STD of 4.1dBA; fountains and demolition, with a mean L_{eq} of 71dBA and STD of 4.2dBA; and demolition only, with a mean L_{eq} of 65.2dBA and STD of 7.7dBA. The high STD with the demolition sounds was mainly caused by the rumbling noise from the diggers.

Figure 3.8a shows the relationships between the measured sound level and the mean subjective evaluation of sound level, with binominal regressions and R. It can be seen that there are strong positive correlations in all the three conditions, and the tendencies of the three regression curves are rather similar. However, the positions of the curves are different, which means that with a given sound level, people have a different perception of the different sounds. Generally, the demolition sounds are perceived as the noisiest, followed by a mixture of the fountains and demolition, and then the fountains only.

For the acoustic comfort evaluation, as shown in Figure 3.8b, the tendencies of the three regression curves are significantly different. The regression is nearly linear for the demolition sounds, which means that the changes in sound level directly contribute to the evaluation of acoustic comfort. For the mixture of the fountains and demolition, however, the regression is a U-shaped curve. When the sound level is lower than around 70dBA, the variation in the acoustic comfort evaluation is almost negligible, possibly due to the masking effect of the fountains, whereas when the sound level is more than 70dBA, the masking effect of the fountains becomes less significant and, thus, the evaluation of acoustic comfort is more affected by sound level changes. For the fountains only, the increase in the sound level has almost no effect on the acoustic comfort evaluation. In addition to the recognition of sound sources, the preference of water sounds rather than demolition sounds might be caused by differences in the spectrum and temporal distribution between the two types of sound (Gifford 1996; Kang and Du 2003). When introducing a pleasant sound in urban open public spaces as a masking sound, there is always a concern regarding its level. From the above results, it can be seen that this level could be 70dBA or even higher.

In the Barkers Pool, as mentioned previously, music could be heard during 35 per cent of the survey period. With music the mean L_{eq} was 61.1dBA (STD 2.2dBA), which is only slightly higher than that without music, 59.7dBA (STD 3.8dBA). In Figure 3.9a correlations between the measured sound level and the mean subjective evaluation of sound level are shown for two conditions, with and without music. A significant difference has been found between the two conditions in terms of the subjective evaluation of sound levels ($p < 0.001$).

(a)

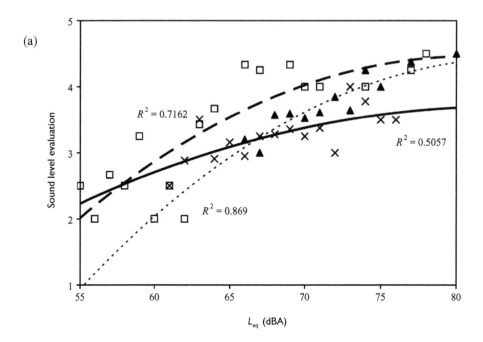

$R^2 = 0.7162$

$R^2 = 0.5057$

$R^2 = 0.869$

Sound level evaluation

L_{eq} (dBA)

(b)

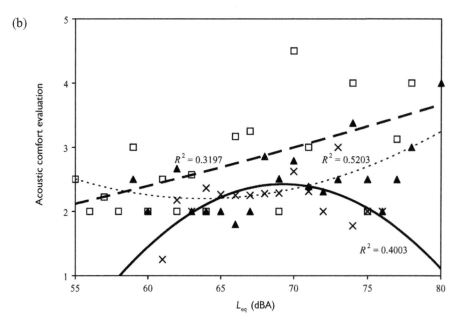

$R^2 = 0.3197$

$R^2 = 0.5203$

$R^2 = 0.4003$

Acoustic comfort evaluation

L_{eq} (dBA)

Figure 3.8 □ and - - - Demolition only
▲ and ·······Fountains and demolition mixture
✕ and— Fountains only
Relationships between (a) the measured sound level and the mean subjective evaluation of sound level; and (b) between the measured sound level and the mean acoustic
comfort evaluation under three source conditions in the Peace Gardens, with
binominal regressions and R^2.

(a)

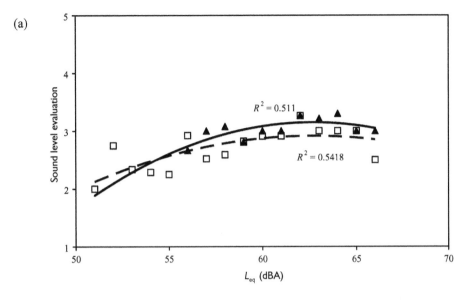

(b)

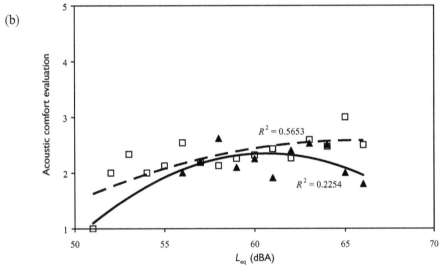

Figure 3.9 □ and ——— Without music
▲ and — — With music
Relationships between (a) the measured sound level and the mean subjective evaluation of sound level, and (b) between the measured sound level and the mean acoustic
comfort evaluation, under two source conditions in the Barkers Pool with binominal
regressions and R^2.

People felt quieter when there was no music, with a given L_{eq}. This suggests that music can
be easily noticed by the users of the square, and thus the perceived sound level is higher.

In Figure 3.9b correlations between the measured sound level and the mean acoustic comfort
evaluations are given. It is seen that the tendencies are rather different in the two conditions.

With music the correlation coefficient is considerably lower than that without music. When there is music, the variation in acoustic comfort evaluation is negligible with the increase of sound level, suggesting that the existence of music can make people feel more acoustically comfortable. It is noted, however, that not all kinds of music are preferred (see Section 3.5).

Overall, the analysis shows that the acoustic comfort evaluation is greatly affected by the sound source type. When a pleasant sound such as music or water dominates the soundscape of an urban open public space, the relationship between the acoustic comfort evaluation and the sound level is considerably weaker than that of other sound sources such as traffic and demolition sounds. In other words, the introduction of a pleasant sound, especially as a masking sound, could considerably improve acoustic comfort, even when its sound level is rather high.

3.4.5   Effects of other physical factors

To analyse the relationship between the overall physical comfort evaluation of an urban open public space and the subjective evaluation of various physical indices, including temperature, sunshine, brightness, wind, view, humidity, as well as sound level, varimax rotated principal component analysis was employed. Based on the data of the 14 sites, with a criterion factor of eigenvalue > 1, three factors were determined. The rotated component matrix is shown in Table 3.3. It can be seen that Factor 1 (22.8 per cent), including temperature, sunshine, brightness and wind, is the most important factor. Factor 2 (17.5 per cent) is associated with visual and auditory senses, showing that the acoustic environment is one of the main factors influencing overall comfort in an urban open public space. Factor 3 (14.8 per cent) is principally related to humidity, including humidity and wind. The above factors only cover 55 per cent of the total variance, which indicates the complexity in evaluating comfort conditions of urban open public spaces. In other words, other aspects, such as social/cultural factors, may also influence the evaluation. The factor analysis has also been carried out for each country, given the considerable variation in their climatic conditions. The results are generally similar to the above.

It is interesting to note that for all countries, visual and auditory aspects are always in the same factor, covering 17–19 per cent of the total variance. Corresponding to the discussion in Section 3.2.4, this again suggests that these two aspects may have interactions, working together as an aesthetic comfort factor, which is important in the design consideration.

Table 3.3 Factor analysis of the overall physical comfort evaluation. Kaiser–Meyer–Olkin measure of sampling adequacy, 0.613; cumulative, 55.1%; extraction method, principal component analysis; rotation method, varimax with Kaiser normalisation; N=9,200.

	Factors		
	1	*2*	*3*
Temperature	0.696	—	—
Sunshine	0.650	—	—
Brightness	0.599	—	—
Wind	−0.532	—	0.521
View	—	0.769	—
Sound level	—	−0.734	—
Humidity	—	—	0.828

3.4.6 Effects of demographic factors

In terms of the subjective evaluation of sound levels, no significant differences were found amongst the different age groups. However, in terms of acoustic comfort, there were significant differences ($p < 0.05$). Teenagers felt most unsatisfied, whereas older people (above 55) were the most satisfied group. No significant difference was found between males and females, both in terms of the subjective evaluation of sound level and acoustic comfort evaluation.

3.5 Sound preferences

The assessment of soundscape is a part of sensory aesthetics research that is concerned with the pleasurableness of the sensations one receives from the environment (Lang 1988). All aesthetic questions involve preference (Prall 1929). Aesthetics also involves the art of discrimination and making judgement. Because of preference, people evaluate the same environment differently and also react differently (Yang and Kang 2003, 2005b).

3.5.1 Preference of sounds

Figure 3.10 shows the results in the two Sheffield squares on the classification of 15 sounds. Corresponding to the results by other researchers (Kariel 1980; Porteous and Mastin 1985;

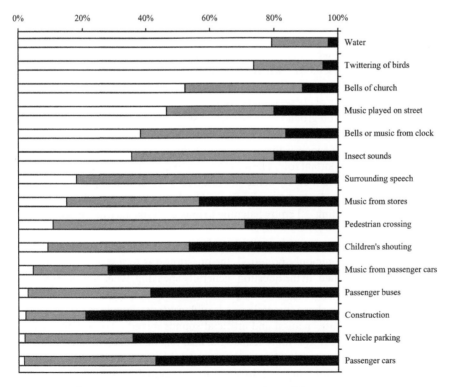

Figure 3.10 ☐ Favourite ▦ Neither favourite nor annoying ■ Annoying
Sound preferences in the two Sheffield case study sites.

Tamura 1998; Carles *et al.* 1999), people showed a very positive attitude towards natural sounds. More than 75 per cent of the interviewees reacted favourably to the sounds of water and birdsong, and only less than 10 per cent of people thought those sounds were 'annoying'. As a university student described: 'Natural sounds tend to be more tranquil and feel less invasive'. For culturally approved sounds, such as church bells, street music, as well as bells or music from clocks, people also showed relatively high levels of preference. For human sounds such as surrounding speech, most people thought they were 'neither favourite nor annoying'. The most unpopular sounds were mechanical sounds, such as construction sounds, music from passenger cars, and vehicle sounds. Like a young man said, 'I cannot think very well if there are too many cars'.

Whilst the above results are somewhat expected, it is interesting to compare the three kinds of music. As can be seen in Figure 3.10, for the music played on street, nearly half of the interviewees chose 'favourite'. For the music from stores, 40 per cent of the interviewees chose 'neither favourite nor annoying', whilst 43 per cent chose 'annoying'. However, more than 70 per cent of the interviewees felt the music from passenger cars was 'annoying'. A possible reason for the different reactions to various music is that the street music involves human activities and thus is less disturbing, whereas the car music always comes with mechanical sounds and high levels of low-frequency sounds, and hence tends to be more annoying.

3.5.2 Sound preferences and the choice of square

Pearson's chi-square test was employed to compare the sound preference between the two case study sites. It is interesting to note that the interviewees at the two squares had significant differences in evaluating some of the sounds. The interviewees at the Peace Gardens were more favourable to the sounds of birds ($p < 0.05$), church bells ($p < 0.01$), water ($p < 0.001$) and children's shouting ($p < 0.001$), whereas the interviewees at the Barkers Pool were more favourable to the music played on streets ($p < 0.05$) and music from stores ($p < 0.001$). As shown in Figure 3.4, most of these sounds were the soundmarks of the square. However, regarding the keynote and foreground sounds, such as surrounding speech, pedestrian crossing, vehicle parking, passenger cars and construction, there was no significant difference between the two squares. This result suggests that when people chose a square to use, their soundscape preferences did play an important role. The appearances of their favourite sounds make people feel more pleasant. As described by some interviewees: 'To feel completely relaxed and comfortable in my surroundings, I like the Peace Gardens – water sounds relaxing', 'The water sounds quite comforting', 'I enjoy listening to music', 'I like hearing birds, but music is nice too'.

Interviewees in Sheffield were asked to select their preferred relaxing sound environments. As natural sounds were commonly preferred, the question was to test how people want the natural sounds to be presented. In the question 'generally speaking, when you want to relax for a short period outside, you prefer____', 56.1 per cent of the interviewees chose 'quiet natural sounds only'; 21.3 per cent chose 'natural sounds with artificial sounds in far distance'; and the other 22.6 per cent chose 'natural and artificial sounds mixed'. To explain the reasons, some interviewees described: 'It is more relaxing to hear natural sounds, when all day you are generally hearing artificial sounds', 'To completely escape the hustle and bustle without background reminders', 'More relaxing and different to my normal experience', 'Want to feel like in a natural space, but don't expect cities to be sanitized', 'Escape from usual busy life more pleasurable environment for eating and reading', 'It's peaceful and makes me appreciate the world around us', 'It gets you away from modern life and its pressures – to make them more interesting, the fountain in the Peace Gardens is a good example'.

From the above analysis it can be seen that soundscape is an important aspect in people's evaluation of urban squares. Some pleasant sound elements could attract people to use a square. The general public is calling for more naturally appealing soundscape in urban squares.

3.5.3   Effects of demographic factors

In Figure 3.10, large variations can be found in interviewees' classification of various sounds. For instance, for the music from stores, the percentages of people who chose 'favourite', 'neither favourite nor annoying', and 'annoying' were 14.9, 43.3 and 41.8 per cent respectively. In order to explore the factors that lead to the lack of agreement in sound preferences, the effects of demographic variables are analysed below.

Age

Distinct differences were found amongst age groups regarding the preference of sounds. It is interesting to note that with the increase of age, people are more favourable to, or tolerant towards, the sounds relating to nature, culture or human activities, for example, birdsongs ($p < 0.001$). In Figure 3.11a it is shown that in Sheffield 93 per cent of the people aged over 65 favour birdsongs, whereas only 46.4 per cent of the age group of 10–17 rated birdsongs as 'favourite' and 14.3 per cent of them even chose 'annoying'. For other sounds significant differences have also been found amongst age groups: church bells ($p < 0.001$), water sound ($p < 0.001$), insect sounds ($p < 0.001$), bells or music from clock ($p < 0.001$), children's shouting ($p < 0.01$), pedestrian crossing ($p < 0.001$), and construction sound ($p < 0.05$).

Younger people, conversely, are more favourable to, or tolerant towards, music and mechanical sounds. Significant differences have also been found amongst age groups for music played on streets ($p < 0.05$), music from passenger cars ($p < 0.001$), vehicle parking ($p < 0.001$) and music from stores ($p < 0.001$). For example, Figure 3.11b shows the differences in preference towards music from stores. It can be seen that for the age group of over 65 only 6.5 per cent classified it as 'favourite', whereas most of them (77 per cent) rated it 'annoying'. By contrast, for the age group of 10–17 the 'annoying' percentage is only 23.6 per cent. Instead, 36.3 per cent of them rated the sound as 'favourite' and 40 per cent of them rated it as 'neither favourite nor annoying'.

For the sounds from passenger cars ($p < 0.01$) and buses ($p < 0.001$), people in the age ranges of 25–44 and 55–64 are the most annoyed groups.

The only sound for which the classifications agree between the various age groups seems to be surrounding speech, as shown in Figure 3.11c. About 70 per cent of the interviewees rated the sound as 'neither favourite nor annoying'.

Although differences exist between various age groups, the difference between the age group of 10–17 and the other groups is particularly significant. For example, in Figures 3.11a and 3.11b it can be seen that from the age group of 10–17 to 18–24 there is usually a considerable change in the curve. Similar results exist for church bells, water sound, insect sounds, and bells or music from clock.

In terms of the preferred relaxing sound environments, comparison is made between different age groups, and a significant difference ($p < 0.001$) was again found. Figure 3.12 shows the preferred relaxing sound environment between the different age groups. It can be seen that most people preferred quiet natural sounds only. With the increase of age, people showed more preference to the quiet natural sound environment, whereas 37.6 per cent of the age group of 10–17 preferred a mixture of natural and artificial sounds.

From the following typical comments the changes in sound preference with increasing age can be clearly seen: 'I like listening to music, the skate park and spotting nice boys' (female,

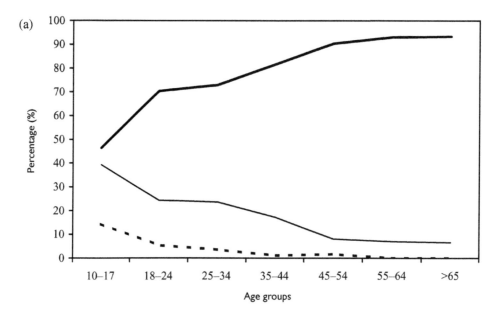

(a)

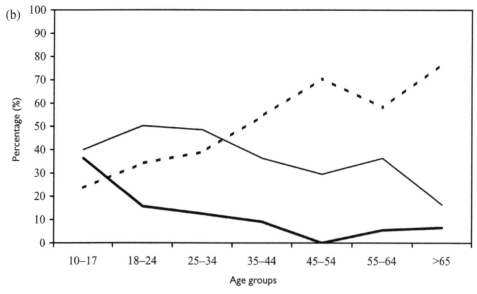

(b)

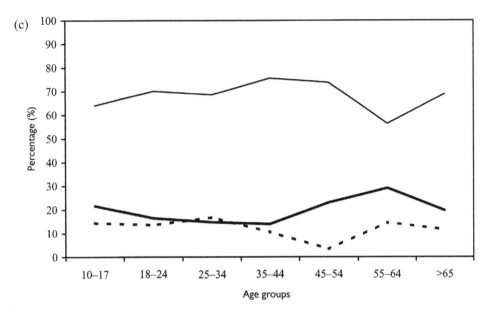

(c)

Figure 3.11 Differences in sound preference amongst age groups in the two Sheffield sites. (a) Bird songs; (b) music from stores; and (c) surrounding speech.
———— Favourite ———— Neither favourite nor annoying ▪ ▪ ▪ Annoying

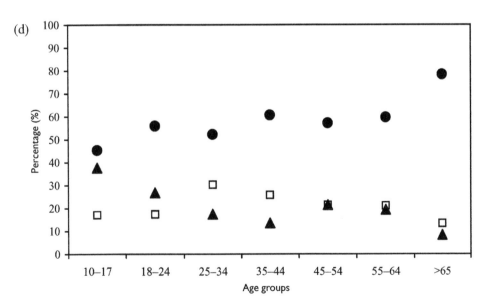

(d)

Figure 3.12 ● Quiet natural sounds only
□ Mainly natural sounds with artificial sounds in far distance
▲ Natural and artificial sounds mixed
Differences amongst age groups in the two Sheffield sites regarding preferred relaxing sound environment.

aged 10–17, preferred mixture of natural and artificial sounds); 'I don't mind having a little of both, it makes the sound environment more interesting with other noises as well as natural' (male, aged 10–17, preferred mainly natural sounds, but artificial sounds in far distance); 'Natural sounds help me to relax, but I like to hear other people getting on with their lives' (male, aged 18–24, preferred mixture of natural and artificial sounds); 'Natural sounds are relaxing, but artificial sounds can be inspiring' (male, aged 18–24, preferred mixture of natural and artificial sounds); 'Because it needs to be quiet, quiet to relax, but I will know I'm not alone' (female, aged 18–24, preferred mainly natural sounds, but artificial sounds in far distance); 'Natural and artificial sounds mix shows that there are many things going on around you and that even though it is relaxing where you are, you are close to a more active social environment' (male, aged 18–24, preferred mixture of natural and artificial sounds); 'I like the peace of my home, but I don't want to be detached from the outer world' (male, aged 18–24, preferred mainly natural sounds, but artificial sounds in far distance); 'I admit to relying upon artificial sounds, for example, CD stereo etc. for stimulation, but natural sounds are very important for peace, tranquillity and reassurance' (male, aged 25–34, preferred mixture of natural and artificial sounds); 'If I want to relax, I want as much peace and quiet as possible' (male, aged 35–44, preferred quiet natural sounds only); 'I want to feel as if I have stepped outside of the hubbub of city life' (male, aged 35–44, preferred quiet natural sounds only); 'As I get older, I prefer peace and quiet' (male, aged 35–44, preferred quiet natural sounds only); 'Noise today is too much of an intrusion in our lives' (male, 55–64, preferred quiet natural sounds only).

A possible reason for the above significant differences amongst age groups is that as people grow older, their sound preferences tend to be shaped by experience. Older people have more emotion when they hear the sound environment. As a result, they may be more appreciative of natural and culturally approved sounds. However, for young people, say aged 10–17, their social lives are just starting, and they may prefer high arousal soundscape in public spaces.

Significant age differences were also found in landscape preference (Lyons 1983). Children under 12 were less discriminating than adults and showed much greater variability in response to landscapes, having little interest in high naturalism in landscape and being less likely to view human intervention in the landscape as detrimental (Zube *et al.* 1983).

Gender

There are also some differences between males and females in sound preference, although less significant than that amongst age groups. Compared to males, female interviewees are more favourable to certain sounds, including church bells ($p < 0.001$), water ($p < 0.001$), music played on the street ($p < 0.05$), bells or music from clock ($p < 0.01$), and children's shouting ($p < 0.05$). It seems that the emotional effect is a common character of these sounds.

Figure 3.13a shows the differences between males and females in rating church bell sounds. More than 60 per cent of the female interviewees classified the church bell as 'favourite', and less than 30 per cent as 'neither favourite nor annoying', whereas these values for males are 44.9 per cent and 43.5 per cent respectively. Figures 3.13b and 3.13c illustrate the differences between males and females in rating the music played on street and children's shouting, respectively. Overall, it can be seen that despite some small differences, male and female interviewees show a similarity in classifying the sounds. It seems that the differences between sounds may exceed gender differences.

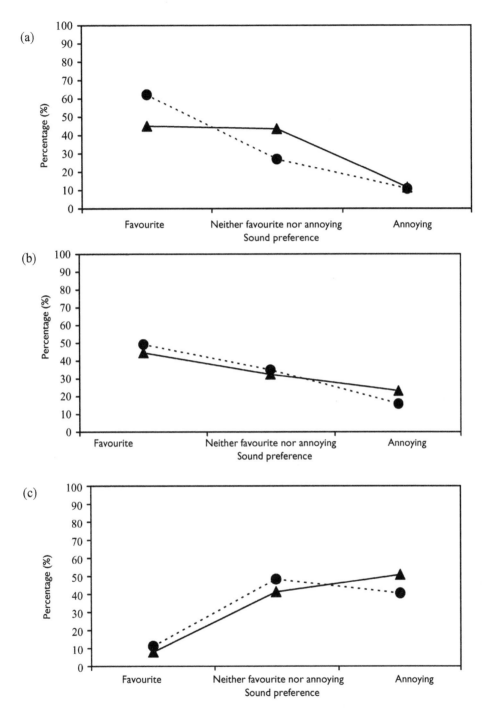

Figure 3.13 ▲ Male ● Female
Differences in sound preference between males and females in the two Sheffield sites.
(a) Church bell; (b) music played on street; and (c) children's shouting.

Other factors

Between local and nonlocal interviewees, it is interesting to note that the difference in sound preference is only significant for surrounding speech ($p < 0.001$). Nonlocal people tend to be more annoyed by this sound.

Between various professions, there is no significant difference, except between students and other professions. However, it is noted that since most of the students are young people, the differences could be related to the differences amongst age groups. A further comparison indicates that there is no significant difference amongst different professions in the same age group.

3.5.4   Effects of cultural factors

Whilst the above analysis is based on the data in Sheffield, it is important to compare the results between various countries and cities (Yang and Kang 2003). Table 3.4 shows the results of sound preferences rating of ten case study sites. Generally speaking, people show positive attitudes towards nature sounds and culture-related sounds. Vehicle sounds and construction sounds are regarded as the most unpopular, whereas those from human activities are normally rated as neutral.

From Table 3.4 it can be seen that interviewees from the same city generally have a similar sound preference rating, whereas the differences among the cities are significant. For example, when classifying water sounds, in Sheffield the results from the two squares are very close, with more than 70 per cent of the interviewees choosing 'favourite', whereas in the IV Novembre Square of Sesto San Giovanni this value is less than 30 per cent. The result in Kassel is similar to that in Sheffield. More than 50 per cent of the interviewees in the Kritis Square of Thessaloniki rated surrounding speech 'annoying', whereas this figure is less than 1 per cent in the two squares in Kassel. In the IV Novembre Square of Sesto San Giovanni about 45 per cent of the interviewees rated surrounding speech as 'favourite'. Probably an important reason for these differences is the cultural factor.

Between the two squares in Thessaloniki, Greece, the sound preference rating is rather different. One possible reason is that the main users of the Makedonomahon Square are not Greeks – the square is used as an open gathering and socialising place for immigrants from Eastern European countries and the former Soviet Union. As a result, the differences between the Makedonomahon Square and the local Greek used Kritis Square might be caused by cultural differences.

Between the two Greek cities there are also some dissimilarities, especially for sounds including surrounding speech and passenger cars. This seems to suggest that although the cultural background is similar, other factors such as differences in city sizes and climatic conditions may cause differences in sound preferences.

3.5.5   Three levels of sound preference

Preference of sounds depends on many more factors than sound level. The results of current and previous studies suggest that the sound preference differences are at three levels (Yang and Kang 2005b). The first can be defined as basic preference. People generally share a common opinion in preferring nature and culture-related sounds rather than artificial sounds. However, the survey across Europe indicates that cultural background and long-term environmental experience play an important role in sound preference. People from different backgrounds may show rather different tendencies in their sound preferences. Thus, there is a second level of sound preference, which can be defined as macropreference. At the third level, within the same cultural background and long-term

Table 3.4 Classifications for various sounds in the case study sites (%), with F, favourite; N, neither favourite nor annoying; and A, annoying.

Source type		Alimos		Thessaloniki		Sesto San Giovanni		Sheffield		Kassel	
		Kara. Square	Seashore	Maked. Square	Kritis Square	Petazzi Square	IV Nov. Square	Peace Gardens	Barkers Pool	Florent. Square	Bahn. Square
Water	F	—	—	—	—	—	27.7	84.0	74.7	80.3	74.5
	N	—	—	—	—	—	66.6	14.8	20.5	17.9	22.4
	A	—	—	—	—	—	5.7	1.2	4.8	1.8	3.1
Insects	F	—	—	—	—	—	—	37.7	33.2	34.0	23.5
	N	—	—	—	—	—	—	43.1	46.1	59.1	75.3
	A	—	—	—	—	—	—	19.2	20.7	6.9	1.2
Bells of church	F	—	—	—	—	31.1	—	56.8	47.9	—	—
	N	—	—	—	—	68.9	—	35.4	37.6	—	—
	A	—	—	—	—	0.0	—	7.8	14.5	—	—
Music played on street	F	—	—	—	—	—	—	44.2	48.8	57.3	88.0
	N	—	—	—	—	—	—	38.3	28.8	27.2	12.0
	A	—	—	—	—	—	—	17.5	22.4	15.5	0
Surrounding speech	F	2.3	7.0	—	32.2	23.5	44.6	17.9	18.0	18.5	15.3
	N	77.8	77.6	—	17.0	69.8	47.2	68.3	69.3	80.5	84.7
	A	19.9	15.4	—	50.8	6.7	8.2	13.8	12.7	1.0	0
Children's shouting	F	20.3	25.5	54.1	29.5	27.4	—	11.7	6.9	1.7	1.0
	N	54.3	50.8	19.9	53.0	53.4	—	48.4	40.3	69.0	54.8
	A	25.4	23.7	26.0	17.5	19.2	—	39.9	52.8	29.3	44.2
Pedestrian crossing	F	5.5	8.0	—	—	—	—	8.6	12.9	7.1	—
	N	89.9	84.7	—	—	—	—	62.0	58.4	17.9	—
	A	4.6	7.3	—	—	—	—	29.4	28.7	75.0	—
Passenger cars	F	0.6	—	3.5	31.3	2.7	1.6	2.4	1.0	—	—
	N	26.0	—	53.0	16.6	59.8	35.4	38.7	43.6	—	—
	A	73.4	—	43.3	52.1	37.5	63.0	58.9	55.4	—	—

Table 3.4 contd.

Passenger buses	F	—	—	3.4	1.3	—	1.6	3.7	2.1	—	—
	N	—	—	52.3	84.3	—	39.2	38.9	37.9	—	—
	A	—	—	44.3	14.4	—	59.2	57.4	60.0	—	—
Vehicle parking	F	—	—	—	—	—	—	2.9	1.0	1.4	2.0
	N	—	—	—	—	—	—	32.2	35.3	57.9	54.7
	A	—	—	—	—	—	—	64.9	63.7	40.7	43.3
Construction	F	—	—	2.1	32.5	—	—	2.2	2.1	—	—
	N	—	—	52.9	11.5	—	—	18.0	19.2	—	—
	A	—	—	45.0	56.0	—	—	79.8	78.7	—	—

environmental experience, personal differences exist. This can be defined as micropreference. In particular, the differences between age groups are more significant than other factors.

3.6   Case studies in urban open public spaces in China

3.6.1   Methodology

During the summer period of 2000, field surveys were carried out in two squares in Beijing, namely XiDan Square and JianGuo Square (Yang 2000; Yang and Kang 2001). A common characteristic of the two squares is that they are located beside at least one busy road, and consequently, unavoidable noise from the nearby traffic is one of the environmental sounds on the sites. There are also considerable differences between the two sites in terms of soundscape. In XiDan Square, no special sounds other than common urban sounds can be heard, whereas in JianGuo Square the users' activities, such as dancing and roller-skating, create a vivid and unique soundscape. Similar to the studies in Europe, both objective measurements and questionnaire surveys were carried out. The questions were generally similar to those described in Section 3.3, with slight variations according to the situations in Beijing. The number of interviewees was 90 in XiDan Square and 74 in JianGuo Square. In 2005, XiDan Square was revisited and another site in Beijing, ChangChunYuan Square, of similar functions/features as JianGuo Square, was also studied, where more than 300 interviews were carried out in each square (Kang and Zhang 2005). Figure 3.14 shows the two squares as well as some typical activities.

3.6.2   Results

A comparative analysis between XiDan Square and JianGuo Square in Beijing and the Peace Gardens in Sheffield has been made, with particular attention on the effects of psychological adaptation as well as cultural difference (Yang and Kang 2001).

The average L_{Aeq} were 63.9dB, 67.4dB and 67.4dB in XiDan Square, JianGuo Square and the Peace Gardens, respectively. It is interesting to note that although the SPL is the lowest in XiDan Square, interviewees in JianGuo Square and the Peace Gardens were more satisfied with the overall acoustic environment. By analysing the results from the subjective and objective surveys in detail, it is noted that psychological adaptation does occur when people perceive the soundscape in those urban open public spaces, which is important in examining the interactions between sound sensitivity and information contained in the soundscape.

The analysis of the relationships between sound sensitivity and the temporal status of the users in JianGuo Square suggests the following tendencies: (1) people using the site for recreation and rest were more satisfied with the acoustic environment than people just passing through the site ($p < 0.005$); (2) people staying for a relatively long period in the site were more satisfied than people staying for a short period ($p < 0.05$); (3) people who visited the site more than 5 times were more satisfied than people who visited the site less than 5 times ($p < 0.05$); and (4) sound sensitivity was reduced with the increase of the group size coming to the site ($p < 0.05$). For the other two sites, some similar tendencies were also noted, but the correlation coefficients failed to reach any significant level. A possible reason for the better correlations in JianGuo Square is that with the sounds from activities psychological adaptation is more marked.

It is noted that the content of the sounds, namely whether they are meaningful or meaningless, is very important in the evaluation process. Psychological adaptation occurs when people

Figure 3.14 (a) XiDan Square and (b) ChangChunYuan Square in Beijing.

demonstrate greater tolerance to a given sound level, because they perceive it to be pleasant. This is especially evident in the case of JianGuo Square. Corresponding to the results in Europe, by comparing the squares in Beijing, it is again suggested that introducing pleasant sounds in urban open public spaces may enhance the acoustic environment although the SPL could be increased.

Moreover, by comparing the sources of different pleasant sounds, this study suggests that the preferred sounds can be divided into two categories: sounds from human activities, defined here as 'active sounds', and sounds from the landscape elements, defined here as 'passive sounds'. Further comparison between the active sounds and passive sounds in the squares indicates that the former affects the depth of the psychological adaptation more than the latter does. This suggests that in the design of urban open public spaces, it would be beneficial to create useful activities.

Overall, soundscape and psychological adaptation should be considered to be complementary, rather than contradictory. Consideration of this duality could improve the use of urban open public spaces, as well as strengthen social interactions between users by encouraging opportunities for such interaction to take place. It is therefore important to consider design issues which would have impact on encouraging successful psychological adaptation.

3.7   Semantic differential analysis

3.7.1   Semantic process

Studying soundscape with semantic processing is becoming more important (Maffiolo *et al.* 1998; Raimbault *et al.* 2001; Guastavino and Cheminée 2004; Guastavino *et al.* 2005). There are two kinds of sounds related to the different ways of processing in terms of the users' listening. One is holistic hearing, which processes the soundscape as a whole without semantic processing. In other words, ambient noise of the city or background noise is considered and no specific event can be isolated. The other is descriptive listening, which is aimed at the identification of acoustic sources or events. It is important that descriptive listening indicates the psychological and social dimensions of a sound source in an urban environment. Based on a cognitive approach of urban soundscape, it was suggested that the meaningful categories of sounds and their properties at linguistic and psychological levels should be identified before describing them in physical dimensions and experimentally manipulating them in psychophysical paradigms (Dubois and David 1999; Dubois 2000). Semantic differential technique, therefore, would be a suitable method to connect users' feeling at both linguistic and psychological levels with sound sources in urban environment.

The semantic differential technique, developed by Osgood *et al.* (1957) in order to identify the emotional meaning of words, has been extended to a large variety of concepts. It has also proved to be a useful method in identifying the most important factors in evaluating sounds. For product sound quality, three main factors – powerful, metallic and pleasant – have been suggested (Kuwano and Namba 2001). Additional factors can be introduced for special sounds, such as dieselness for diesel cars (Patsouras *et al.* 2001). For urban environment sounds, the semantic differential technique has been used to analyse connotative and denotative meanings. A study has suggested that evaluation, timbre, power and temporal change were four essential factors for general environmental sounds (Zeitler and Hellbrück 2001). For residential areas, research in Sweden showed that the soundscape was characterised in four dimensions – adverse, reposing, affective and expressionless (Berglund *et al.* 2001).

3.7.2   Case study methodology in urban open public spaces

A case study through semantic differential analysis was carried out for urban open public spaces in 2002–2005, with the aim of identifying factors that characterise the soundscape (Kang and Zhang 2002, 2005). The study is based on a three-stage field survey. Stage one, as a pilot study, was a soundscape walk in four typical urban open public spaces. Stage two included more detailed interviews in two selected sites, with a much larger sample size from the general public. Stage three consisted of several soundscape walks with architectural students, examining differences between the general public and designers.

Stage one

Soundscape walks are frequently used in environmental acoustics research. The general purpose is to encourage the participants to listen carefully and make judgements about the sonic environment and sounds they are experiencing. Listening is one of the psychological functions through which people perceive the world. As discussed previously, the evaluation of sound effects on people is primarily a subjective issue, rather than being merely based on objective parameters.

The soundscape walk was carried out with 48 university students aged 18–25. There were 30 males and 18 females, all audiologically normal. The walk started from a square in front of the Sheffield University Student Union building. The square is close to a major roundabout and under a busy traffic bridge. It is used by students to distribute information flyers, meet friends and take breaks between classes. Sound sources included heavy traffic above the square, loud music, birds, pedestrian/bike crossings, and conversations. The second site, Devonshire Green, is a large green space surrounded by low buildings and small roads. It is a place for lunch breaks and leisure activities such as cycling. In a corner of the site there is a skate park. Sound sources included skating and shouting of children/teenagers, traffic in distance, birds, pedestrian/bike crossings and construction noise near the site. The third and fourth sites, the Peace Gardens and the Barkers Pool, are described in Section 3.3.

During the soundscape walk, the subjects were asked to list the sounds they heard, evaluate the overall soundscape as well as three main individual sounds, and give further comments for each site. Table 3.5 shows the evaluation form. The study used 28 indices with a 7-point bipolar rating scale. Some of the indices were based on previous research related to urban soundscape as well as product sound quality (Zeitler and Hellbrück 2001; Hashimoto and Hatano 2001; Schulte-Fortkamp 2001), and some were compiled specifically for this study according to the actual situations, including close–far, social–unsocial, safe–unsafe, friendly–unfriendly, happy–sad, and like–dislike.

Both connotative meanings of urban environment sounds, such as calming–agitating, interesting–boring, and like–dislike, and denotative meanings such as quiet–noisy, sharp–flat, and smooth–rough were included. The indices also covered various aspects of soundscape, for example,

- satisfaction: comfort–discomfort, quiet–noisy, pleasant–unpleasant, interesting–boring, like–dislike, calming–agitating, happy–sad, and beautiful–ugly;
- strength: gentle–harsh, high–low, hard–soft, light–heavy, and strong–weak;
- fluctuation: sharp–flat, directional–everywhere, varied–simple, fast–slow, echoed–deadly, far–close, smooth–rough, pure–impure, and steady–unsteady; and
- social aspect: meaningful–meaningless, bright–dark, friendly–unfriendly, safe–unsafe, and social–unsocial.

By analysing the soundscape walk results, it was found that some indices were seldom selected/evaluated, so that in the second stage of the study only 18 indices were selected, which are boldfaced in Table 3.5.

Stage two

The characteristics of sound sources are vital for soundscape evaluation. The Peace Gardens and the Barkers Pool were selected for the second stage of study since they were relatively more representative of the typical soundscape in urban open public spaces, including continuous and intermittent sounds, human-made and natural sounds, meaningful and meaningless sounds, and pitched and varied sounds. There were also activity related sounds as well as soundmarks such as fountains in the Peace Gardens and music in the Barkers Pool.

As shown in Table 3.6, 491 users were interviewed in two seasonal periods, autumn/winter and spring/summer at the two selected sites. The interviewees were randomly sampled from

Table 3.5 Soundscape evaluation form used in the soundscape walk in Sheffield. Boldfaced indices are those used in the second stage survey in the Peace Gardens and the Barkers Pool.

	Very	Fairly	Little	Neutral	Little	Fairly	Very	
Agitating	3	2	1	0	-1	-2	-3	**Calming**
Comfortable	3	2	1	0	-1	-2	-3	**Uncomfortable**
Directional	3	2	1	0	-1	-2	-3	**Everywhere**
Echoed	3	2	1	0	-1	-2	-3	**Deadly**
Far	3	2	1	0	-1	-2	-3	**Close**
Fast	3	2	1	0	-1	-2	-3	**Slow**
Gentle	3	2	1	0	-1	-2	-3	**Harsh**
Hard	3	2	1	0	-1	-2	-3	**Soft**
Interesting	3	2	1	0	-1	-2	-3	**Boring**
Like	3	2	1	0	-1	-2	-3	**Dislike**
Meaningful	3	2	1	0	-1	-2	-3	**Meaningless**
Natural	3	2	1	0	-1	-2	-3	**Artificial**
Pleasant	3	2	1	0	-1	-2	-3	**Unpleasant**
Quiet	3	2	1	0	-1	-2	-3	**Noisy**
Rough	3	2	1	0	-1	-2	-3	**Smooth**
Sharp	3	2	1	0	-1	-2	-3	**Flat**
Social	3	2	1	0	-1	-2	-3	**Unsocial**
Varied	3	2	1	0	-1	-2	-3	**Simple**
Beautiful	3	2	1	0	-1	-2	-3	Ugly
Bright	3	2	1	0	-1	-2	-3	Dark
Friendly	3	2	1	0	-1	-2	-3	Unfriendly
Happy	3	2	1	0	-1	-2	-3	Sad
High	3	2	1	0	-1	-2	-3	Low
Impure	3	2	1	0	-1	-2	-3	Pure
Light	3	2	1	0	-1	-2	-3	Heavy
Safe	3	2	1	0	-1	-2	-3	Unsafe
Steady	3	2	1	0	-1	-2	-3	Unsteady
Strong	3	2	1	0	-1	-2	-3	Weak

the sites. Each interviewee was asked to fill an evaluation form (see Table 3.5) for the semantic differential analysis.

Demographic factors are also important for soundscape evaluation. In terms of gender, there were generally more males than females in both sites, as can be seen in Table 3.6. Figure 3.15 shows the age distribution of the interviewees at the two sites, and the education and occupation profiles of the interviewees are shown in Figure 3.16. Between the two sites, the demographic profiles are generally similar, including gender, age, education and occupation.

Table 3.6 Number of interviewees in the Peace Gardens and the Barkers Pool at the second stage of study.

	Peace Gardens			Barkers Pool			
	Male	Female	Sum	Male	Female	Sum	Total
Autumn/winter	69	36	105	56	40	95	200
Spring/summer	75	71	146	72	73	145	291
Total	144	107	251	128	113	240	491

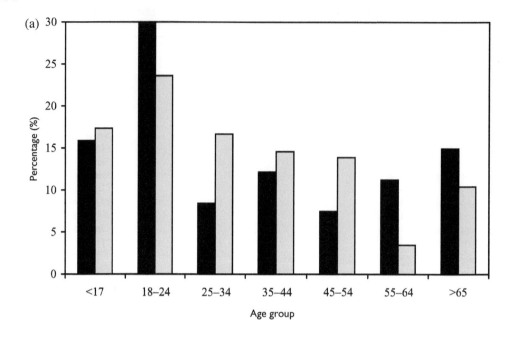

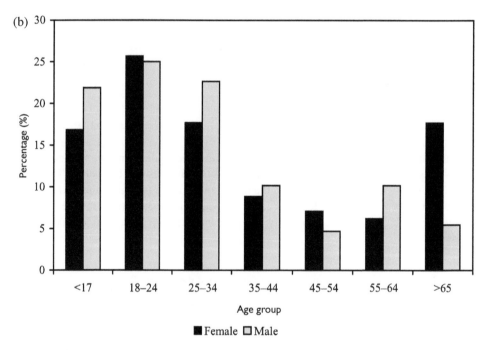

Figure 3.15 Age group distribution at (a) the Peace Gardens and (b) the Barkers Pool. Average of two seasonal periods.

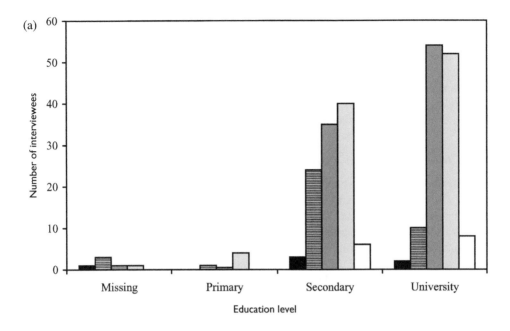

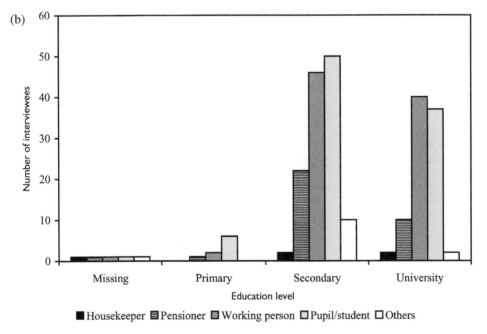

Figure 3.16 Education and occupation profiles of the interviewees at (a) the Peace Gardens and (b) the Barkers Pool. Average of two seasonal periods.

Stage three

A further soundscape evaluation was made with 223 architectural students, examining how future architectural/urban designers value the urban soundscape they experience everyday, and how they would like to design urban soundscapes (Zhang and Kang 2004). The four urban open public spaces at stage one were again used and the evaluation was for both the overall soundscape and for individual sounds, with the 18 indices shown in Table 3.5. In addition, they were asked to evaluate the importance of acoustics compared with other physical factors, to indicate preferred sounds, and to give design suggestions.

3.7.3   Main factors for soundscape evaluation

The analysis in this section is based on the second stage of the study. Semantic differential analysis of the results was carried out using SPSS (Field 2000) to identify main factors for soundscape evaluation in urban open public spaces. Factor analysis was first performed using all the data in the Peace Gardens and the Barkers Pool, from both autumn/winter and spring/summer periods. Varimax rotated principal component analysis was employed to extract the orthogonal factor underlying the 18 adjective indices. With a criterion factor of eigenvalue > 1, four factors were determined, as shown in Table 3.7. It can be seen that factor 1 (26 per cent) is mainly associated with relaxation, including comfort–discomfort, quiet–noisy, pleasant–unpleasant, natural–artificial, like–dislike and gentle–harsh. Factor 2 (12 per cent) is generally associated with communication, including social–unsocial, meaningful–meaningless, calming–agitating, and smooth–rough. Factor 3 (8 per cent) is mostly

Table 3.7 Factor analysis of the soundscape evaluation – overall results of the Peace Gardens and the Barkers Pool in the two seasonal periods. Kaiser–Meyer–Olkin measure of sampling adequacy, 0.798; cumulative 53%; extraction method, principal component analysis; rotation method, varimax with Kaiser normalisation; $N=491$.

Indices	Factors			
	1 (26%)	2 (12%)	3 (8%)	4 (7%)
Comfort–discomfort	0.701	0.164	0.138	—
Quiet–noisy	0.774	—	—	—
Pleasant–unpleasant	0.784	0.258	0.157	—
Interesting–boring	0.435	0.272	0.274	0.103
Natural–artificial	0.532	0.102	0.240	—
Like–dislike	0.519	0.575	0.247	0.151
Gentle–harsh	0.502	0.531	0.123	—
Hard–soft	—	—	—	0.812
Fast–slow	—	—	—	0.827
Sharp–flat	0.220	—	0.345	0.488
Directional–everywhere	0.234	—	0.441	0.267
Varied–simple	0.115	—	0.674	0.167
Echoed–deadly	0.204	—	0.531	—
Far–close	—	—	0.550	—
Social–unsocial	—	0.672	0.462	—
Meaningful–meaningless	0.126	0.585	0.469	—
Calming–agitating	-0.143	0.708	0.286	—
Smooth–rough	—	0.683	0.396	—

associated with spatiality, including varied–simple, echoed–deadly and far–close. Factor 4 (7 per cent) is principally related to dynamics, including hard–soft and fast–slow.

It is interesting to note that these four factors cover the main facets of designing the acoustics of an urban open public space: function (relaxation and communication), space, and time. It is also noted that the four factors cover 53 per cent of the total variance. It is lower than most results in product sound quality studies and general environmental noise evaluation. This is perhaps due to the significant variations in urban open public spaces, in terms of sound source number and type, as well as their characteristics. Another possible reason is that some indices, although well evaluated by the university students in the pilot study, were not well under-stood/evaluated by the interviewees in the second stage of survey.

3.7.4 Effects of season, site, sample size and special sounds

To examine the difference between the two seasonal periods, factor analysis was carried out based on the autumn/winter and spring/summer data separately. For the spring/summer data, including 291 subjects at the two sites, five factors cover 58 per cent variance. Factor 1 (24 per cent), relaxation; factor 2 (12 per cent), communication; factor 3 (8 per cent), spatiality; and factor 5 (6 per cent), dynamics, are similar to those in the overall results, except that in factor 3 sharp–flat is added. Factor 4 (8 per cent), including calming–agitating and smooth–rough, can be related to the communication factor in the overall result. Factors based on the autumn/winter data have similar tendencies.

The data of the two sites, the Peace Gardens and the Barker Pool, were also analysed sepa-rately. The results in the Peace Gardens show that factor 1 (25 per cent) is related to relaxation, factor 2 (14 per cent) is related to communication, factor 4 (8 per cent) is concerned with spatiality, and factor 5 (6 per cent) is associated with dynamics. Factor 3 (9 per cent) includes interesting–boring and meaningful–meaningless, which might be contributed to by the fore-ground sounds from fountains and demolishing work. The result at the Barkers Pool is generally similar to that at the Peace Gardens, and also similar to the overall result of the two sites, although there are some slight differences in the order of the factors.

Further analysis was made using the data of each site for each seasonal period. It has been shown that the number of factors usually increases with a decreasing sample size. A comparison between various sample sizes suggests that a sample size of 100 to 150 is generally acceptable for evaluating soundscape in urban open public spaces.

It is also noted from the analysis that when there is a special/dominant sound source, the results of factor analysis can be considerably affected. For example, with the high level of demolition noise at the Peace Gardens in the autumn/winter period, the factor analysis result is rather different from other situations, suggesting that attention must be paid to some special sources, especially unpleasant ones.

Overall, although the situation in urban open public spaces is rather complicated compared to product sound quality and general environmental noise evaluation, it is still possible to identify several major factors, which include relaxation, communication, spatiality and dynamics.

3.7.5 Comparisons between users and designers

Based on the results at stage three, semantic differential analysis is made for the architectural student group and then compared to the stage two results of general public. Generally speaking, there is no significant difference between the two groups, but with the architectural students there are slightly more factors and also, more variation between the two sites. In other words, it seems that the evaluations of the architectural students are more diverse than that of the general public (Zhang and Kang 2004, 2006b).

Table 3.8 shows the average evaluation score of several typical indices including comfortable–uncomfortable, pleasant–unpleasant, quiet–noisy and social–unsocial. It is interesting to note that in the two squares with natural sounds and considerable green spaces, namely the Peace Gardens and Devonshire Green, the scores for the first three indices are systematically higher than those in the other two squares, indicating the preference of the architectural students. A series of paired samples t-tests show that those differences are significant ($p < 0.05$). Correspondingly, the difference between the four squares in terms of the distribution at each scale is shown in Figure 3.17, using the comfortable–uncomfortable index as an example. It is also noted that when the architectural students were asked to select a number of preferred sounds when considering urban square design, natural sounds were strongly preferred. In terms of the social–unsocial index, however, the results are rather different from the above, with the Devonshire Green having a significantly lower score ($p < 0.05$) than the other three, suggesting the importance of considering other facets too.

In Table 3.8 the result of the architectural students is also compared with that of the general public based on stage two. It is interesting to note that for the Peace Gardens the scores of the architectural students are significantly higher, according to a series of independent samples t-tests ($p < 0.05$), for all four indices, showing their strong preference for such kinds of urban open public spaces. For the Barkers Pool, the difference between architectural students and the general public is much less, and generally there is no statistically significant difference.

Since the architectural students and the general public are of different age ranges, only the results of the latter relating to the 18–24 age group are also shown in Table 3.8. Generally speaking, there is no significant difference between the general public of all age groups and of

Table 3.8 Average evaluation score of typical indices in the four squares, by the architectural student group (stage three) and by the general public (stage two).

	Comfortable (3) – uncomfortable (-3)	Pleasant (3) – unpleasant (-3)	Quiet (3) – noisy (-3)	Social (3)– unsocial (-3)
Peace Gardens: architectural students	0.77	1.58	0.10	1.14
Barkers Pool: architectural students	0.31	0.37	-0.56	0.91
Devonshire Green: architectural students	0.88	0.93	0.55	0.42
Union Square: architectural students	0.37	0.26	-0.48	1.67
Peace Gardens: general public (all)	0.45	0.32	-0.74	0.63
Peace Gardens: general public (18–24 years)	0.54	0.54	-0.72	0.75
Barkers Pool: general public (all)	0.39	0.27	0.01	0.37
Barkers Pool general public (18–24 years)	0.31	0.23	0.19	0.24

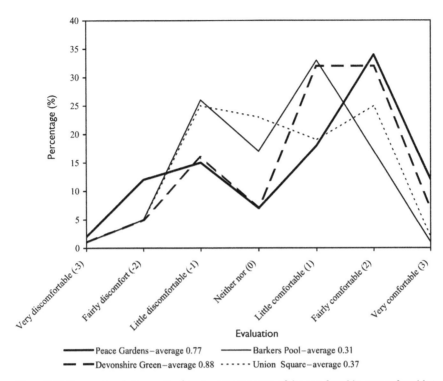

Figure 3.17 Comparison between the four squares in terms of the comfortable–uncomfortable index.

the 18–24 age group, according to a series of independent samples *t*-tests ($p < 0.05$). Further statistical analysis shows that the difference between architectural students and the general public is similar to the above when only the 18–24 age group is considered for the latter.

The architectural students were asked to select one or more important environmental factors from a given list. The results are shown in Figure 3.18. It can be seen that in general sound is a rather important consideration, ranked as second/third on the list. It is important to note that there are considerable variations between different squares, especially for some factors such as view, temperature, air quality and sun, suggesting that for different types of squares different evaluation and weighting systems should be used. A further question was also asked: when you choose a location to sit/relax in this square, is the acoustic environment an important factor? The percentage answering 'yes' is 75, 54, 87 and 63 per cent in the Peace Gardens, Barkers Pool, Devonshire Green, and Union Square, respectively. This again demonstrates the importance of sound environment in the opinion of those future designers.

3.8 Soundscape description in urban open public spaces

To investigate the acoustic environment of an urban open space or to design a soundscape, it is important to develop an appropriate description system. Raimbault *et al.* (2003) suggested that three categories of analysis should be considered for urban open public spaces: activities such as human presence or transport, spatial attributes like location, and time history including moment or period.

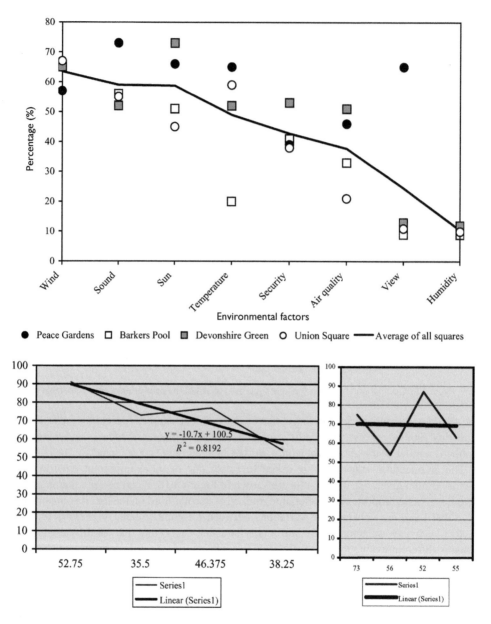

Figure 3.18 Percentage of the architectural students who selected a given environmental factor as 'important'.

Based on a literature review and the studies described in the previous sections, a model for describing the soundscape of urban open spaces is proposed (Kang 2003b; Zhang and Kang 2006a), as shown in Figure 3.19. This includes four facets: characteristics of each sound source, acoustic effects of the space, social aspect of the users, and other aspects of the physical conditions. Since at different locations of an urban open space the

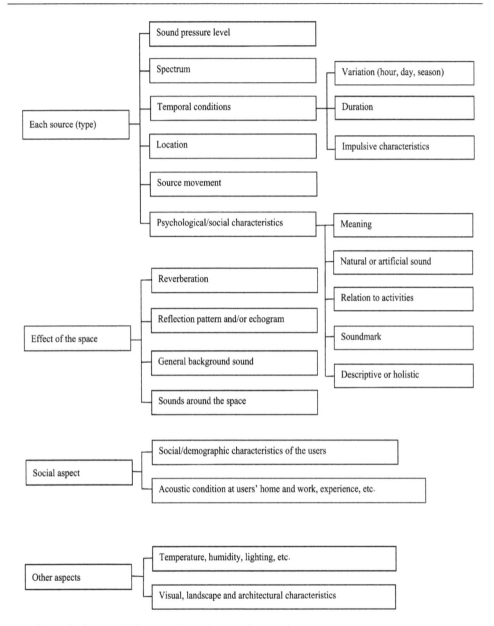

Figure 3.19 A model for describing the soundscape of urban open spaces.

soundscape could be rather different, the description should consider a number of typical receivers.

For each sound source, the SPL, spectrum, temporal conditions, source location, source movement, and the psychological and social characteristics should be considered. In terms of sound level, both the steady-state SPL and statistical SPL should be taken into account. In terms of frequency, if tonal components are noted, it would be useful to consider a narrowband spectrum.

The temporal effect may be related to the dynamic characteristics of hearing. It has been demonstrated that when the temporal pattern of a sound is systematically varied, the sound which has the high level portion at the beginning is perceived to be louder, which might be caused by the overshoot at the onset of the sound (Kuwano and Namba 2001). The description should include rate and pattern of the sound occurrence, sound sequences which tell a story in sound, and passage of time such as acoustic actions of starting/stopping, adding/subtracting, and expanding/contracting. The perception of a sound also varies according to its duration. The shorter the duration, the sharper the sound is judged (Kuwano and Namba 2001). Impulsive characteristics, including peak level as well as rise and fall time, should be taken into account too. Considerable research has been carried out on the determination of the loudness of impulse sounds (Meunier *et al.* 2001; Scharf 1978).

The location and movement of sound sources are of particular importance for the soundscape in urban open spaces. People have a natural ability to isolate sounds in relation to their approximate positions: whether sounds are behind, to the side, above, below, or in front of the head (Wenzel 1992). The auditory system is also capable of detecting detailed information about the distance of the sound source, its velocity, direction of its movement, and even its size and weight from a variety of acoustic events.

Another aspect of a sound source in an urban open space is its psychological and social characteristics, as discussed in Section 2.1 and the previous sections of this chapter. Sound figures can be natural in occurrence or selected by the will of the listener.

The acoustic effects of an urban space should be considered. More detailed analyses are made in Chapters 4 and 7. Relevant factors include the shape of the space, boundary materials, street/square furniture and landscape elements. In addition to sound level distribution and reverberation, reflection pattern and/or echogram, possible acoustic defects such as echoes and focus effects should be checked for.

It is also important to describe the general background sound and any special sound sources around an urban open space. The subjective evaluation of an urban open space can be affected by the surrounding acoustic environment (Yang and Kang 2001).

Social/demographic aspects of the users in an urban open space are vital, as demonstrated in the previous sections, and thus relevant information should be taken into account. This includes gender, age group, place of living, that is, local resident or from other cities, as well as their cultural and education background. The acoustic experience of the users is also important, so is the acoustic environment at their home and working places.

In addition to the description of soundscape, it is essential to describe other aspects of physical conditions, including temperature, humidity, wind, sun, luminosity, and glare. The visual environment as well as landscape and architectural features of an urban open space should also be considered.

3.9 Soundscape evaluation using artificial neural networks

Following the description system as discussed in the above section, it would be useful to develop a tool to predict subjective evaluation at the design stage, using known design conditions such as the physical features of a space, acoustic environmental variables, and demographic characteristics of the users to aid planners/designers in making decisions. For more general cases, an environment similarity index to compare two different environments was proposed (Hiramatsu *et al.* 2001). This index is calculated on the basis of physical properties of, and community responses to, sound species existing in the two environments. However, given the complexity of soundscape evaluation in urban open spaces as well as the effects of other physical factors, it may not always be appropriate to rank the soundscape of different types of urban open spaces. A possible way to integrate various factors is to use the artificial neural networks (ANN) method, as explored in the following (Yu 2003; Yu and Kang 2005a, 2005b).

3.9.1 Artificial neural networks

For a complex system involving multiple disciplines and factors, artificial intelligence (AI) technique is found to be appropriate. The technique, introduced as ANN, is a simplified model of the central nervous system that consists of networks of highly interconnected neural computing elements that have the ability to respond to input stimuli and learn to adapt to the environment (Patterson 1996).

The ANN technique is based on our understanding of the brain's working patterns. It introduces an idea of using the silicon logic gates of the microprocessors in personal computers as structural constituents of the brain. The basic computing element in biological systems is the neuron. A biological neuron is a small cell that receives electrochemical stimuli from multiple sources and responds by generating electrical impulses that are transmitted to other neurons. About 10 per cent of the neurons are input and output receivers, and 90 per cent are interconnected with other neurons, which can store information or transform the signals being propagated through the network. Neuron connections are made through two synapses. Neuronal activity is related to the creation of an internal electric potential. This potential may be increased or decreased by the input activity received from other neurons through the synapses (Patterson 1996).

The same function is achieved in an ANN model through learning algorithms that develop weights between its processing elements (artificial neurons). The weights stress the strength of the response, and in the whole training process, the weights are constantly adjusted to reduce the difference between desired and actual responses (Hecht-Nielsen 1990). The adjustment process is continued until there is no further significant change. The network has then learnt from the training examples by input–output target mappings.

In the biological world, subjective evaluations come from inter-reactions between biological neurons. It is the same way for people to obtain soundscape perception. No matter how many complex variables are involved, a biological brain can give individual perception. ANN technique simulates this and can thus be applied to the study of soundscape evaluation, which is a multidisciplinary research relating to various social and environmental aspects that requires building multiple and nonlinear links among various factors.

3.9.2 Framework of using ANN for soundscape evaluation

The basic process of ANN is to (1) use as many input variables as possible to design initial models; (2) utilise existing data to train models; (3) reconstruct the model architecture according to the analysis of training results; and (4) use well-trained models to make predictions.

ANN learning comes from input–output target mappings, so that the difference between targets (output nodes) and network predictions (output from learning process) is a key point to judge the model performance. Several analysis tools can be used to trace the training process, among them rms errors plot and target/output plot are commonly used. The former presents how the training network reaches its minimum value of error, and the latter measures how well the network makes predictions compared to the original targets.

A large number of input variables are considered corresponding to Figure 3.19. According to these variables, a model framework has been established. The framework works as a tree system, as shown in Figure 3.20. The main model is soundscape evaluation, and in addition there are two kinds of submodels: a series of models for other physical factors, such as for thermal, lighting and visual comfort; and a series of foreground sound models, one for each main sound source. The input data for other physical aspects include relevant physical parameters as well as social and demographic factors, whereas the input variables for a foreground sound model include SPL, sound spectrum and other physical and psychological characteristics of the source. The outputs from the submodels act as input variables for the main model of soundscape evaluation, as can be seen in Figure 3.20. General social and demographic factors of the users form another part of the input variables for the main model, together with some other aspects such as general background sound. Note in Figure 3.20 variables are expressed as nodes, as commonly used in ANN modelling.

The software package Qnet (Vesta Services 2004) was used to build the above models (Yu and Kang 2005b).

3.9.3   Database construction and initial analysis

The ANN models need to be trained through many cases, which means that a large amount of data is needed from surveying existing urban open public spaces. Based on the results described in the previous sections, a database suitable for ANN modelling has been established. Results from further field surveys and laboratory experiments can also be integrated for a refined model system.

Initial results suggest that there is a good convergence of using ANN to predict people's perception of soundscape in urban open spaces. With a sample size of 2,000, in the acoustic comfort evaluation model, by using 11 input variables including SPL, individual social and behaviour aspects, and spatial and temporal information of users, the correlation coefficient is 0.670 for the training set and 0.600 for the test set. This demonstrates the feasibility of using ANN for the prediction of acoustic comfort, although improvements can still be made, in terms of number of variables, for example. Similar results have been obtained for the submodels for lighting, thermal, humidity and wind comfort, where the correlation coefficient is 0.664, 0.900, 0.658 and 0.717 for the training set; 0.658, 0.620, 0.314 and 0.537 for the test set; and the number of variables is 16, 19, 17 and 16, respectively.

It is worth noting that despite the fact that ANN can be used to build intelligent prediction models, it is often difficult to use it to make direct analysis between various input and output factors, since ANN largely operates as a black box. To better understand the relationships between various factors, statistical analysis is still necessary.

3.10   Soundscape design in urban open public spaces

Whilst most investigations have focused on the study of soundscape as a passive perception factor, it is important to put soundscape into the intentional design process comparable to landscape, and to introduce the theories of soundscape into the design process of urban public spaces. In this section some issues relating to the soundscape design in urban open spaces are discussed.

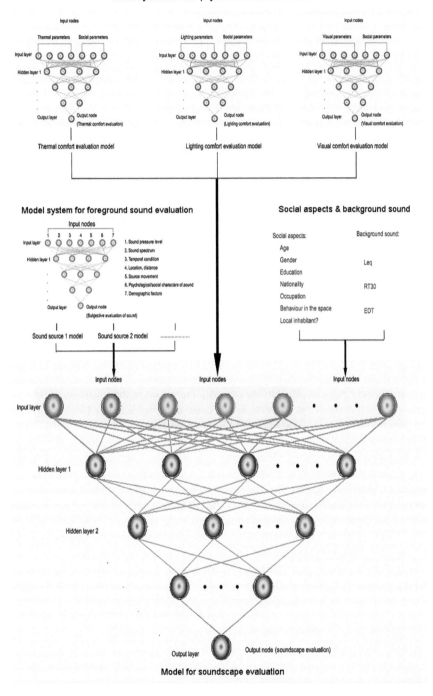

Figure 3.20 Framework for using ANN for soundscape evaluation. A colour representation of this figure can be found in the plate section.

3.10.1   Sound

First, it would be important to consider soundmarks, reflecting traditional and cultural characteristics. From Section 3.4 it is interesting to note that in perceiving the sound elements in a given soundscape, the first noticed sounds do not have to be the loudest. People always mention the soundmarks as their first noticed sounds. Moreover, the preferences of soundscape elements are proved to influence people's choice of using an urban square. Therefore, the soundscape identity is important for a designated space. A more aesthetically appealing soundscape would attract more users to a square.

Another important aspect of the soundscape design of urban open spaces is spectrum analysis, for both individual sounds and the overall acoustic environment. An interesting phenomenon is that it seems animal and insect vocalisations tend to occupy small bands of frequencies leaving spectral niches into which the vocalisations of other animals, birds or insects can fit. As urban areas spread, the accompanying noise might block or mask spectral niches and, if mating calls go unheard, a species might die out (Krause 1993). When introducing natural sounds into urban open public spaces, it is important to consider this phenomenon. Spectrum analysis is also important when using psychoacoustic magnitudes.

The design of soundscape in an urban open space should be considered as a dynamic process. The soundscape variation with seasons, days and different times of typical days should be taken into account, as well as differences in soundscape between the designed space and the surrounding acoustic environment. It is useful to relate design with sound excursion of the urban open space or the city, using a series of typical listening points (Westerkamp 2000; Dietze 2000).

As discussed in Section 3.6.2, sound sources in an urban open space can be divided into active sounds and passive sounds. The former relates to sounds from the activities in the space and the latter relates to the sounds from the landscape elements. As a typical active sound, live music is always very popular. People are not only interested in the music itself, but are also attracted by the activities of the players. In this case, the type of music (for example, classic music or pop music) is not a very important issue. However, when music is from a store or played through a public address (PA) system, the type of music as well as the sound level needs to be carefully considered. Most people do not like loud music played from loudspeakers, whatever the music type is (see Section 3.5.1). In terms of spectrum characteristics, case studies in Sheffield suggest that the low-frequency components in music are often not loud enough to mask traffic sound, whereas the high-frequency components can bring the music sound out from other background sounds and make the soundscape more pleasant.

Because there are many factors influencing participation in human activities, the presence of active sounds is uncertain in some conditions. It is often useful to introduce passive sounds. Many kinds of design features with favourable sounds can be applied, both for functional and aesthetical purposes. A typical passive soundmark, water, in the form of fountains, springs or cascades, is often used as a landscape element in open public spaces, and has been proved to have endless effects in colouring the soundscape. As Kaplan (1987) stated, in the visual aesthetic field, there are contents called 'primary landscape qualities', which have a special effect on preference, and water and foliage were two of the contents first identified. Similarly, water sound can be defined as a 'primary soundscape quality'. In the case studies described in the previous sections, water sound was classified as 'favourite' by the majority of the interviewees, and the introduction of water elements has dramatically improved soundscape quality.

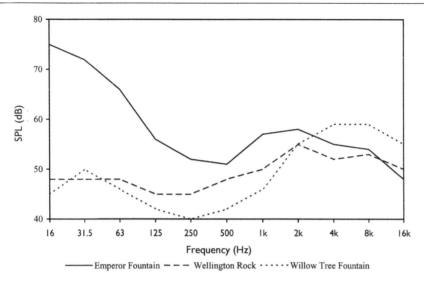

Figure 3.21 Comparison of the spectra of three water features in the Chatsworth Garden, England.

Yang (2005) analysed the characteristics of a series of water features from the viewpoints of their sound masking effect, spectrum enrichment function, spiritual life, fun making, as well as contemporary design tendencies. From detailed spectrum and dynamic range analysis, it was found that most water sounds have significant high-frequency components around 2–8kHz and some of them also have notable low-frequency components. This result can probably explain why water sound is always distinctive from the background.

The spectrum of water features is designable. Different flow methods result in different sound frequencies. Generally speaking, high-frequency components come from the water splash itself, whereas when a large flow of water is raised to a very high level and then dropped to a water body or hard surface, notable low-frequency components can be generated. In Figure 3.21 the spectra of three water features in the Chatsworth Garden in England are compared based on near field measurements, demonstrating the diversity in spectrum.

3.10.2 Space

If there are several acoustic zones in an urban open space, a suitable aural space or the source–listener distance for each zone should be designed. The scale of an aural space changes with time and place. For example, pre-industrial soundscapes and sounds emanating from a listener's own community may be heard at a considerable distance, reinforcing a sense of space and position and maintaining a relationship with home (Wrightson 2000). Nowadays, it is common that one's aural space is reduced to less than that of human proportions (Truax 2001).

An urban open space can be designed to encourage activities which generate active soundmarks. Green spaces, hard spaces, as well as thoroughfares should be arranged well in a square. A green space may enhance the natural appeal of a square, attract wild animals' activities such as bird singing, and improve the microclimate conditions and sound level distribution. Hard spaces are useful for generating many activities, especially among young people, such as dancing and skateboarding. From field surveys it has been shown that some patterns of design are more suitable for certain activities (Kang and Yang 2002), for example, defined edges, such as walls, colonnades, or shrub planting, often encourage activities to take place.

As analysed in Section 3.4, if the overall sound level is higher than a certain value, say 70dBA, people will feel annoyed, whatever the type of sound is. In this case it is important to reduce sound levels, for example, by designing square forms and boundaries, landscape elements, vegetation, urban furniture and noise barriers. Effects of architectural changes and urban design options on the sound field of urban open spaces are discussed in Chapters 4–7.

Landscape elements can also be used to create certain sound fields. For example, experimental research in woods has shown that tree trunks can scatter sound with different time delays, so that the conditions for the sensations of spaciousness and envelopment are created (Ruspa 2001).

Given the aural–visual interaction described in Section 3.2.4, integrative consideration of landscape and soundscape elements is important. For example, with the same traffic, the soundscape could differ significantly with the highway view, vegetation view and noise barrier view (Nathanail and Guyot 2001). A more direct connection between landscape and soundscape is sonic sculptures (Harvey 2000).

3.11 From outdoor soundscape to indoor acoustic comfort

Outdoor and indoor soundscapes are naturally connected and there are many interactions between them. Recently, the acoustic comfort in a series of buildings/spaces was studied (Lin 2000; Keeling-Roberts 2001; Chen 2002; Du 2002; Christophers 2003; Stepan 2003; Chung 2004; Lee 2004; Bai 2005; Ip 2005; Bradley 2005; Di Carlo 2005), and some of those studies are reviewed in this section, aiming at exploring the relationships between the characteristics of sound fields and perceptions of acoustic comfort, as well as ways to create comfortable acoustic environments (Kang 2003a).

3.11.1 Shopping mall atriums

A case study was carried out on the acoustic comfort in Sheffield's Meadowhall, one of the largest indoor shopping malls in the United Kingdom (Chen 2002; Chen and Kang 2003). Three aspects were considered: measurements of objective acoustic indices including sound level and reverberation, surveys among customers and staff on the levels of acoustic comfort, and correlations between the two aspects. During the study 90 customers and 80 staff members were interviewed. Three main atriums were studied: the Oasis, a multifunctional atrium containing stores, restaurants, cafes, cinemas and a games room; the Lower High Street, a long shopping atrium consisting of stores, booths, resting spaces and plants; and the Upper Central Dome, an open atrium linking the main pedestrian axes.

There are some special features in such spaces in terms of objective acoustic indices. The reverberation is generally long, and the longest RT occurs at middle frequencies at 2–3s. The decay curves are mostly concave, which means that the EDT is shorter than RT. Figure 3.22 shows measured temporal sound fluctuation and typical spectra. It can be seen that the SPL fluctuates considerably at different times of the day and week, and this is related to the number of customers, space features, as well as background music. The spectra in such atriums typically show peaks at middle frequencies, and considerable drops at high frequencies.

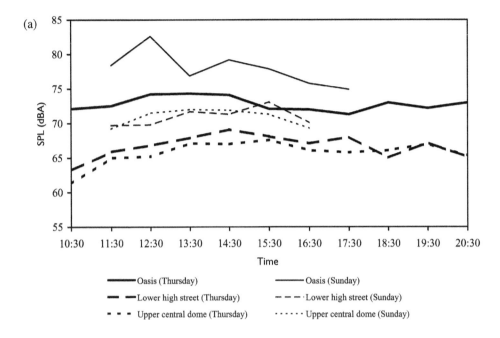

(a)

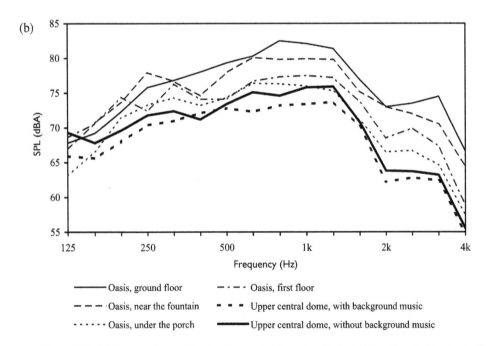

(b)

Figure 3.22 (a) Temporal sound fluctuations and (b) spectra (at typical locations in Meadowhall).

Generally speaking, people are not satisfied with the current acoustic environment in Meadowhall. In terms of demographic factors, no significant difference is found between age groups, and there is also no significant correlation between the acoustic evaluation and the acoustic condition at interviewees' home. On the other hand, the results suggest that the acoustic evaluation is affected by people's duration of stay and their activities. They may feel acoustically uncomfortable just after arriving, but after a short period they get used to it. After a longer period, they may feel uncomfortable again as they become tired with the continuous high noise level.

As expected, significant differences have been found between the acoustic perceptions of different sounds – sounds from fountains are considered to be the most pleasant and sounds from nearby people are the most annoying.

There is a tendency that the overall acoustic comfort evaluation becomes less satisfactory with increasing SPL, but the correlation coefficient is rather low, only 0.6 ($p < 0.01$), due to the complicated features of the various sound sources. It has also been found that for a given SPL, the annoyance scores are usually higher than, or the same as, those for loudness, showing people's tolerance.

In terms of speech intelligibility, the survey results suggest that there is a significant correlation between the communication quality and the EDT. In general, people feel more satisfied with the communication quality than with the overall acoustic comfort. It is also interesting to note that the staff group are more tolerant in terms of communication comfort than customers.

3.11.2 Library reading rooms

Acoustics in library reading rooms were investigated through a case study at the Sheffield University Main Library (Du 2002; Kang and Du 2003). Measurements in the architectural reading room (AR), which was 22.5m long, 7.2m wide and 2.4m high, showed that the SPL attenuation with distance was considerable, over 20dB across the room; the RT was rather short, 0.3–0.5s across the frequency range from 125Hz to 8kHz; and the general background noise, 37–45dBA, was not high. However, the acoustic comfort evaluation, with 67 students, was only at a satisfaction level of medium or lower, and it seemed that there was no significant correlation between the sound level and acoustic comfort evaluation. This reveals the contradiction in designing the acoustic environment in such spaces – balance between privacy and annoyance. In the main reading room (MR) the results were similar.

The study then compared natural and artificial sounds as background in the AR. Four sounds were played back with the same level of 50dBA, with four loudspeakers positioned at the room corners, creating a rather even sound field in the room. The sounds, pre-recorded on-site, included rain and wind in a small forest, rain hitting the ground, running water in a small stream, and noise from the library ventilation system. The spectra and temporal characteristics of the sounds are shown in Figure 3.23. The acoustic comfort under various conditions was then evaluated with 40 students, through a series of carefully designed questions.

It is important to note that the mean evaluation scores for the running water sound are generally significantly higher than those for other sounds ($p < 0.05$). A possible reason is that in comparison with other sounds, the running water sound has rather weak low-frequency components, as can be seen in Figure 3.23. Rain/wind and rain sounds, although also from

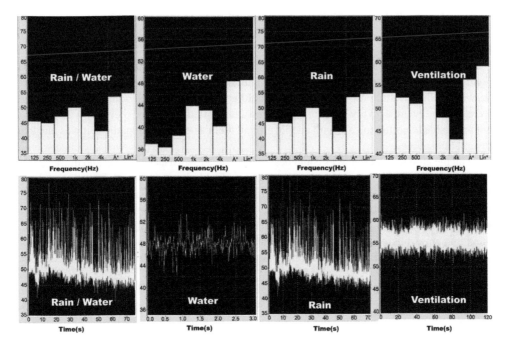

Figure 3.23 Spectra and temporal characteristics of the sounds played in the architectural reading room (AR) at the Sheffield University Main Library. A colour representation of this figure can be found in the plate section.

nature, received similar scores to ventilation noise. This is probably due to their notable low-frequency components and, more importantly, their large dynamic range.

Subjects were asked to identify the sounds when they were played back. It is interesting to note that although various sounds caused rather different sensations, they were mostly incorrectly identified.

3.11.3   Football stadiums

The atmosphere inside a football stadium is of fundamental importance to the performance of the team, and thus the prosperity of the club. In six typical football grounds, including McAlpine Stadium, Huddersfield; Ewen Fields, Hyde; Valley Parade, Bradford; Edgeley Park, Stockport; Pride Park, Derby; and Maine Road, Manchester, SPL measurements and subjective surveys were carried out relating to the acoustic atmosphere (Keeling-Roberts 2001). The measured average SPL was 77–98dBA, and the maximum SPL was 102–120dBA. In the subjective survey five linear scales were used. For example, for 'how well can you hear sounds from the pitch', the scales were 1, very well; 2, quite well; 3, ok; 4, not very well; and 5, not at all. At least 30 fans were interviewed in each stadium. The general aim was to find out what exactly makes a good acoustic atmosphere, and what architectural features of a football stadium combine to create this effect.

Fans at all stadiums except Ewen Fields wanted to hear sounds from the pitch better than they could – mostly by about one point on the scale. The mean answers for all interviewees were 2.99 for how well they could currently hear sounds from the pitch, and 2.14, or 'quite well' for how well they would like to be able to. A mean answer of 2.06 shows the sounds from other parts of the stadium to be slightly more important to fans than hearing sounds from the pitch.

All the stadiums had very audible PA systems. Whilst they are of great importance regarding safety, they do not seem to contribute to a good atmosphere. Although the mean answers to the question 'how well can you hold a conversation with someone near to you' were invariably either 'very well' or 'quite well', fans often suggested that they would prefer not to be able to communicate with people around them as easily, especially if it was due to a better atmosphere.

At all the grounds, interviewees could hear external noise, such as wind, rain and traffic, better than they would like to. It could be heard best at Ewen Fields, but it was also tolerated most there. Perhaps this is because the fans are used to the noise, or the noise is of a more tolerable variety.

For five of the stadiums, the responses to two questions regarding quality of atmosphere and loudness of the stadium were very similar. Further analysis suggests that most fans do think that atmosphere is very, if not totally, dependent on sound volume. For the question of 'how important is the acoustic atmosphere to you', the mean rating for all grounds is 2.21, or 'important'.

Overall, the subjective analysis suggests several strategies for good acoustic atmosphere in a stadium: a large capacity; a high attendance–capacity ratio; huge, multi-tiered stands; standing areas; large proportion of capacity for away fans; and seats close to and all around the pitch.

3.11.4   Swimming spaces

Subjective surveys were carried out at three typical swimming spaces in Sheffield, including the Cofield swimming pool at Sheffield University, Ponds Forge sport centre, and Hillsborough leisure centre (Lin 2000). The number of interviewees were 51, 52 and 90 respectively. The questionnaire included three sections: demographic information of the users and the use of the spaces, general satisfaction of the physical environment and facilities, and the acoustic comfort. The surveys were mainly carried out in swimming and audience areas, but other areas including restaurants and changing areas were also considered. The acoustic questions included evaluation of overall acoustic environment as well as of various sound sources. During the survey, the SPL was measured in a number of typical positions. Figure 3.24 shows the SPL variation with time in the three swimming spaces. It can be seen that the level varies considerably in different pools due to different activities, and the SPL range is about 60–80dBA, which is rather high. The middle frequency RT is 2.8, 3 and 2.1s, and the L_{eq} is 62, 67 and 77.3dBA, in the Cofield, Ponds Forge and Hillsborough spaces, respectively.

The correlations between the RT and SPL and the overall acoustic comfort of the swimming area, where the comfort rating is 1, poor; 2, average; 3, good; and 4, excellent, suggest that in terms of acoustic comfort, people prefer long reverberation, but not high SPL. The correlations between RT and subjective rating of liveliness and reverberation seem to suggest that people are generally satisfied with the current situation, although the RT varies considerably in the three spaces. In the subjective rating three scales were used: 1, not lively enough and should be more lively (reverberant); 2, fine; and 3, too lively (reverberant). The survey results also indicate that in the three spaces studied, there is no significant correlation between

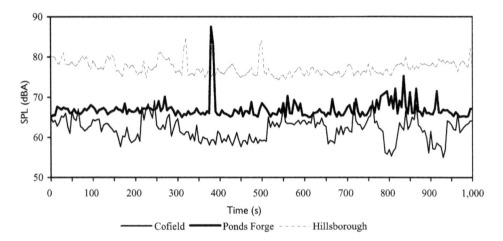

Figure 3.24 Typical SPL distribution with time in the main swimming areas.

RT and subjective evaluation of speech intelligibility, for long and short distance communication as well as for the PA system.

On average, in the three swimming spaces, 50 per cent of the interviewees believe acoustics is an important or very important issue in swimming spaces; 75 per cent of the interviewees indicate that the major noise source is children's shouting; and 32 per cent of people feel acoustically uncomfortable after swimming, and another 33 per cent sometimes have such a feeling.

3.11.5 Churches

Objective measurements and questionnaire surveys were carried out in five churches in Shef-field, including the Buddhist Centre (St Josephs Church), Walkley; St Marks Church, Broomhill; Wesley Hall, Crookes; Christ Church, Fulwood; and Sheffield Cathedral (Christophers 2003; Stepan 2003; Stepan *et al.* 2003). Again, the measurements included sound level and reverberation, and the questionnaire included people's general feeling about acoustic comfort, and evaluation regarding various usages. In each church 30–35 interviews were conducted.

Statistical analysis of the survey results suggests that there is no significant correlation between the acoustic comfort of a church and the RT. Generally, an RT value of 1.8–3.3s at middle frequencies corresponds to a 'good' and 'satisfactory' level. For speech intelligibility, conversely, the subjective rating score tends to become less favourable with increasing RT. For the quality of choir and musical instruments, people tend to prefer longer reverbera-tion. Overall, it seems that there are considerable differences between the evaluation of acoustic functions and the acoustic comfort.

Further analysis of the results suggests the importance for designers to consider whether people come to church for the sole purpose of hearing the priest's sermon and then praying or whether they also come to share in a collective atmospheric experience. If going to church is all about community and spirituality and if the acoustics serve only to facilitate the basic func-tions but dampen the communal atmosphere then the building is failing in its deeper purpose.

It seems clear that there is an important differentiation to be made between how well a church is performing its acoustic functions and its level of acoustic comfort. Acoustic comfort is less easily definable than function; it requires the designer to think of space as an acoustic environment rather than a facilitator of events. It is to do with creating a feeling, rather than fulfilling a function. This is also true in many other architectural and urban spaces.

Chapter 4

Microscale acoustic modelling

As most acoustic problems cannot be resolved by purely analytical procedures, computer simulation has been continuously developed since the application of computers in the 1960s. The models for calculating sound distribution in urban areas can be roughly divided into two groups: microscale and macroscale (Kang 2005b). The former, often using simulation techniques, is used for accurately calculating the sound field for small- and medium-scale urban areas such as a street or square. The latter, normally involving statistical methods and simplified algorithms, is for describing the sound distribution in a relatively large urban area. For both kinds of models, the main acoustic index is the distribution of SPL, but reverberation has also been given attention, especially for microscale urban areas.

This chapter presents typical microscale acoustic models as well as related acoustic theories, including energy-based image source methods for street canyons and urban squares with geometrically (specularly) reflecting boundaries (Section 4.1), image source method considering interference (Section 4.2), ray tracing (Section 4.3), radiosity model for diffusely reflecting boundaries (Section 4.4), transport theory (Section 4.5), the equivalent source method (Section 4.6), and other models (Section 4.7). In Section 4.8 techniques for acoustic animation are discussed, mainly for urban squares. Section 4.9 briefly introduces/reviews physical scale modelling techniques for urban acoustics, whereas Section 4.10 presents some actual measurements; both are useful for validating the simulation models.

Where buildings have a constant height in a street/square, the term street/square height is used in the following sections for convenience.

4.1 Energy-based image source method

The image source method treats a flat surface as a mirror and creates an image source. In other words, the boundaries are regarded as geometrically reflective. The reflected sound is then modelled with a sound path directly from the image source to a receiver. Multiple reflections are achieved by considering further images of the image source. When the wavelengths are short compared to the space/boundary dimensions, the energy-based image source method can be used, namely ignoring the physical wave nature of sound and treating a sound wave as a sound ray.

With the image source method the situation of a source in an urban space is replaced by a set of mirror sources in a free field visible from the receiver considered. The acoustic indices at the receiver are determined by summing the contribution from all the image sources. For each reflection the strength of the image source is reduced due to the surface absorption.

A disadvantage of the image source method is that it slows exponentially with increasing orders of reflection as the number of images increases. In addition, validity and visibility tests are required for image sources.

4.1.1 Street canyons

Consider an idealised rectangular urban street with buildings along both sides of a constant height, where the façades are called A and B, and the ground is called G (Kang 2000d). The street width is W, the street height is H, and a point source S is at (S_x, S_y, S_z). Figure 4.1 illustrates the distribution of image sources in the street. It can be seen that there are two lines of image sources. For calculation convenience, the image sources are divided into four groups, namely $A1$, $A2$, $B1$ and $B2$. Groups $A1$ and $A2$ correspond to the reflections between two façades, and groups $B1$ and $B2$ include the reflection from the street ground.

With reference to Figure 4.1, the energy from an image source to a receiver, R, at (R_x, R_y, R_z) can be determined. First consider an image source i $(i = 1, \ldots, \infty)$ in group $A1$. For odd values of i the energy to the receiver is

$$E_i(t)_{A1} = \frac{1}{4\pi d_i^2}(1-\alpha_A)^{(i+1)/2}(1-\alpha_B)^{(i-1)/2}e^{-Md_i} \qquad \left(t = \frac{d_i}{c}\right) \tag{4.1}$$

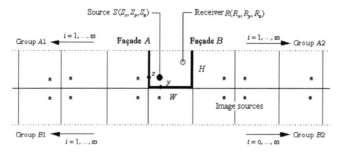

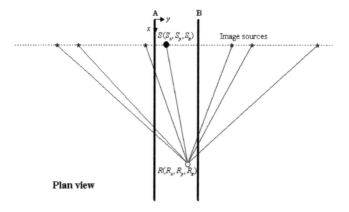

Plan view

Figure 4.1 Distribution of image sources in an idealised street canyon.

where α_A and α_B are the absorption coefficient of façades A and B, respectively. M (Np/m) is the intensity-related attenuation constant in air. t $(t = 1,\ldots,\infty)$ is the time and $t = 0$ represents the moment at which the source generates an impulse. d_i is the distance from the image source i to the receiver:

$$d_i^2 = (S_x - R_x)^2 + [(i-1)W + S_y + R_y]^2 + (S_z - R_z)^2 \tag{4.2}$$

For even i

$$E_i(t)_{A1} = \frac{1}{4\pi d_i^2}(1-\alpha_A)^{i/2}(1-\alpha_B)^{i/2}e^{-Md_i} \quad \left(t = \frac{d_i}{c}\right) \tag{4.3}$$

with

$$d_i^2 = (S_x - R_x)^2 + (iW - S_y + R_y)^2 + (S_z - R_z)^2 \tag{4.4}$$

For an image source i $(i = 1,\ldots,\infty)$ in group $A2$, for odd values of i the sound energy to the receiver is

$$E_i(t)_{A2} = \frac{1}{4\pi d_i^2}(1-\alpha_A)^{(i-1)/2}(1-\alpha_B)^{(i+1)/2}e^{-Md_i} \quad \left(t = \frac{d_i}{c}\right) \tag{4.5}$$

with

$$d_i^2 = (S_x - R_x)^2 + [(i+1)W - S_y - R_y]^2 + (S_z - R_z)^2 \tag{4.6}$$

For even i,

$$E_i(t)_{A2} = \frac{1}{4\pi d_i^2}(1-\alpha_A)^{i/2}(1-\alpha_B)^{i/2}e^{-Md_i} \quad \left(t = \frac{d_i}{c}\right) \tag{4.7}$$

with

$$d_i^2 = (S_x - R_x)^2 + (iW + S_y - R_y)^2 + (S_z - R_z)^2 \tag{4.8}$$

For groups $B1$ and $B2$, the energy from the image sources to the receiver can be determined using Equations (4.1)–(4.8) but replacing the term $S_z - R_z$ with $S_z + R_z$ and also, considering the ground absorption, α_G.

By summing the energy from all the image sources in groups $A1$, $A2$, $B1$ and $B2$, and taking direct sound transfer into account, the energy response at receiver R can be given by

$$L(t) = 10\log\left[E_d(t) + \sum_{i=1}^{\infty} E_i(t)_{A1} + \sum_{i=1}^{\infty} E_i(t)_{A2} + \sum_{i=1}^{\infty} E_i(t)_{B1} + \sum_{i=0}^{\infty} E_i(t)_{B2}\right] - L_{ref} \tag{4.9}$$

where L_{ref} is the reference level. The direct energy can be calculated by

$$E_d(t) = \frac{1}{4\pi d_r^2} e^{-Md_r} \qquad \left(t = \frac{d_r}{c} \right)$$

(4.10)

where d_r is the source–receiver distance. By introducing a term, d_r / c, to translate the arrival time of direct sound to zero, the decay curve can be obtained by the reverse-time integration of $L(t)$. Consequently, the RT and EDT can be determined. The steady-state SPL at receiver R can be calculated by

$$L = 10\log \sum_{\Delta t} 10^{\frac{L(t)}{10}}$$

(4.11)

4.1.2 Urban squares

Consider an idealised rectangular square as shown in Figure 4.2 (Kang 2005a). Similar to the situation in urban streets, by assuming the boundaries to be geometrically reflective, a series of image sources can be created, as illustrated in Figure 4.3. Again, for the convenience of calculation, the image sources in Figure 4.3 are divided into a number of groups. Consider an image source $(j,k)(j=1,\ldots,\infty; k=1,\ldots,\infty)$ in group I-iii, for example, the energy from an image source to a receiver R at (R_x, R_y, R_z) can be determined by

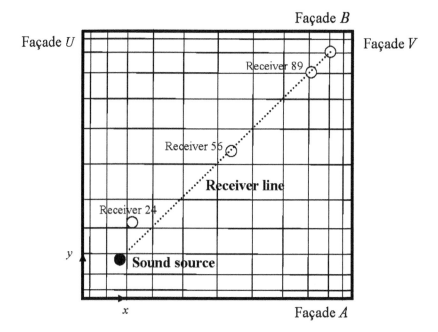

Figure 4.2 Plan view of an idealised urban square. The grid lines show the division of patches (see Section 4.4). The source and receiver positions used in a parametric study are also shown (see Section 7.3), where for the 50 x 50m square, the source is at (10m,10m), and the positions of four typical receivers are 24 (12.5m, 17.5m), 56 (27.5m, 27.5m), 89 (42.5m, 42.5m), and 100 (47.5m, 47.5m), corresponding to source–receiver distances of 8, 25, 46 and 53m, respectively.

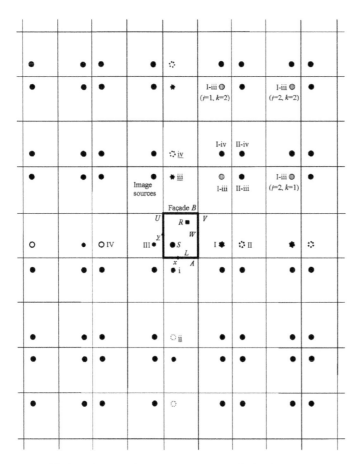

Figure 4.3 Distribution of image sources in an idealised square, plan view.

$$E_{j,k}(t) = \frac{1}{4\pi d_{j,k}^2}(1-\alpha_A)^{k-1}(1-\alpha_B)^k(1-\alpha_U)^{j-1}(1-\alpha_v)^j e^{-Md_{j,k}} \quad \left(t = \frac{d_{j,k}}{c}\right) \qquad (4.12)$$

where α_A, α_B, α_U and α_v are the absorption coefficient of façades $A, B, U,$ and V, respectively, and $d_{j,k}$ is the distance from the image source (j,k) to the receiver:

$$d_{j,k}^2 = (2jL - S_x - R_x)^2 + (2kW - S_y - R_y)^2 + (S_z - R_z)^2 \qquad (4.13)$$

where L and W are the square length and width respectively.

By considering the ground reflection, an image source plane similar to that in Figure 4.3 can be obtained. The energy from those image sources to receiver R can be determined in a similar manner as above, but the term $S_z - R_z$ must be replaced by $S_z + R_z$ and the ground absorption α_G must be taken into account.

4.2   Image source method considering interference

For relatively narrow streets and/or low frequencies, interference effects due to multiple reflections from building façades and ground are important. At each reflection of a sound wave from a boundary, a complex reflection coefficient should be used to take the amplitude and phase change into account. In other words, a coherent model rather than an energy-based, incoherent model should be used.

Iu and Li (2002) addressed the problem of predicting sound propagation in narrow street canyons, assuming that the height of the buildings is much greater than the street width. A point source is considered, simulating, for example, air conditioners installed on building façades and powered mechanical equipment for repair and construction work. Starting from the Helmholtz equation, the solution at a receiver is derived, where the sound fields due to the point source and its images are summed coherently such that mutual interference effects between contributing rays can be included in the analysis:

$$
p = \frac{1}{4\pi} \sum_{l=0}^{\infty} \Bigg[Q^l(d_{l1},\theta_{l1},\beta_L)Q^l(d_{l1},\theta_{l1},\beta_R)\frac{e^{ikd_{l1}}}{d_{l1}}
$$

$$
+ Q^{l+1}(d_{l2},\theta_{l2},\beta_L)Q^l(d_{l2},\theta_{l2},\beta_R)\frac{e^{ikd_{l2}}}{d_{l2}}
$$

$$
+ Q^l(d_{l3},\theta_{l3},\beta_L)Q^{l+1}(d_{l3},\theta_{l3},\beta_R)\frac{e^{ikd_{l3}}}{d_{l3}}
$$

$$
+ Q^{l+1}(d_{l4},\theta_{l4},\beta_L)Q^{l+1}(d_{l4},\theta_{l4},\beta_R)\frac{e^{ikd_{l4}}}{d_{l4}}
$$

$$
+ Q^l(\bar{d}_{l1},\bar{\theta}_{l1},\beta_L)Q^l(\bar{d}_{l1},\bar{\theta}_{l1},\beta_R)Q(\bar{d}_{l1},\bar{\Theta}_{l1},\beta_G)\frac{e^{ik\bar{d}_{l1}}}{\bar{d}_{l1}}
$$

$$
+ Q^{l+1}(\bar{d}_{l2},\bar{\theta}_{l2},\beta_L)Q^l(\bar{d}_{l2},\bar{\theta}_{l2},\beta_R)Q(\bar{d}_{l2},\bar{\Theta}_{l2},\beta_G)\frac{e^{ik\bar{d}_{l2}}}{\bar{d}_{l2}}
$$

$$
+ Q^l(\bar{d}_{l3},\bar{\theta}_{l3},\beta_L)Q^{l+1}(\bar{d}_{l3},\bar{\theta}_{l3},\beta_R)Q(\bar{d}_{l3},\bar{\Theta}_{l3},\beta_G)\frac{e^{ik\bar{d}_{l3}}}{\bar{d}_{l3}}
$$

$$
+ Q^{l+1}(\bar{d}_{l4},\bar{\theta}_{l4},\beta_L)Q^{l+1}(\bar{d}_{l4},\bar{\theta}_{l4},\beta_R)Q(\bar{d}_{l4},\bar{\Theta}_{l4},\beta_G)\frac{e^{ik\bar{d}_{l4}}}{\bar{d}_{l4}} \Bigg] \tag{4.14}
$$

where the first four terms are the contributions due to the reflections from the façades only, and the last four terms represent a set of image sources due to the presence of a reflecting ground. k is the wave number of the source. $d_{l1}, d_{l2}, d_{l3}, d_{l4}$ and $\bar{d}_{l1}, \bar{d}_{l2}, \bar{d}_{l3}, \bar{d}_{l4}$ are path lengths of the image sources. $\theta_{l1}, \theta_{l2}, \theta_{l3}, \theta_{l4}, \bar{\theta}_{l1}, \bar{\theta}_{l2}, \bar{\theta}_{l3}, \bar{\theta}_{l4}, \Theta_{l1}, \Theta_{l2}, \Theta_{l3}, \Theta_{l4}$ are the angles of incidence of the reflected waves measured from the normal of the reflecting plane. The general term $Q(d,\theta,\beta)$ is the spherical wave reflection coefficient that can be determined for a given separation of an image source and receiver, d; the angle of incidence of the reflected wave, θ; and the normalised admittance of the boundary, β, including left façade β_L, right side façade β_R and ground β_G. According to Attenborough (1988),

$$Q(d,\theta,\beta) = R_p + (1 - R_p)F(w)$$ (4.15)

$$R_p = \frac{\cos\theta - \beta}{\cos\theta + \beta}$$ (4.16)

$$F(w) = 1 + i\sqrt{\pi}we^{-w^2}\operatorname{erfc}(-iw)$$ (4.17)

$$w = +\sqrt{\frac{1}{2}ikd}(\cos\theta + \beta)$$ (4.18)

where $\operatorname{erfc}()$ represents the complementary error function.

The direct sound with the consideration of phase can be calculated by

$$P_d = \frac{1}{4\pi}\left(\frac{e^{ikd_r}}{d_r}\right)$$ (4.19)

where d_r is the source–receiver distance.

A comparison has shown that the coherent and incoherent models give rather similar results when the width of a street canyon is greater than about 10m, whereas when the street width becomes narrower, the differences are more significant. This is also proved by the measurements in a 1:10 scale model and in an actual side lane of 1.55m wide (Iu and Li 2002).

It is noted, however, when the boundaries reflect sound diffusely, energy-based and incoherent models, such as those presented in Sections 4.4 and 4.5, can also be applicable to narrower streets, since the effects of phase cancellation and addition tend to be averaged. Measurements in a street where the width is 7.9m and the boundaries are relatively diffuse show that within frequency range from 250Hz to 5kHz the effect of frequency on sound attenuation is insignificant (Picaut *et al.* 2005).

4.3   Ray tracing

4.3.1   General principles

Ray tracing creates a dense spread of rays, which are subsequently reflected around a space and tested for intersection with a detector (receiver) such as a sphere or a cube (Krokstad *et al.* 1968; Kulowski 1984). A sound ray can be regarded as a small portion of a spherical wave with a vanishing aperture, which originates from a certain point. An echogram at a receiver can be constructed using the energy attenuation of the intersecting rays and distances travelled.

Particle tracing uses similar algorithms to ray tracing, but the method of detection is different. With the particle model, the longer a particle stays in the detector, the higher is its contribution to the energy density.

Beams are rays with a nonvanishing cross section (Drumm and Lam 2000). The beams may be cones with a circular cross section or pyramids with a polygonal cross section. By using beams, a point detector can be used, instead of a sphere or a cube. Beams are reflected around a space and tested for illumination of the detector.

The above methods are generally used for geometrically reflecting boundaries. Exact simulation of diffusely reflecting boundaries could be extremely time consuming after several reflections since a large number of rays/particles/beams need to be traced. A commonly used approximation is to generate a random number to determine the direction of reflection, rather than tracing reflected sound to many directions.

A number of models based on ray tracing, particle tracing or beam tracing have been developed for urban areas, such as models for calculating sound distribution in interconnected streets (Thomas 2000), for considering strategic design options in a single street (Rahim 2001), and predicting dynamic traffic noise distribution (De Coensel *et al.* 2005; see also Section 5.4.3). An advantage of such models is that they can be used for relatively complicated urban configurations.

4.3.2   A model for urban squares

Aiming at real-time simulation, a ray tracing model has been developed for microscale urban environments, especially urban squares, as a key part of an animation/auralization tool (Meng *et al.* 2005; see also Section 4.9). The core program is based on a multiple loop, tracing the current ray around the environment until it hits the receiver, gets lost or exceeds the predefined ray number or reflection order. All the rays are arranged according to their arrival, and subsequently, an impulse response can be produced.

In addition to the basic ray tracing algorithms, some special methods relating to microscale urban environments have been implemented in the model in order to reduce the computation time. The total number of rays that hit a receiver can be predefined, so that the resolution of the impulse response can be controlled. A simple transformation algorithm is applied based on the input configuration to define the range of calculation area and simplify the geometry.

Another important algorithm is the optimisation of rays. For example, since urban open spaces have a totally absorbent 'ceiling', the model does not generate any rays to that area. Similarly, any ray, after one or more reflections, will be stopped once it hits the 'ceiling'.

The QuickSort algorithm (Weiss 1998), an effective method of arranging a large number of data in an array, is adopted in the model to arrange the time and energy of the rays for creating impulse responses. This algorithm has been compared with various methods including Bubble, Selection and Merge, and the results show that it is the quickest method of arranging elements for the situation of urban squares. The model outputs *.sim files, which are required by the animation/auralization tool (see Section 4.9).

4.4   Radiosity model

The image source method and ray tracing are usually applied for geometrically reflecting boundaries. Starting with a brief discussion about the importance of taking boundary diffusion into account, this section presents a radiosity model, considering urban streets and squares with diffusely reflecting boundaries according to the Lambert cosine law.

4.4.1   Role of diffusion

Since there are always some irregularities on building or ground surfaces, it is necessary to consider diffuse reflections (Lyon 1974). The back-diffusion effect of reverberation has been demonstrated through measurements in a street canyon (Picaut *et al.* 2005). Another investigation on the role of sound reflection from building façades using physical scale models (Ismail and Oldham 2005) suggests that the scattering coefficient is about 0.09–0.13 for façades with surface irregularities similar to those found on real building façades and the coefficient is not very sensitive to the degree of typical surface irregularities. However, although the coefficient appears to be small, the effect of multiple reflections is to make the diffuse reflection mechanism dominant at higher orders of reflection. A progression of the reflected sound field, from being dominated by geometrical reflections to domination by diffuse reflections in a similar way to pure diffuse reflections, has been demonstrated.

Several ways of considering diffuse reflections have been proposed. Bullen and Fricke (1976) considered the effects of scattering from objects and protrusions in streets by analysing the sound field in terms of its propagating modes. In a model suggested by Davies (1978), the sound field was assumed to be the sum of a multiple-geometrically reflected field and a diffuse field that was fed from scattering at boundaries at each reflection of the geometrical field. Wu and Kittinger (1995) developed a model using the method proposed by Chien and Carroll (1980) for describing a surface by a mixed reflection law with an absorptivity or reflectivity parameter and a smoothness or roughness parameter. Heutschi (1995) suggested modelling sound propagation by a continuous energy exchange within a network of predefined points located on individual plane surfaces, where it was possible to define any directivity pattern for the reflections.

4.4.2 General principles of the radiosity model

The radiosity method is an effective way of considering diffusely reflecting boundaries. The method, also called radiation balance, radiation exchange or radiant interchange, was first developed in the nineteenth century for the study of radiant heat transfer in simple configurations (Siegel and Howell 1981). Computer implementations of the method have been substantially developed in computer graphics research (Foley *et al.* 1990; Sillion and Puech 1994), where radiosity is predominantly used to calculate light energy. By considering relatively high frequencies, the method can also be used in the field of room (Moore 1984; Lewers 1993; Kang 2002e, 2002f, 2004b) and environmental acoustics. A significant feature of this application is that when the reverberation is taken into account the computation time is considerably increased.

The radiosity method divides boundaries in a space such as an urban street and square into a number of patches (elements) and replaces the patches and receivers with nodes in a network. The sound propagation in the space can then be simulated by energy exchange between the nodes. The energy moving between pairs of patches depends on a form factor, which is the fraction of the sound energy diffusely emitted from one patch which arrives at the other by direct energy transport.

4.4.3 Patch division

Consider an idealised street canyon with a length L, width W and height H, as illustrated in Figure 4.4 (Kang 2000d). The first step of the model is to divide each boundary into a number of patches. Similar to Section 4.1, the boundaries are defined as: G, ground; A, building façade at $y = 0$; and B, building façade at $y = W$. Also, the patches along the length, width and height are defined as l $(l = 1,\ldots,N_x)$, m, $(m = 1,\ldots,N_y)$ and n $(n = 1,\ldots,N_z)$, respectively.

The model is more accurate with finer patch parameterisation, but there is a square-law increase of calculation time with the number of patches. To reduce the patch number, the

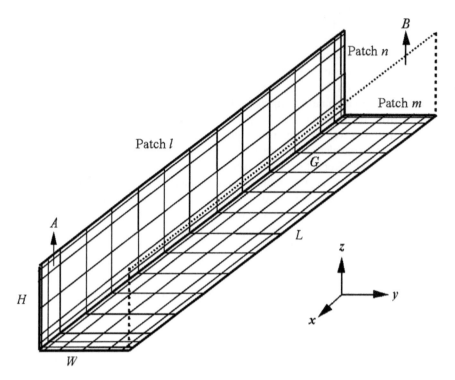

Figure 4.4 Three-dimensional projection of an idealised street showing an example of patch division.

boundaries are so divided that a patch is smaller when it is closer to an edge. This is because, for a given patch size, form factor calculations become less accurate as the patch moves closer to an edge. For the convenience of computation, the division of boundaries is made using geometrical series (see Figure 4.4).

If a dimension is not large, for example, along the width, the patch size dd_m increases from the edges to the centre, namely

$$dd_m = \begin{array}{c} k_m q_y^{m-1} \\ k_m q_y^{N_Y-m} \end{array} \begin{pmatrix} 1 \le m \le \dfrac{N_Y}{2} \\ \dfrac{N_Y}{2} < m \le N_Y \end{pmatrix} \tag{4.20}$$

where q_y $(q_y > 1)$ is the ratio between two adjacent patches, N_Y should be an even number, and

$$k_m = \frac{W}{2} \frac{1-q_y}{1-q_y^{\frac{N_Y}{2}}} \tag{4.21}$$

For a relatively large dimension, for example, along the length, the patch size dd_l increases from $l = 1$ to $N_X/4$, decreases from $l = 3N_X/4 + 1$ to N_X, and is constant between $l = N_X/4 + 1$ and $3N_X/4$. In this way extreme differences in patch size can be avoided. The patch sizes can be determined by

$$dd_l = \begin{cases} k_l q_x^{l-1} & \left(1 \leq l \leq \dfrac{N_X}{4}\right) \\[3mm] k_l q_x^{\frac{N_X}{4}-1} & \left(\dfrac{N_X}{4} < l \leq \dfrac{3N_X}{4}\right) \\[3mm] k_l q_x^{N_X-l} & \left(\dfrac{3N_X}{4} < l \leq N_X\right) \end{cases} \tag{4.22}$$

where q_x $(q_x > 1)$ is the ratio between two adjacent patches, N_X should be divisible by 4, and

$$k_l = \frac{L}{2} \frac{1}{(1-q_x^{N_X/4})/(1-q_x)+(N_X/4)q_x^{(N_X/4)-1}} \tag{4.23}$$

Correspondingly, the coordinates of the centre of a patch, for example, $G_{l,m}$, can be determined by

$$d_l = -\frac{1}{2}dd_l + \sum_{l=1}^{l} dd_l \quad \text{and} \quad d_m = -\frac{1}{2}dd_m + \sum_{m=1}^{m} dd_m \tag{4.24}$$

4.4.4   First-order patch sources

The next step of the model is to distribute the sound energy of an impulse source to the patches. The basic principle of the source energy distribution is that the energy fraction at each patch is the same as the ratio of the solid angle subtended by the patch at the source to the total solid angle.

Consider a point source S at (S_x, S_y, S_z). A first-order patch source, $G_1(t)_{l,m}$, for example, as shown in Figure 4.5, can be calculated by

$$G_1(t)_{l,m} = K(1-\alpha_{G_{l,m}})e^{-MS_{l,m}}\frac{1}{4\pi}\int_0^{\beta_{l,m}}\int_{\varphi_{l,m}}^{\varphi_{l,m}+\Delta\varphi_{l,m}}\cos\varphi\, d\varphi\, d\beta$$

$$= K(1-\alpha_{G_{l,m}})e^{-MS_{l,m}}\frac{1}{4\pi}\left|\sin(\varphi_{l,m}+\Delta\varphi_{l,m})-\sin\varphi_{l,m}\right|\beta_{l,m} \quad (t=\frac{S_{l,m}}{c}) \tag{4.25}$$

where K is a constant in relation to the sound power of the source. $\alpha_{G_{l,m}}$ is the angle-independent absorption coefficient of patch $G_{l,m}$. To consider buildings with a lower height than H, the absorption coefficient can be given as 1. $S_{l,m}$ is the mean beam length between the source and patch $G_{l,m}$, which can be approximated by the distance from the source to the centre of the patch. $\varphi_{l,m}$, $\Delta\varphi_{l,m}$, and $\beta_{l,m}$ are the angles determining the location of patch $G_{l,m}$ with reference to the source, which can be calculated by

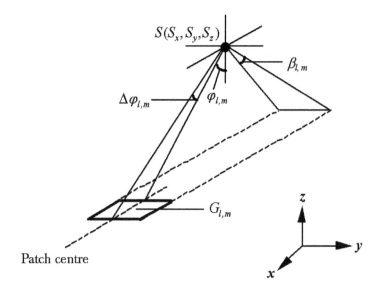

Figure 4.5 Distribution of the energy from a point source to a patch on the ground.

$$\sin(\varphi_{l,m} + \Delta\varphi_{l,m}) = \frac{d_l + (1/2)dd_l - S_x}{\sqrt{(d_l + (1/2)dd_l - S_x)^2 + (d_m - S_y)^2 + S_z^2}} \qquad (4.26)$$

$$\sin\varphi_{l,m} = k_\varphi \frac{d_l - (1/2)dd_l - S_x}{\sqrt{(d_l - (1/2)dd_l - S_x)^2 + (d_m - S_y)^2 + S_z^2}} \qquad (4.27)$$

and

$$\beta_{l,m} = \left|\arctan\left|\frac{d_m + (1/2)dd_m - S_y}{S_z}\right| - k_\beta \arctan\left|\frac{d_m - (1/2)dd_m - S_y}{S_z}\right|\right| \qquad (4.28)$$

In Equation (4.27) k_φ is used to consider the source position which is between the two sides of patch $G_{l,m}$. That is, $k_\varphi = -1$ when $d_l - dd_l/2 \le S_x \le d_l + dd_l/2$, otherwise $k_\varphi = 1$. Similarly, in Equation (4.28) $k_\beta = -1$ when $d_m - dd_m/2 \le S_y \le d_m + dd_m/2$, and $k_\beta = 1$ otherwise.

If a source is directional, a term in Equation (4.25) representing the radiation strength of the source in the direction of patch $G_{l,m}$ should be added.

The patches are now regarded as sound sources, and are subsequently referred to as first-order patch sources.

4.4.5   Form factors

In an idealised street canyon, as illustrated in Figure 4.4, the relative location of any two patches is either orthogonal or parallel. For orthogonal patches, the form factor can be calculated using Nusselt's method (Foley *et al.* 1990), where computing form factor is equivalent to projecting the receiving patch onto a unit hemisphere centred on the radiation patch, projecting this area orthographically down onto the hemisphere's unit circle base, and dividing by the area of the circle. As an example, Figure 4.6 illustrates the calculation from emitter $A_{l',n'}$ ($l'=1,\ldots,N_X$, $n'=1,\ldots,N_Z$) to receiver $G_{l,m}$. By considering the absorption of patch $A_{l',n'}$ and air absorption, the energy emitted from $A_{l',n'}$ to $G_{l,m}$, $AG_{(l',n'),(l,m)}$, can be calculated by

$$AG_{(l',n'),(l,m)} = (1-\alpha_{A_{l',n'}})e^{-Md_{(l',n'),(l,m)}}$$

$$\frac{1}{2\pi}\left|\cos^2\gamma_{(l',n'),(l,m)} - \cos^2\left(\gamma_{(l',n'),(l,m)} + \Delta\gamma_{(l',n'),(l,m)}\right)\right|\vartheta_{(l',n'),(l,m)} \qquad (4.29)$$

where $d_{(l',n'),(l,m)}$ is the mean beam length between patches $A_{l',n'}$ and $G_{l,m}$, which can be approximated using the distance between the centres of the two patches.

$\gamma_{(l',n'),(l,m)}$, $\Delta\gamma_{(l',n'),(l,m)}$ and $\vartheta_{(l',n'),(l,m)}$ are the angles for determining the relative location of the two patches:

$$\cos\gamma_{(l',n'),(l,m)} = \frac{d_m - (1/2)dd_m}{\sqrt{(d_l - d_{l'})^2 + (d_m - (1/2)dd_m)^2 + d_{n'}^2}} \qquad (4.30)$$

$$\cos\left(\gamma_{(l',n'),(l,m)} + \Delta\gamma_{(l',n'),(l,m)}\right) = \frac{d_m + (1/2)dd_m}{\sqrt{(d_l - d_{l'})^2 + (d_m + (1/2)dd_m)^2 + d_{n'}^2}} \qquad (4.31)$$

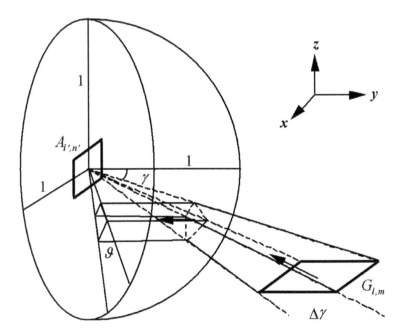

Figure 4.6 Determination of the form factor from emitter $A_{l',n'}$ to an orthogonal patch $G_{l,m}$.

$$\vartheta_{(l',n'),(l,m)} = \left| \arctan\frac{d_l - (1/2)dd_l - d_{l'}}{d_{n'}} - k_\vartheta \arctan\frac{d_l + (1/2)dd_l - d_{l'}}{d_{n'}} \right|$$ (4.32)

In Equation (4.32) K_ϑ is used to consider the case where the two patches have the same coordinate in the length direction, namely $k_\vartheta = -1$ when $l = l'$, otherwise $k_\vartheta = 1$.

For parallel patches, the form factor can be calculated by a method developed by Cohen and Greenberg (1985), which projects the receiving patch onto the upper half of a cube centred on the radiation patch. Consider patches $A_{l',n'}$ ($l' = 1, \ldots, N_X$, $n' = 1, \ldots, N_Z$) and $B_{l,n}$, for example. The energy emitted from $A_{l',n'}$ to $B_{l,n}$, $AB_{(l'n'),(l,n)}$, can be calculated by

$$AB_{(l',n'),(l,n)} = \left(1 - \alpha_{A_{l',n'}}\right) e^{-Md_{(l',n'),(l,n)}} \frac{W^2 dd_l dd_n}{\pi[(d_l - d_{l'})^2 + (d_n - d_{n'})^2 + W^2]^2}$$ (4.33)

where $d_{(l',n'),(l,n)}$ is the mean beam length between the two patches, which can be approximated using the distance between the two patch centres.

4.4.6 Energy exchange between patches

Using the form factors obtained above, the sound energy of each first-order patch source can be redistributed to other patches and consequently, the second-order patch sources can be generated. Continue this process and the kth-order patch sources can be obtained ($k = 1, \ldots, \infty$). Note the energy exchange between patches depends only on the form factors and the patch sources of preceding order, which is significantly different from the process in ray tracing type models. The calculation of a kth-order patch source can be made by summing the contribution from all the $(k-1)$th-order patch sources, except those which are on the same boundary as the kth-order patch source considered. For example, to calculate a kth-order patch source on the ground, $G_k(t)_{l,m}$, the contribution from the patches on the two façades should be summed:

$$G_k(t)_{l,m} = \sum_{l'=1}^{N_X} \sum_{n'=1}^{N_Z} AG_{(l',n'),(l,m)} A_{k-1}\left(t - \frac{d_{(l',n'),(l,m)}}{c}\right)_{l'n'}$$

$$+ \sum_{l'=1}^{N_X} \sum_{n'=1}^{N_Z} BG_{(l',n'),(l,m)} B_{k-1}\left(t - \frac{d_{(l',n'),(l,m)}}{c}\right)_{l'n'} \quad \left(t - \frac{d_{(l',n'),(l,m)}}{c} \geq 0\right)$$ (4.34)

4.4.7 Energy from patches to receiver

Consider a receiver R at (R_x, R_y, R_z). The energy response at the receiver can be determined by taking all orders of patch sources into account. For the kth-order patch sources, the energy at receiver R at time t can be written as

$$E_k(t) = E_k(t)_G + E_k(t)_A + E_k(t)_B$$ (4.35)

where, for example, the contribution from the patch sources on boundary G is $E_k(t)_G$, which can be determined by

$$E_k(t)_G = \sum_{l=1}^{N_X} \sum_{m=1}^{N_Y} \left[\frac{G_k\left(t - \dfrac{R_{l,m}}{c}\right)_{l,m}}{\pi R_{l,m}^2} \cos\left(\xi_{l,m}\right) \right] e^{-MR_{l,m}} \quad \left(t - \frac{R_{(l,m)}}{c} \geq 0\right) \tag{4.36}$$

where $\xi_{l,m}$ is the angle between the normal of patch $G_{l,m}$ and the line joining the receiver and the patch:

$$\cos\left(\xi_{l,m}\right) = \frac{R_z}{\sqrt{(d_l - R_x)^2 + (d_m - R_y)^2 + R_z^2}} \tag{4.37}$$

In Equation (4.36) $R_{l,m}$ is the mean beam length between receiver R and patch $G_{l,m}$, which can be approximated using the distance between the receiver and the patch centre:

$$R_{l,m} = \sqrt{(d_l - R_x)^2 + (d_m - R_y)^2 + R_z^2} \tag{4.38}$$

A more accurate calculation of $R_{l,m}$ can be made by subdividing patch $G_{l,m}$ into N_l by N_m ($N_l, N_m \geq 1$) equal elements and then calculating their average distance to the receiver:

$$R_{l,m} = \frac{1}{N_l N_m} \sum_{i=1}^{N_l} \sum_{j=1}^{N_m} \left[\left[d_l - \frac{1}{2} dd_l + \frac{dd_l}{N_l}(i - \frac{1}{2}) - R_x \right]^2 \right.$$
$$\left. + \left[d_m - \frac{1}{2} dd_m + \frac{dd_m}{N_m}(j - \frac{1}{2}) - R_y \right]^2 + R_z^2 \right]^{1/2} \tag{4.39}$$

A similar method can also be used for the calculations of source–patch (Section 4.4.4) and patch–patch (Section 4.4.5) distances. By considering all orders of patch sources as well as the direct energy transport from source to receiver, the energy response at receiver R and consequently, various acoustic indices including RT, EDT and the steady-state SPL, can be obtained.

4.4.8 Geometrically reflecting ground

In the above discussion it is assumed that the ground is diffusely reflective, but the formulation can be modified to consider geometrically reflecting ground (Kang 2002d). A geometrically reflecting ground can be treated as a mirror and the sound source S and the patch sources will have their images, as illustrated in Figure 4.7.

The initial energy in the patch sources on the façades is from the sound source S as well as its image, S'. For the latter, the absorption coefficient of the ground is taken into account. During the energy exchange process, the energy in a patch source on a façade, say A, is calculated by summing the contribution from all the patch sources on façade B and its image, B'.

At receiver R, for each order of energy exchange between patches, the sound energy contributed from each patch source and its images is summed. In addition to the multiple reflections considered above, the direct sound and the first reflection from source to receiver through the ground should also be included.

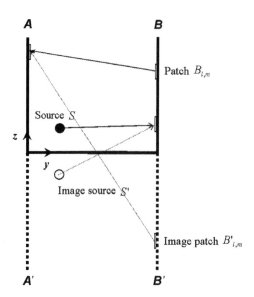

Figure 4.7 Cross section of an idealised rectangular street with diffusely reflecting facades and geometrically reflecting ground, showing the distribution of source energy and energy exchange between patches.

4.4.9   Cross streets and urban squares

Radiosity models have also been developed for cross streets (Kang 2001) as well as urban squares (Kang 2005a). The basic principles and algorithms are similar to those for a single street canyon. The cross street configuration is illustrated in Figure 7.1b, where 11 boundaries are taken into account. Again, there are two kinds of relative locations between patches, either parallel or orthogonal. If there is no line of sight between a pair of patches, the form factor is zero.

In an idealised rectangular square, as illustrated in Figure 4.2, five boundaries including four façades A, B, U, V and ground G are considered. In the square all the form factors are nonzero since there is always a line of sight between any pair of patches. In Figure 4.2 an example of patch division on the ground is also shown, in a similar way to that in street canyons.

4.4.10   Numerical simulation and validation of the algorithms

A series of computer programs have been developed based on the above algorithms. In the calculation there is an initial stage of determining patch division, including patch numbers and ratios between adjacent patches, so that a required accuracy in calculating form factors and the source energy distribution can be achieved. The accuracy in calculating form factors can be evaluated by the fact that the sum of the form factors from any patch to all the other patches should be unity. The accuracy in distributing the source energy to patch sources can be similarly evaluated.

A typical street configuration is used to test the algorithms. The street length, width and height are 96, 8 and 12m, respectively, and a point source is positioned at (16m, 4m, 2m). The patch numbers are $N_X = 60$ and $N_Y = N_Z = 12$. Along the length the patch size increases from $l = 1$ to 15, decreases from $l = 46$ to 60, and is constant between $l = 16$ and 45. For the varied patch sizes, the ratio between two adjacent patches is $q_x = 1.05$. Along the width and height the patch size increases from the edges to the centre with a ratio of $q_y = q_z = 1.1$. Using these parameters the model calculates the form factors and the source energy distribution on patches accurate to three decimal places.

In the numerical simulation, calculation stops when the total energy reduces to a certain amount. When reverberation is considered this is typically 10^{-6} of the source energy, whereas when only steady-state SPL is considered, this amount can be much greater, which can increase the calculation speed significantly.

To calculate energy response and consequently reverberation times, the time step/interval is proportional to the value of reverberation time, typically 3–5ms. When only sound distribution is considered, there is no need to arrange/divide the sound energy according to their arriving time, so that the calculation time can be much shorter.

Similar algorithms can also be used for rectangular enclosures. Such a model has been shown to correctly calculate the acoustic characteristics of long, flat and regularly shaped enclosures with various distributions of boundary absorption (Kang 2002a, 2002d, 2002g, 2002h; Kang and Neubauer 2001). This can be regarded as a further validation of the algorithms.

4.4.11   Comparison with measurements

A comparison has been made (Kang 2002d) in a single street between the radiosity prediction and the measurements carried out by Picaut *et al.* (1999) in a 1:50 scale model. The street length, width and height are 96, 8 and 12m, respectively. The model façades follow a statistic distribution extracted from an actual Haussmann building façade (Picaut and Simon 2001). This type of façade is common in European cities and they are considered to be rather diffusely reflective. The measured absorption coefficient of the façades is about 0.05 at middle frequencies. The ground is acoustically smooth and highly reflective, so that an absorption coefficient of 0.01 is used in the calculation. The measurement data are based on the average in the frequency range 400Hz to 1.6kHz, so that the calculation is also in this range. A point source is used in the calculation, which corresponds to the spark source used in the measurement. The source is positioned at (16m, 4m, 2m), and the receivers are along line (20–95m, 4m, 2m) with an interval of 2.5m.

A comparison of sound distribution between calculation and measurement is shown in Figure 4.8a, where the SPLs are normalised with respect to the value at $x = 20$m, and in the calculation the patch divisions correspond to those discussed in Section 4.4.10. In the calculation the sound power level of the source is set as 0dB. It can be seen that the agreement is very good, generally within ±1.5dB accuracy.

The calculated SPL distribution based on diffusely reflecting ground is also shown in Figure 4.8a. It can be seen that the result is generally rather close to that with geometrically reflecting ground, although the difference between the two kinds of grounds increases with increasing source–receiver distance. An important reason for the difference is that with more diffusely reflecting boundaries the sound path length generally becomes longer. At $x = 95$m, namely 79m from the source, the SPL with diffusely reflecting ground is about 3dB lower than that with geometrically reflecting ground.

The comparison in RT between calculation and measurement is shown in Figure 4.8b. The agreement is very good, with an average difference of 6 per cent. The calculated RT by assuming the ground as diffusely reflective is also shown in Figure 4.8b. It can be seen that the results are rather close for both kinds of grounds. This is also the case in terms of decay curve, as shown in Figure 4.9. The reason is that reverberation is mainly determined by multiple reflections, with which the diffuse reflection mechanism becomes dominant, as discussed in Section 4.4.1.

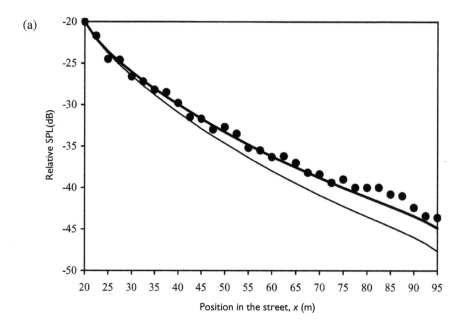

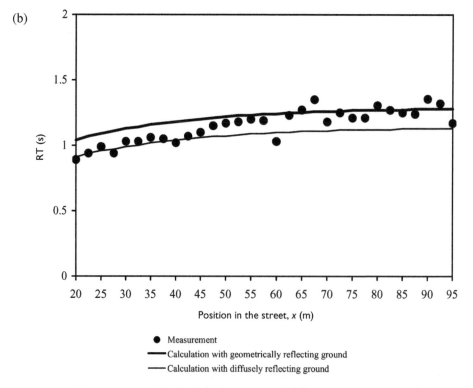

● Measurement
——Calculation with geometrically reflecting ground
——Calculation with diffusely reflecting ground

Figure 4.8 Comparison of (a) SPL attenuation and (b) RT along the length between calculation and measurement.

4.5 Transport theory

Sound propagation may be simulated by a ray beam that represents the path of a sound particle or phonon. The phonon obeys classic mechanics laws according to the Hamilton stationary action principle. Based on the concept of sound particles and the application of the classic theory of particle transport, a model has been developed to predict the temporal and spatial sound distribution in urban areas (Picaut *et al.* 1999; Le Pollès *et al.* 2004, 2005). In the model a particle undergoes a straight line until it meets an obstacle, and interactions and collisions between particles are neglected. It is assumed that the effects of phase cancellation and addition are averaged and the sound sources are not correlated. The model can consider partially diffusely reflecting building façades, scattering by urban objects, atmospheric attenuation and wind effects.

4.5.1 General equation

Since there are a large number of particles in a street, the description of the N particles system can be reduced to the knowledge of an artificial single particle system in a probabilistic way (Williams 1971). A sound particle is defined by its elementary energy e, position $\mathbf{X}$, and velocity $\mathbf{V}$, the norm of which is equal to the sound velocity c. A six-dimensional phase space Γ is used, involving the three usual space and velocity coordinates $(\mathbf{X}, \mathbf{V})$. A probability density, named the single particle distribution function (SPDF) $f(\mathbf{X}, \mathbf{V}, t)$, is introduced, representing the amount of particles, at time t, with velocity $\mathbf{V}$ to within about $d\mathbf{v}$, in an elementary volume $d\mathbf{x}$ located at $\mathbf{X}$:

$$\iint f(\mathbf{x}, \mathbf{v}, t) d\mathbf{x} d\mathbf{v} = 1 \tag{4.40}$$

When collisions of phonons only take place on the boundaries, the evolution of the sound particle density in urban areas is similar to that of the molecular density in a rarefied gas or Knudsen gas. The main equation of the model can be derived using the transport equation of the free molecular flow to describe the spatial and temporal evolution of the sound particle density and consequently of the sound field energy

$$\frac{\partial f}{\partial t} + \mathbf{v} \nabla_x f = 0 \quad (\mathbf{x} \in X, \mathbf{v} \in V) \tag{4.41}$$

where $\nabla_x f$ represents the spatial derivation. The equation could be generalised to take into account more phenomena, such as the atmospheric attenuation, the wind effect, and the scattering by urban objects, by introducing an absorption term σ, a strength of transport $\mathbf{F}$ and a scattering term, respectively

$$\frac{\partial f}{\partial t} + \mathbf{v} \nabla_x f - \sigma f + \mathbf{F} \nabla_v f = \left. \frac{\partial f}{\partial f} \right|_{\text{scattering}} \quad (\mathbf{x} \in X, \mathbf{v} \in V) \tag{4.42}$$

To consider boundary reflection, the outward unit normal $\mathbf{n}$ on the boundary is introduced, and $\Gamma^{\pm}$ is used to represent points $(\mathbf{x}, \mathbf{v})$ in the phase space Γ. Γ^+ and Γ^- represent the incident and reflected sound particles on the building façades, and the restrictions of the SPDF to Γ^+ and Γ^- are f^+ and f^-, respectively.

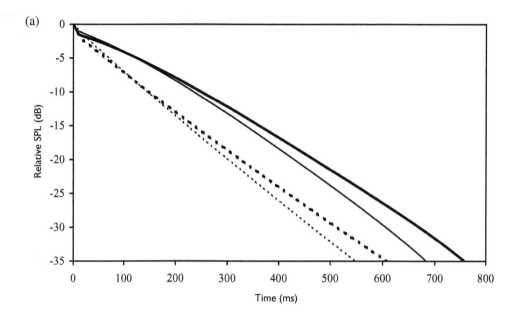

(a)

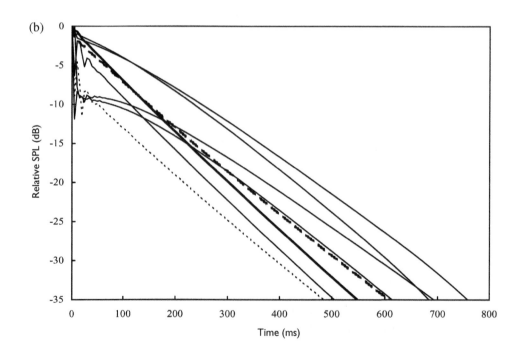

- - - - Source–receiver distance 9m, diffusely reflecting gound
■ - ■ - ■ Source–receiver distance 9m, geometrically reflecting gound
——— Source–receiver distance 59m, diffusely reflecting gound
━━━ Source–receiver distance 59m, geometrically reflecting gound

(b)

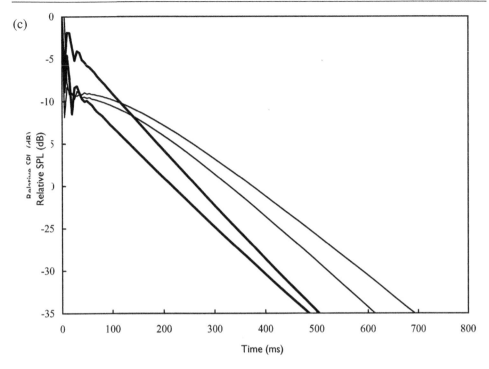

(c)

Figure 4.9 Comparison of decay curves between geometrically and diffusely reflecting ground at two typical receivers.

The boundary absorption is expressed by considering the probability $\alpha(\mathbf{x})$ that a sound particle hitting a boundary at position $\mathbf{x}$ is absorbed. It would be possible to consider angle dependent absorption, although in the model this is assumed to be angle independent.

The degree of boundary diffusion is expressed by an accommodation coefficient, $d(\mathbf{x})$, with $d(\mathbf{x}) = 0$ representing pure diffuse reflection. In a general way, the nongeometrical reflection can be taken into account by considering a probabilistic approach. A positive, integrable, and smooth function $\Re(\mathbf{x}, \mathbf{v}, \mathbf{v}')$, defined on $\Gamma^{\pm}$, can be introduced, representing the probability that an incident sound particle with a velocity $\mathbf{v}'$ leaves the boundary, at position $\mathbf{x}$ after reflection, with a velocity $\mathbf{v}$. The reflection law is normalised as

$$\int_{\Gamma^+} \Re(\mathbf{x}, \mathbf{v}, \mathbf{v}')d\mathbf{v}' = 1 \qquad\qquad (\mathbf{x} \in \partial X, \mathbf{v} \in V) \qquad\qquad (4.43)$$

$$\int_{\Gamma^-} \Re(\mathbf{x}, \mathbf{v}, \mathbf{v}')d\mathbf{v}' = 1$$

The boundary conditions express the flow of reflected particles as a function of the incident particle flow:

$$|\mathbf{n} \cdot \mathbf{v}| f^-(\mathbf{x}, \mathbf{v}, t) = (1 - \alpha(\mathbf{x}))[d(\mathbf{x})|\mathbf{n} \cdot \mathbf{v}^*| f^+(\mathbf{x}, \mathbf{v}^*, t)$$

$$+ (1 - d(\mathbf{x}))\int_{\Gamma^+} \Re(\mathbf{x}, \mathbf{v}, v')|\mathbf{n} \cdot \mathbf{v}| f^+(\mathbf{x}, \mathbf{v}', t)d\mathbf{v}'] \quad (\mathbf{x} \in \partial X, \mathbf{v} \in \Gamma^-) \qquad (4.44)$$

where $\mathbf{v}^*$ is the incident velocity leading to a geometrical reflection in the direction of $\mathbf{v}$. The left side of the equation represents the reflected flow, and the first and second term of the right side expresses geometrical and nongeometrical flow respectively, both weighted by the reflection coefficient.

4.5.2   An empty street

The above model is applied to an empty street canyon with partially diffusely reflecting surfaces characterised by Lambert's Law (Le Pollès *et al.* 2004). Since there is no exact analytical solution for such a system, an asymptotic solution (Börgers *et al.* 1992) is explored, showing that the transport equation with the appropriate boundary conditions may be reduced to a diffusion equation. It is assumed that the street width is much smaller than the length and height. If the sound source is located on the ground, the sound propagation in a street canyon is similar to the propagation between two parallel planes. The distribution function $f(\mathbf{x}, \mathbf{v}, t) = f(x, y, z, u, v, w, t)$ is expressed as a product of two functions $q(x, y, t)$ and $\phi(z, u, v, w)$, where $\mathbf{x} = (x, y, z)$ with $(x, y) \in X, z \in Z$, and $\mathbf{v} = (u, v, w)$ with $(u, v, w) \in [-c, c] \times [-c, c] \times [-c, c]$ and $u^2 + v^2 + w^2 = c^2$.

For $\phi(z, u, v, w)$, two changes are considered in the transport equation in order to reach a diffusion approximation: t is rescaled to t / ε, so that a long time is needed to observe a mixture of the sound particles between the planes; and z is replaced by εz, so that the two planes are brought closer in order to have more changes of the sound particle velocity direction, by increasing the frequency collision on the planes. Le Pollès *et al.* (2004) then derived that $\phi(z, u, v, w)$ depends only on the street width W :

$$\phi(z, u, v, w) = \frac{1}{4\pi W c^2} \tag{4.45}$$

$q(x, y, t)$ expresses the spatial and temporal distribution of sound particles in a plane parallel to the building façades, namely between $z = 0$ and W. First consider the path of a sound particle between two successive Lambert's reflections. The distribution of the horizontal distance of propagation after a Lambert's reflection is

$$W\widetilde{G}(r) = W \frac{2r}{(1 + r^2)^2} \tag{4.46}$$

The flight time of propagation between two collisions in spherical coordinates is

$$WE(\widetilde{\tau}) = \frac{2W}{c} \tag{4.47}$$

Then consider the path of a sound particle between two Lambert's reflections including n geometrical reflections. The probability $P(n = k)$ that k geometrical reflections occur between two successive Lambert's reflections, is

$$P(n = k) = (1 - d)d^k \tag{4.48}$$

The probability density associated to the horizontal distribution of propagation is

$$G(r) = \sum_{k=1}^{+\infty} (1-d)d^{k-1} \frac{1}{k} \tilde{G}\left(\frac{r}{k}\right) \tag{4.49}$$

The flight time of propagation is

$$E(W\tau) = \frac{2W}{c\,(1-d)} \tag{4.50}$$

The probability density G describes the spatial and temporal distribution of sound particles in the horizontal plane $z = 0$:

$$G(x,y,t) = \frac{1}{2\pi t\sigma^2} \exp\left[-\frac{x^2 + y^2}{2t\sigma^2}\right] \tag{4.51}$$

where

$$\sigma^2 = \frac{(1+d)}{(1-d)} \frac{c}{2} \tag{4.52}$$

By construction, G is equal to $q(x,y,t)$, and it is the solution of a diffusion equation

$$\frac{\partial}{\partial t} q(x,y,t) - \kappa \frac{\partial^2}{\partial x^2} q(x,y,t) - \kappa \frac{\partial^2}{\partial y^2} q(x,y,t) = 0 \tag{4.53}$$

where

$$\kappa = \frac{(1+d)}{(1-d)} \frac{Wc}{4} \tag{4.54}$$

It is thus shown that for street canyons, with Lambert's reflection law, the transport equation may be reduced to a diffusion equation for the sound energy, defined by only one parameter κ.

The boundary absorption is neglected in the above. Absorption at the openings, which depends only on the reflection law and the speed of sound c, can be taken into account by introducing an exchange coefficient

$$\eta_{open} = \frac{c}{2} \tag{4.55}$$

The ground absorption α_p is considered using an exchange coefficient

$$\eta_p = \alpha_p \frac{c}{2} \tag{4.56}$$

The absorption by building façades α_f can also be introduced, in a simplified way, in the exchange coefficient of the ground

$$\eta'_p = \left[\alpha_p + 2\frac{H}{W}\alpha_f\right]\frac{c}{2} \tag{4.57}$$

Finally, the solution of $q(x,y,t)$ is obtained (Le Pollès et al. 2004; Picaut et al. 1999):

$$q(x,y,t) = \sum_{n=1}^{\infty}\sum_{m=1}^{\infty}\frac{a_n}{u_n}\frac{b_m}{v_m}\left[u_n \cos\left(\frac{u_n x}{L}\right) + B_{\text{open}} \sin\left(\frac{u_n x}{L}\right)\right]$$

$$\left[v_m \cos\left(\frac{v_m y}{H}\right) + B_P \sin\left(\frac{v_m y}{H}\right)\right]\exp\left[-\left(\frac{u_n^2}{L^2} + \frac{v_m^2}{H^2}\right)\kappa t\right] \tag{4.58}$$

where L and H are street length and height, $B_{\text{open}} = \eta_{\text{open}} L/\kappa$, and $B_p = \eta'_p H/\kappa$. a_n, b_m, u_n and v_m verify $(u_n^2 - B_{\text{open}}^2)\tan u_n = 2B_{\text{open}} u_n$, and $(v_m^2 - B_p B_{\text{open}})\tan v_m = (B_p + B_{\text{open}})v_m$. The ratio a_n/u_n and b_m/v_m are given by the initial conditions. For a point source at (x_0, y_0),

$$\frac{a_n}{u_n} = \frac{2}{L}\frac{u_n \cos(u_n x_0/L) + B_{\text{open}} \sin(u_n x_0/L)}{u_n^2 + B_{\text{open}}^2 + 2B_{\text{open}}} \tag{4.59}$$

$$\frac{b_m}{v_m} = \frac{2}{H}\frac{v_m \cos(v_m y_0/H) + B_P \sin(v_m y_0/H)}{v_m^2 + B_P^2 + 2B_P + \cos^2 v_m (v_m^2 + B_P^2)(B_{\text{open}} - B_P)/(v_m^2 - B_P B_{\text{open}})} \tag{4.60}$$

From the above equations the steady-state SPL and the reverberation times at position (x,y) in a plane parallel to the building façades can be determined.

Comparison between calculation and measurement (Picaut et al. 2005; see also Section 4.11) has been made in a pedestrian street. A low absorption coefficient around 0.05 is assumed. Since no information is available for the accommodation coefficient of the building façades, the calculation is made for several values of d. For RT the model is in agreement with experimental data if $d = 0.4–0.7$. For SPL, any accommodation coefficient leads to a good agreement with the experimental results near the sound source, whereas when the source–receiver distance increases, the best agreement is obtained for $d = 0.5–0.8$, which is approximately in agreement with the values for RT. It is noted that the street has traditional buildings with highly modulated façades, which explains why the accommodation coefficient should be relatively small, namely rather diffuse.

4.6   Wave-based models

With the development of more powerful computers, a number of models based on numerically solving wave equations have been developed and applied in urban situations, including acoustic finite element method (FEM) and boundary element method (BEM), equivalent sources method (ESM), finite difference time domain method (FDTD), and parabolic equation (PE) method.

4.6.1 Finite element method and boundary element method

Acoustic FEM and BEM are based on the approximation to the wave equation. The methods can model resonances in the frequency domain and wave reflections in the time domain. They have been successfully applied in acoustic simulation of small spaces where the wavelength is larger or at least of the same order as the room dimensions (Wright 1995). The application range has also been extended to relatively large spaces including urban streets.

A two-dimensional boundary element numerical model was used to study the sound field in the region of balconies in a tall building close to a roadway (Hothersall *et al.* 1996). This is a typical situation where energy-based models are less appropriate since the wavelengths are not small compared to the dimensions of the balcony spaces and building elements. It was found that treatment of the ceiling or the rear wall of the balcony is the most efficient in terms of noise reduction.

4.6.2 Equivalent sources method for parallel street canyons

The basic idea of the equivalent sources method is to reduce a problem to a simplified geometry with boundary conditions that are easy to handle. On boundaries with different conditions, virtual sources are placed, with their strengths adjusted by solving an equation system so that the boundary conditions are fulfilled everywhere. The method has been developed in various situations (Cummings 1992; Ochmann 1995; Johnson *et al.* 1998; Bérillon and Kropp 2000), including sound propagation in two parallel street canyons by Ögren and Kropp (2004), as summarised below.

A two-dimensional configuration was considered. For the source canyon, the geometry is divided into two parts, the domain inside the canyon and the half space above. The problem can be handled by considering radiation into a half space by a Rayleigh integral and a sound field in a rigid cavity by a modal approach. In the street canyon the actual source can be considered by taking into account the vehicle flow density, percentage of heavy vehicles and velocity. The coupling between the half space and the cavity is obtained by a set of equivalent sources at the opening of the street canyon which correct the field impedance along the boundary. The opening (boundary) is divided into equally sized elements of one-tenth of the wavelength, and the corresponding equivalent sources are approximated with a piecewise constant complex source strength. At the opening the pressure and the velocity fields must be continuous. The loss factors from the street boundaries and from air absorption are also considered. Patches with absorbers can be included in the boundary conditions as impedances.

The source strengths calculated are seen as sources on a rigid plane for the receiving canyon and accordingly, the pressure at receiving points can be calculated. It is assumed there is no reflected wave from the receiving canyon back to the source canyon.

Comparison between calculation and measurement has been made for the SPL in courtyards in an area in Stockholm. The area consists of buildings 5–7 stories high, and the streets close to the courtyards of interest are 11 to 20m wide. The courtyards are of different shapes and sizes. The results show that the calculation using only the two closest streets underestimates the SPL substantially, whereas with more source canyons the agreement between calculation and measurement is rather good.

Atmospheric turbulence could cause an increase in SPL compared to a homogeneous case. With the ESM it is possible to model a turbulent atmosphere by accounting for the loss in coherence between different sound paths to a single receiver, although this only considers a nonrefracting atmosphere. The level increase is estimated using a von Kármán turbulence model and the mutual coherences of all equivalent sources' contributions. A comparison with a ray-based model shows a reasonably good agreement (Ögren and Forssen 2004).

Compared with the standard BEM or FDTD (see Section 4.6.3), the ESM is relatively less computationally heavy since it only discretises the opening of the canyon and the impedance patches. The BEM is slightly more flexible but requires the whole canyon to be discretised, whereas with FDTD the whole domain must be discretised, including an area above the canyon, as described in the next section.

4.6.3   Finite difference time domain method and parabolic equation method

The FDTD model is based on the numerical integration of the linearised Euler equations in the time domain (Botteldooren 1994; Blumrich and Heimann 2002; Van Renterghem and Botteldooren 2003; Ostashev *et al.* 2005). It solves the moving-medium sound propagation equations, taking into account the combined effect of multiple reflections, multiple diffractions, inhomogeneous absorbing and partly diffusely reflecting surfaces. An advantage of FDTD compared with ESM and BEM is its applicability to a moving, inhomogeneous and turbulent atmosphere, namely the consideration of the effects of refraction. Hence, the FDTD can be considered as a complete model.

The PE model is based on a one-way wave equation in the frequency domain. It is suitable for long-range sound propagation over flat ground, but less suitable in situations with several reflecting obstacles and arbitrary wind fields (Gilbert and Di 1993; Salomons 1998).

Since the computational resources needed for FDTD simulations are large, Van Renterghem *et al.* (2005, 2006) used a coupled FDTD–PE model, where the FDTD is applied in the complex source region and the PE is used for propagation over flat ground to a distant receiver. The coupling of the two models occurs at a vertical array of intermediate receivers located at the boundary of the source region. The FDTD results are used to generate starting functions for PE. A two-dimensional idealised configuration as shown in Figure 4.10 is studied using the coupled FDTD–PE model. To reduce calculation time, the

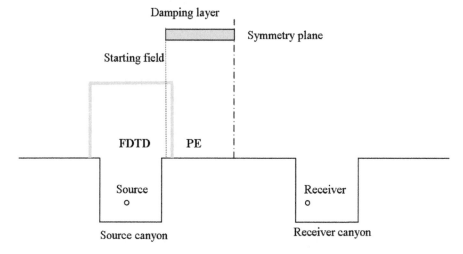

Figure 4.10 Set-up of the coupled FDTD-PE model, redrawn based on Van Renterghem *et al.* (2006).

source and receiver canyons are geometrically identical, so that simulations are only needed in half of the sound propagation domain. A single FDTD calculation is performed in the source canyon (grey area) using a broadband source. At a short distance from the canyon edge, time signals are recorded on a vertical array to generate starting functions for the PE, using a transition from the time domain to the frequency domain by means of fast Fourier transform (FFT). Finally, PE calculations are performed up to a receiver at the symmetry plane for the frequencies of interest. In the calculation Green's function PE (GFPE) model is applied (Salomons 1998). A very good agreement has been obtained between the coupled FDTD–PE model and reference calculations, namely with the FDTD applied completely from source to receiver.

4.7 Empirical formulae

For urban designers, it would be useful at the design stage to use relatively simple formulae to estimate the sound propagation in microscale urban areas. Based on both analytic theory and regression of data obtained using computer simulation models, a series of formulae have been developed for calculating the RT, EDT and SPL under various boundary conditions (Kang 2004a).

Some formulae are given below for urban squares with diffusely reflecting boundaries (Kang *et al.* 2003a), based on simulation using the radiosity method, with a range of urban square configurations: length $L = 20$–200m, width $W = 20$–200m, height $H = 5$–100m, and square area 400–40,000m^2. The length/width ratio is 1:1 to 4:1, and the side/height ratio $\sqrt{LW} / H$ is =0.5–40. Buildings are considered to be along two, three or four sides of a square, with an absorption coefficient of 0.1–0.9.

It has been shown that for an urban square surrounded by buildings with low side/height ratio and low boundary absorption coefficient, the average RT using the radiosity simulation is rather close to that calculated by the well-known Eyring formula, as described in Section 1.6. 2. However, with the increase of side/height ratio and boundary absorption the Eyring formula becomes increasingly inaccurate. A modified formula is consequently developed by introducing several correction items

$$RT = \frac{0.16V}{-S_0 Ln\,(1-\overline{\alpha}) + 4MV}\left(88.6 + 49\alpha_b + 2.7\frac{\sqrt{LW}}{H}\right) \qquad (4.61)$$

where S_0 is the total surface area and $\overline{\alpha}$ is the average absorption coefficient, both including an imaginary square ceiling. $V = LWH$, and α_b is the average absorption coefficient of boundaries, that is, façades and ground only. It has been demonstrated that within the configuration range described above, calculations using Equation (4.61) and the radiosity model are very close, generally within an accuracy of 10–15 per cent.

A formula for the SPL distribution in a square with diffusely reflecting boundaries has also been derived:

$$L = L_W + 10\log\left(\frac{Q}{4\pi d_r^{\,2}} + \frac{3H}{W+L}\frac{4}{R}\right) \qquad (4.62)$$

where $R = S_0 \alpha_T / (1 - \alpha_T)$ and $\alpha_T = \overline{\alpha} + 4MV / S_0$. L_W is the sound power level of the source, Q is the directivity factor of the source, and d_r is the source–receiver distance. The difference between calculations using Equation (4.62) and using the radiosity model is generally within 2dB. It is noted that Equation (4.62) is inapplicable to long rectangular spaces where the SPL attenuation is considerably greater than that in quasi-square spaces.

4.8 Other models

Whilst many models concentrate on the sound propagation, attention has also been paid to the effects of traffic sources. A series of models developed in Poland (Walerian *et al.* 2001a, 2001b; Janczur *et al.* 2001a, 2001b) consider different representations of equivalent point sources for various classes of vehicles, and the time-average sound level within an urban system, especially street canyons. The change of average vehicle speeds and vehicle speed limits are also taken into account.

Kuttruff (1975) derived formulae for the average value and the variance of noise level in urban streets. The reverberation in urban streets is considered in two ways: (1) the sound field is regarded as diffuse and the average absorption coefficient is calculated with a totally absorbent ceiling, hard walls and hard ground, and (2) the image source model is used and the side walls are considered to have an absorption coefficient.

Combinations of various models have also been proposed. For example, Ismail and Oldham (2005) suggested that a possible approach to the modelling of urban noise propagation might be to employ geometrical models in the near field, the radiosity method for mid-range propagation, and models based upon classic diffusion for far field propagation.

4.9 Acoustic animation

To aid urban soundscape design as well as for public participation, it would be useful to present the three-dimensional visual environment with an acoustic animation tool, where consideration should be given to various urban sound sources, such as traffic, fountains, street music, construction, human voice and bird singing; to the dynamic characteristics of the sources, such as variation of traffic in a day; and to the movements of sources and receivers. The calculation speed should be reasonably fast, so that a designer can adjust the design and then immediately listen to the difference.

Some noise mapping programs (see Section 5.1) can provide acoustic animations, but the source conditions are rather simple and also, they are only monaural SPL since reverberation is not considered. In the virtual reality sector, although digital sound rendering techniques are being rapidly developed (Begault 1990; Gomes and Gerges 2001), practically only simple acoustic effects are considered, for example, by accurately simulating a small number of discrete echoes, and generating uncorrelated decay to simulate later-arriving reverberation tail. Modelling techniques for room acoustics can produce rather accurate acoustic indices and good auralisation functions, but they become inappropriate and time-consuming when multiple sources with dynamic characteristics are taken into account, since a high order of reflections and a large number of rays are needed.

A key issue of achieving fast acoustic animation/auralization for urban soundscape is to simplify the simulation algorithms, whilst retaining reasonable accuracy. This is feasible since compared to room acoustics the requirements for urban soundscape animation are relatively low. In this section such simplifications in urban squares are explored from the viewpoints of objective indices and subjection perception, respectively, and then an animation prototype is described.

4.9.1 Simplification through parameter studies

This section examines possible simplifications when the ray/beam tracing method is used for urban squares, especially the effects of two key parameters, reflection order and ray number, on the simulation accuracy and calculation speed. A parametric study has been carried out (Meng and Kang 2004) using a beam tracing software package Raynoise (LMS 2005a). The objective indices considered include SPL, RT, and EDT at 1kHz octave band.

Three idealised rectangular squares are considered, namely 25 × 25m, 50 × 50m and 100 × 100m, as shown in Figure 4.11. The square height is 20m, and the gaps between the building blocks are 5, 10 and 15m for the three squares, respectively. The depth of all the building blocks is 10m. A point source is located at a corner position, namely (28.2m, 29.8m), (48m, 44m), and (99m, 87m) for the three squares respectively. For each square the results presented below are based on a typical receiver in a relatively far field, at (12.8m, 18.3m) for the 25 × 25m square, (15.6m, 26.7m) for the 50 × 50m square, and (31m, 52m) for the 100 × 100m square (receiver 1). The source and receiver heights are all 1m. For the 50 × 50m square a receiver in the near field is also considered, at (43.3m, 43.3m) (receiver 2). It is assumed that the absorption coefficient of all the façades and ground is 0.09 (Chourmouziadou and Kang 2003). In terms of the diffusion coefficient of the boundaries, three values are considered, dc = 0, 0.1 and 0.2.

Figure 4.12 shows the variation of SPL, RT and EDT with increasing ray number from 5k to 120k, where the reflection order is 50. It can be seen that the SPL is rather stable and a ray

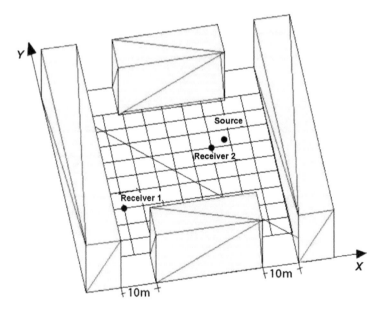

Figure 4.11 Typical square configuration (50 × 50m) used in the parametric study, showing the source and receiver positions.

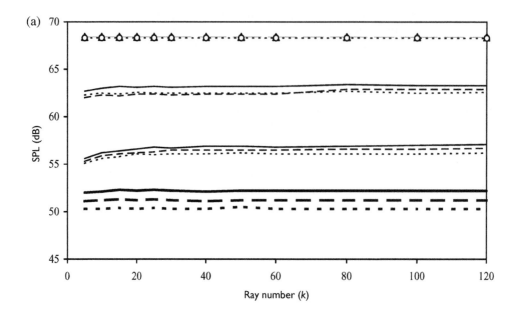

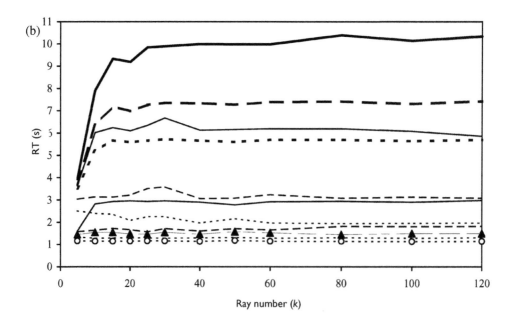

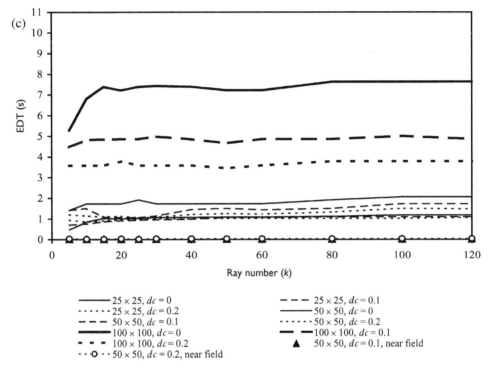

(c)

EDT (s)

Ray number (k)

——— 25 × 25, *dc* = 0   – – – 25 × 25, *dc* = 0.1
· · · · · · 25 × 25, *dc* = 0.2   ——— 50 × 50, *dc* = 0
– – – 50 × 50, *dc* = 0.1   · · · · · 50 × 50, *dc* = 0.2
——— 100 × 100, *dc* = 0   — —100 × 100, *dc* = 0.1
■ ■ ■ 100 × 100, *dc* = 0.2   ▲   50 × 50, *dc* = 0.1, near field
· · O · · 50 × 50, *dc* = 0.2, near field

Figure 4.12 Variation in SPL, RT and EDT with increasing ray number (1 kHz and 50 reflection order).

number of 5k is acceptable if only SPL is considered. For RT and EDT, for the 25 × 25m and 50 × 50m squares the computation results become approximately stable when the ray number is about 10k, whereas for the 100 × 100m square this value is increased to about 15k, as expected. With increasing diffusion coefficient, lower ray numbers are required to reach a stable RT and EDT, about 5k for the 25 × 25m and 50 × 50m squares and 10k for the 100 × 100m square. In the near field, differences caused by ray number are relatively less, probably because the direct sound plays an important role.

The effect of reflection order is shown in Figure 4.13, where the ray number is 100k. It can be seen that the SPL increases consistently with increasing reflection order until about 5, and then becomes rather stable. For RT, when the diffusion coefficient is 0, a reflection order of 20 is required for the three squares, whereas when the diffusion coefficient is increased to 0.2, the reflection order required reduces to about 10. For EDT, the required reflection number is similar to, or slightly less than, that for the RT. At the receiver in the near field, similar results are found. In terms of square size, a higher reflection order is generally required for a larger square.

The changes in SPL, RT and EDT with various combinations of reflection order and ray number have also been systematically examined, where the diffusion coefficient is 0.1 and the square size is 50 × 50m (Meng and Kang 2004). It is shown that for SPL, a reflection order of 5 and a ray number of 5k is already acceptable, with an error within 2dB. For RT, a reflection order of 20 and a ray number of 5k is a good combination with an accuracy of less than 10–15 per cent approximately. In terms of EDT, the required reflection order and ray number are much higher. It seems that a good combination is a reflection order of 20 and a ray number of 40k, with an accuracy of 20 per cent.

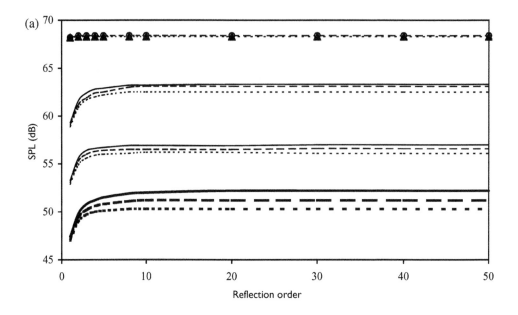

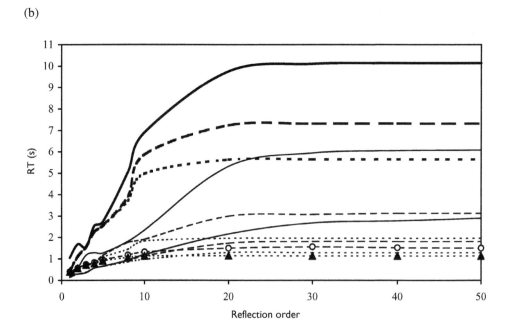

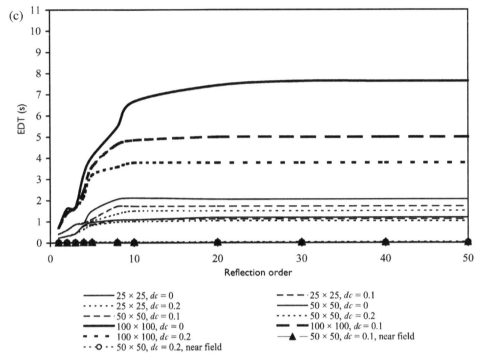

(c)

Figure 4.13 Variation of SPL, RT and EDT with increasing reflection order (1kHz and 100k ray number).

Further calculations show that with buildings on one or two sides a lower ray number and reflection order are generally required.

4.9.2   Simplification through subjective tests

From the above results it can be seen that for urban squares, it is possible to reduce reflection order and ray number compared to the parameters conventionally used in room acoustics. However, the computation time may still not be practically acceptable when generating impulse responses for acoustic animation. On the other hand, human sensitivity to a particular sound source might be reduced within a complex sound environment with multiple and moving sources. Therefore, to provide a fast urban acoustic animation, further simplification of calculation parameters has been explored through a series of subjective experiments, with particular attention paid to reverberation and reflection order (Meng *et al.* 2005).

Experimental method

Based on the 50 × 50m square as described in Section 4.9.1, a series of impulse responses were generated using Raynoise at receiver 1. Four reflection orders, 0, 5, 20 and 50, were considered. The ray number was 200k. Four boundary diffusion coefficients, 0, 0.1, 0.2 and 0.9, were also considered. The impulse responses were two-channel stereo 16-bit .wav file and were convolved with dry signals for listening tests. Four dry sound signals were used, including human voice (male speech), music (flute/guitar), car, and the fountain at the Peace Gardens in Sheffield. The sound signals generated with a reflection order of 50 were regarded as the standard signal, namely an accurate simulation of the sound environment of the square.

Three stages of experiment were carried out. In Experiment I eight subjects were asked to listen to the above-mentioned binaural sound clips via headphones connected to a standard PC. Each subject was asked to compare a series of the paired signals – one was the standard signal with a reflection order of 50 and the other was a signal based on simplified simulation, namely with a reduced reflection order. The length of each signal was 4–6s. The comparison was made by answering the following question: compared with the first signal, do you feel the second signal is: 1, much less reverberant; 2, less reverberant; 0, same; –1, more reverberant; and –2, much more reverberant. In addition, a number of pairs both with the standard signal were given to test the repeatability. In order to study the effect of background noise, a car noise was mixed with the human voice, with a range of S/N ratios. Similar tests were also carried out for music.

Experiment II was carried out using an array of eight loudspeakers forming a cube in an immersive virtual reality environment, RAVE (MultiGen-Paradigm 2005), also known as the REFLEX studio, although no image or video was shown in the experiment. The sound system was controlled via the Huron audio workstation (Mcgrath 1995). With the audio workstation the sound source movement was also simulated, using the ambisonic technique, from receiver 1 towards the source (see Figure 4.11) at a speed of 0.5m/s. The duration of the movement was 15s and the moving distance was 7.5m. The tests were all made with a background noise binaurally recorded at a park in Sheffield using a dummy head. The background noise included distant traffic and some natural sounds such as birdsong. With a background SPL of 61dBA, a range of S/N ratios were tested. Experiment III was similar to Experiment II, but two different square sizes, 25 × 25m and 100 × 100m, were considered. In Experiment II and III, two additional questions were asked: (1) compared with the first signal, do you feel the space corresponding to the second signal is: 1, much larger; 2, larger; 0, same; –1, smaller; and –2, much smaller; and (2) compared with the first signal, do you feel the distance corresponding to the second signal is: 1, much farther; 2, farther; 0, same; –1, closer; and –2, much closer. Again, eight subjects were used.

Results

The reliability of the experimental results was examined through the mean value and STD of the paired comparisons between two standards signals, as well as by carrying out the one-sample t-test (Field 2000) with a target value of 0, namely, 'same'. It was shown that for all the tests there is no significant difference, and the mean values are all within 0.5, suggesting that the test results are reliable. On the other hand, it was noted that even for the same signals, the STD values were rather high, from 0.46 to 1.19, indicating the range of sensitivity of human perception in this aspect.

Based on Experiment I, comparisons between various reflection orders were made for human voice, music and fountain. Paired and independent samples t-tests were carried out, typically with $p < 0.05$. Compared with reflection orders of 5, 20 and 50, with 0 reflection order, namely direct sound only, there were usually significant differences. Correspondingly, the mean values were rather high at 1.9 in average. In other words, listeners could easily distinguish the difference between signals with and without reverberation. There were several exceptions, mainly for fountain sounds and for signals with background noise, suggesting that, in such cases, people are less sensitive to reverberation.

Between reflection orders 20 and 50 there is generally no significant difference. This corresponds to the results in terms of objective indices, as described in Section 4.9.1. It is interesting to note that between reflection orders 5 and 20, as well as 5 and 50, there is also no significant difference for music and fountain sounds. For human voice, the difference between reflection orders 5 and 20/50 tends to be significant. Nevertheless, a reflection order of 5 is acceptable in the situation with background noise.

Since with diffusely reflecting boundaries a square may have a considerably shorter reverberation than that with geometrically reflecting boundaries (see Section 7.2), a comparison between diffusion coefficients 0, 0.1, 0.2 and 0.9 was made. It was shown that with diffusion coefficients 0, 0.1 and 0.2 the differences between various reflection orders and sounds were generally similar to the above – again, a reflection order 5 is acceptable for music and fountain sound, but not for human voice, whereas with a diffusion coefficient of 0.9 the result seems to be slightly different – a reflection order of 5 is also acceptable for human voice. For car sound, the results are generally similar to those of fountain sound.

In Experiment II, similarly to the previous discussion, there is no significant difference for music and fountain sounds between reflection orders 5 and 20, as well as between 5 and 50. For human voice, the difference between reflection orders 5 and 20/50 tends to be significant. Nevertheless, a reflection order of 5 is acceptable with background noise. Comparison was made between the two S/N ratios using 'independent samples' t-tests, for music and fountain sounds. No significant difference was found.

For the other two square sizes, 25×25m and 100×100m, the results were generally similar in terms of the differences between various reflection orders and different sounds. It seems that with the increase of square size, it was slightly more acceptable to consider a lower reflection order, especially for human voice. This is different from the objective result, as shown in Figure 4.13.

In terms of the question relating to space, the result is rather similar to that of reverberation. A reflection order of 5 is acceptable, especially for music and fountain. In terms of the question about distance, it seems that the differences between various sounds and reflection orders are much less significant than those for reverberation and space, suggesting simplifications can be made in this aspect.

In summary, the subjective tests suggest that a reflection order of 5 is generally acceptable in urban squares for many typical sound sources such as music, fountain and car. For certain sounds, such as human voice, more reflection orders are needed. This rule is largely applicable for squares with different diffusion coefficients and sizes, although with a greater diffusion coefficient and/or a larger square it tends to be possible to consider a lower reflection order.

It is expected that, with interactive visual and aural stimuli, the acoustic simulation could be further simplified to reduce the calculation time.

4.9.3 Prototype

A prototype of the animation/auralisation tool for urban soundscape has been developed for cross streets and squares (Kang *et al.* 2003a). It aims to aid design process, rather than simply being used for the final presentation. During the design process, a designer can change the configuration, and then listen to the acoustic environment in real time. The program allows users to input a design graphically, or to import drawing files from programs commonly used in architectural and urban design. There is a database with dry signals of a number of urban sounds from which multiple sources can be selected and they can be put at any position in a street or square.

The impulse responses are generated using the ray tracing program described in Section 4.3.2. By integrating the simplifications described in Sections 4.9.1 and 4.9.2, the computation is rather fast. Where appropriate, the calculation speed can be further increased by integrating a series of formulae of calculating acoustic indices in urban squares (see Section 4.7), or even databases with pre-computed acoustic indices for a range of configurations.

The system then produces sound files based on the digital audio processing. The source unit reads the dry signal and passes it to the reverb unit. The reverb unit renders the sound via audio buffers and the sink unit converts those back into audio files (Kahrs 1998; Steiglitz 1996). Huron software can also be applied and integrated with the RAVE system (Mcgrath 1995).

4.10   Physical scale modelling

4.10.1   General principles

Acoustic physical scale modelling has been used in simulating environmental sound propagation for a number of years (Delany et al. 1978; Mulholland 1979; Jones et al. 1980; Kerber and Makarewicz 1981). A notable advantage, compared with computer simulation, is that some complex acoustic phenomena can be considered more accurately, such as the diffraction behaviour of sound when it meets obstacles. Compared with real measurements in urban areas, a significant advantage of scale modelling is that the geometry, source and receiver condition, as well as background noise are relatively easy to control.

In a 1:n scale model, the measured time factors should be enlarged n times, and the frequency should be n times higher. This is because the speed of sound in air is constant, and the sound behaviour when it hits an obstacle is determined by the relationship between the size of the object and the wavelength of sound. The sound levels are not subject to any ratio scaling. As to the model size, a wide range of scale factors from 1:2 to 1:100 have been used in environmental acoustic modelling. The choice of scale factor is often a compromise between the characteristics of the measuring equipment and the size of available test space. The scale model is normally put in an anechoic or semi-anechoic chamber, so that reflections from the boundaries of the test chamber can be avoided.

In terms of model materials, although it is ideal to accurately simulate the boundary impedance, in practice it is often sufficient to model the absorption coefficient. The absorption coefficients of model materials at model frequencies can be determined in a similar way as at full scale, for example, using a model reverberation chamber. Considerable data have been published on the absorption characteristics of various model materials, based on both room and environmental acoustic modelling.

Air absorption becomes very large in the ultrasonic range, which causes a significant problem in scale modelling. The absorption from air, due to both molecules of oxygen (O_2)

and moisture (H_2O), depends mainly on sound path length, temperature and humidity. At certain scales, by filling a scale model with dry air of relative humidity 2–3 per cent or with oxygen-free nitrogen, the air absorption can be directly simulated (Barron 1983; Hodgson and Orlowski 1987; Kang 1995). If the measurement data were processed using a computer, the excessive air absorption could be numerically compensated (ANSI 1999a).

The size of sound sources and receivers should also be scaled down. For example, spark pulses or small loudspeakers are often used to simulate actual sound sources, and small condenser microphones are normally used as receivers in scale models.

Subjective tests can be undertaken in relatively large models of 1:8 or 1:10 scale, although this technique has mainly been used in room acoustics rather than in environmental acoustics. Dry sound signals recorded in an anechoic chamber are played back in a scale model at an increased tape velocity and the recorded signal in the model is slowed down to be listened to over headphones.

Water and light models have also been used to simulate acoustic phenomena since there are some common properties between sound waves, water waves and light waves. Water models are useful for demonstration because the wave velocity is relatively slow. In place of sound, ripples on a shallow water basin are used. The practical use of water models, however, is restricted because they are only two-dimensional and the wavelength range to be handled is rather narrow. The use of light models in acoustics is limited to very high frequencies since the wavelengths of light, unlike sound, are very small compared to room dimensions. They are useful for examining first- or second-order reflections, as well as for investigating sound intensity distribution using luminous intensity, but with light models it is difficult to obtain information about the arrival time of reflections and reverberation since the speed of light is much faster than that of sound. Sound absorption can be simulated by light absorption, and diffusely reflecting boundaries can also be modelled. Lasers are often used to simulate sound rays.

4.10.2 Applications

In this section the dimensions and frequencies refer to full scale, except where indicated.

Iu and Li (2002) used a 1:10 model to study the effects of interference in a street canyon of 24m long with two parallel façades of 18m high. Three street widths, 2, 5 and 8m were considered.

A varnished plywood board of 18mm thickness was used to simulate an acoustically hard surface. A carpet, laid on the varnished plywood board, was used to simulate an impedance ground surface. Attenborough's (1992) two-parameter model was applied to describe the impedance of the ground surface, where the two parameters, the effective flow resistivity and the effective rate of change of porosity with depth, were the best fit values based on the measurements of sound propagation over the ground.

A Tannoy driver with a tube of internal diameter of 3cm and length of 1.5m was used as a point source. The model source was first located at 0.065m (model scale) above the ground, simulating a realistic urban noise source such as a typical engine. Elevated noise sources, such as air conditioners and cooling towers installed on building façades, were then simulated at 0.05m (model scale) from a model façade.

A PC-based maximum length sequence (MLS) system analyser (MLSSA) was used both as the signal generator for the source and as the analyser for subsequent data processing. This is useful to improve the S/N ratio as compared to the conventional stationary excitation technique. MLSSA operates in the time domain where the impulse response is measured.

Ismail and Oldham (2005) investigated the role of sound reflection from building façades using scale modelling technique. Based on the size of the anechoic chamber and the frequency range of readily available instruments, a 1:15 scale model was built, simulating an idealised street section of 18 × 18 × 18m.

The building façades were modelled using varnished MDF sheets. The absorption coefficient of the sheets was measured using the MLSSA system by comparing the intensity spectrum of direct and reflected pulses after correcting for path difference. In the test a sample was placed vertically on the floor of a reverberation chamber, and the sound source and microphone were placed in a parallel plane to the sample at the same level above the floor. Since the path length was rather short, the effect of air absorption was ignored.

To consider diffusion effects, three degrees of surface irregularity were created by sticking three different arrangements of varnished MDF blocks of size 10 × 10 × 1.2cm (model scale) to the façades.

If a point source and a point receiver are located at the same position at the centre of the street and on a perfectly reflecting ground, the maximum possible interval between successive geometrical reflections is ensured and also, the reflections from one façade are identical to those from the opposite façade. To approximate this configuration, a ½ inch B&K type 4165 condenser microphone was used as the source. With this small size the directivity is uniform over most of the frequency range studied. The receiver was a microphone of the same type located immediately next to the source.

The scale model measurements were also made using the MLSSA system. By using a delay in the signal processing procedure, it was possible to block the direct sound at the receiver microphone which would have otherwise dominated the recorded data. For excessive air absorption, numerical compensation was applied. With the measurement data various indices can be derived, such as the ratio of total energy in an order of reflection to the total energy in the following order of reflection.

Picaut and Simon (2001) built a 1:50 scale model of a street canyon, where the façades were simulated by a series of varnished wooden cubes. Their disposition side by side followed a statistic distribution extracted from a real Haussmann building façade. Various façade elements such as windows, carpentries and balconies could also be modelled. The street height was 0.24m (12m full scale), and the length and width were adjustable.

The absorption coefficient of the wooden cubes, measured by an ultrasonic method, ranged from 0.06 at 50kHz (1kHz full scale) to 0.19 at 125kHz (2.5kHz full scale), for incident angles of 15° to 60° from the normal. Road and pavement were simulated by a varnished wooden panel.

For the sound source, a three-electrode spark was specially designed, aiming at good directivity, reproducibility and high acoustic energy level. Tungsten electrodes of 0.5mm diameter were used, controlled by a system allowing to exactly adjust the gap between electrodes. Except for the 500Hz octave band (full scale), where the forward radiation is maximum, the sound radiation is rather uniform around the spark source at ±2dB. To avoid the problems of small fluctuations and occasional bad discharges of the spark source, a coherent average of impulse responses was made at each microphone location, which also increased the S/N ratio (Papoulis 1991). For each receiver 100 discharges were recorded, and about 95 per cent of them were used for the average. The excessive air absorption was numerically corrected.

The receiver was a B&K type 4138 ⅛ inch capacitor microphone connected to a preamplifier and then a measurement amplifier of B&K type 2633 and 2670. The measured frequency range was up to 3150Hz one-third octave band at full scale. The microphone was positioned vertically in order to minimise the variation of microphone sensibility with the incident angle. Movements of the microphone were controlled by a Charlyrobot 3D-robot via a D/A 16-bit converter.

Horoshenkov *et al.* (1999) used a 1:20 scale model to study the sound propagation in a street canyon. A number of ultrasonic sources were used to simulate noise from the traffic on two lanes. The sources were supplied with air from pipes beneath the floor of the model. Two source heights were considered in the model, 0.025 and 0.05m (model scale) above the ground, simulating light and heavy vehicles, respectively.

Pneumatic noise sources are often used in urban scale models to simulate traffic noise, including a single source and a traffic flow (Yamashita and Yamamoto 1990). They are useful to consider parameters such as the density of vehicles and the composition of traffic, but they are not suitable for estimating impulse responses.

4.11 Site measurements

Validation against measurements is very important for computer models. Although scale models are useful for validation, it is often necessary to carry out full scale measurements, since most scale models are designed for certain specific studies and the data are not always applicable to other works. This section briefly reviews/discusses the methods for site measurements.

The first measurements of sound propagation in urban areas were carried out in the 1940s and 1950s for optimal positioning of air raid sirens (Ball 1942; Volkmann and Graham 1942; Jones 1946). In terms of the sound propagation in streets, the first measurements were made by Wiener *et al.* (1965) in the 1960s. Both noise level and reverberation were measured. Sound field in built-up areas was also measured by other researchers (Schröder 1973; Yeow 1976, 1977; Donavan 1976; Steenackers *et al.* 1978; Ko and Tang 1978; Sergeev 1979), and the variation in reverberation along the street length was noticed.

Picaut *et al.* (2005) recently carried out a systematic measurement in a street canyon in downtown Nantes, France. This is a pedestrian street 210m long, 18m high and 7.9m wide, with only one junction. An alarm pistol was used as the sound source. An array of nine microphones was arranged in a cross section of the street, and the array was moved along the street from 6–50m to the source at an interval of 2m. Two source heights, 0.52 and 1.65m, and two locations of the sound source, at the beginning and in the middle of the street, were used. For each configuration, a reference microphone was located at 2m from the sound source, and this was also used to activate the data acquisition. Five pistol shots were recorded for each sound source position and each location of the microphone array. The sound attenuation along the street and the decay curves at each receiver were obtained. Meteorological data were also measured, including temperature, pressure, humidity and wind speed.

Iu and Li (2002) carried out a measurement in a side lane 1.55m wide in Hong Kong. The two parallel façades were marble stone and ceramic tiles. The ground was concrete with a small drain channel at the edge of the side lane. A Roland type KC–300 amplifier and loudspeaker unit was used to generate broadband white noise. The source was positioned at the central line of the side lane at 0.5m above the ground. The SPL at various receiver locations was measured in one-third octave bands using a precision sound level meter. The results were used to validate a theoretical model considering interference effects (see Section 4.2).

Chapter 5

Macroscale acoustic modelling

Although the microscale simulation techniques described in Chapter 4 can give a rather accurate prediction of urban sound propagation, it is normally impractical to apply these algorithms on a large scale such as a whole city. Based on a series of simplified algorithms, a number of software packages have been developed for large area noise mapping. This chapter first describes main algorithms of noise mapping techniques (Section 5.1). It then discusses the accuracy, efficiency, and strategic application of noise mapping techniques through case studies (Section 5.2). This is followed by a brief description of some noise mapping applications (Section 5.3). Finally, this chapter describes some other recently developed models for medium- and large-scale urban areas, typically a series of urban elements such as street canyons (Section 5.4).

5.1 Noise mapping techniques

A noise map, typically in the form of interpolated isocontours, is a way of presenting geographical distribution of noise exposure, either in terms of measured or calculated levels. In this book noise mapping is computing-based except where indicated.

Noise mapping is based on a series of algorithms specified in various standards, internationally (ISO 1993) and nationally, for various noise sources including aircraft (ECAC 1997), road (UK DfT 1988; Jonasson and Storeheier 2001), railway (UK DfT 1995) and industry (DAL, 1987). The calculation procedures and indices are often different in different standards. Comparisons of the traffic noise levels predicted by the German and French procedures applied to the same scenario show that a difference of up to 7dB could occur even for nominally similar favourable, that is, downwind, meteorological conditions (Kang *et al.* 2001a). Whilst efforts are made to improve the prediction accuracy (Daigle and Stinson 2005), especially across EU countries, for example, through the Harmonoise and IMAGINE projects (Watts 2005), various national standards are still being widely applied in noise mapping software (AEAT 2004).

Following the basic principles and theories of outdoor sound propagation described in Section 1.5, noise mapping algorithms, including the source models for road traffic based on the Harmonoise project, and the propagation models based on ISO 9613 (ISO 1993) are outlined in this section. The project Harmonoise includes the development of road and railway source models, reference propagation models which are precise for considering different atmospheric conditions but require intensive computer resources, an engineering/simpler model for use in noise mapping, and validations against measured results. A follow-up project, IMAGINE, refines the prediction models developed in Harmonoise and also considers industrial sources and aircraft. The scope of ISO 9613 extends to the calculation of noise levels from road, rail, industrial, construction and other sources, but not aircraft in flight, military operations and blast waves that are associated with mining type activities.

5.1.1   Source model for road traffic

The source models for road traffic developed in Harmonoise entail the description of sound power of various categories of vehicle (Jonasson *et al.* 2004; Watts 2005). The sources on the vehicles are simplified into two point sources: the lower source at 0.01m above the road surface, which is mainly due to tyre/road noise; and the higher source, which is mainly propulsion noise with its height depending on the vehicle category. Vehicles are divided into five main categories including light vehicles such as cars and vans; medium heavy vehicles (with two axles) such as buses, light/medium heavy trucks and heavy vans; heavy vehicles (with more than two axles) such as heavy trucks and buses; other heavy vehicles such as construction trucks and agricultural tractors; and two-wheelers such as mopeds, scooters and motorcycles. In each main category there are a number of sub-categories.

The rolling noise for the reference condition is calculated by

$$L_{WR}(f) = a_R(f) + b_R(f)\log\left(\frac{v}{v_{ref}}\right) \tag{5.1}$$

where the coefficients a_R and b_R are given for each vehicle category in one-third octave band, v is the vehicle speed, and $v_{ref} = 70$km/h is the reference speed. The rolling noise is assumed to radiate 80 per cent from the lower source and 20 per cent from the higher source.

The propulsion noise is described as a linear function of speed:

$$L_{WR}(f) = a_P(f) + b_P(f)\log\left(\frac{v - v_{ref}}{v_{ref}}\right) \tag{5.2}$$

where the coefficients a_P and b_P depend on the vehicle category and frequency, and the reference speed v_{ref} is again 70km/h. For propulsion noise 20 per cent is assumed to radiate from the lower source and 80 per cent from the higher source.

A number of corrections are made to the basic sound power levels, including for the road surface texture and condition, for directivity both in the horizontal and vertical plane, and for tyres. Propulsion noise increases during acceleration and decreases during deceleration and, thus, a correction $\Delta L_{acc} = Ca\,(-2\text{m/s}^2 < a < 2\text{m/s}^2)$ is given, where a is the acceleration/deceleration and the coefficient C depends on vehicle category.

It is important to consider the influence of variation in vehicle speed. For example, the additional noise produced by relatively fast moving traffic could cancel out the benefits of lower speed vehicles to a certain extent. Speed may also vary along a section of road where there is a junction, pedestrian crossing or traffic calm device.

5.1.2   General calculation procedure

The following calculations are based on point sources which can be moving or stationary. Extended noise sources are divided into cells each with their own characteristics. For example, a road segment can be composed of a series of point sources of different types depending on the percentage of vehicles of various categories and their speed. For noise mapping purposes, 5° is a reasonable value for the angle of view – the maximum angle each segment subtends at the receiver position. A smaller angle can be used where greater precision is required (Watts 2005). Octave or one-third octave band algorithms are used in calculating various attenuations.

In ISO 9613 the equivalent continuous downwind octave band SPL at a receiver position is determined from

$$L_{fT}(DW) = L_W + D_C - A \tag{5.3}$$

where L_W is the sound power level of the source, D_C (dB) is a directivity correction, and A is the attenuation between source and receiver and includes the following five attenuation mechanisms: geometrical divergence A_{div}, atmospheric absorption A_{atm}, ground effects A_{gr}, barrier effects A_{bar}, and miscellaneous effects A_{misc}:

$$A = A_{div} + A_{atm} + A_{gr} + A_{bar} + A_{misc} \tag{5.4}$$

Considering eight octave bands from 63Hz to 8kHz, the equivalent continuous A-weighted downwind SPL is

$$L_{AT}(DW) = 10\log\left\{\sum_{i=1}^{n}\left[\sum_{j=1}^{8} 10^{0.1[L_{fT}(ij)+A_f(j)]}\right]\right\} \tag{5.5}$$

where n is the number of contributions considering sources and paths, and A_f is the A-weighting factor for each frequency (see Section 1.2.3). By subtracting the effect of the meteorological conditions, C_{met} (see Section 5.1.8), the long-term averaged A-weighted SPL can be obtained

$$L_{AT}(LT) = L_{AT}(DW) - C_{met} \tag{5.6}$$

5.1.3 Geometrical divergence

Geometrical divergence A_{div} is due to free-field spherical spreading of sound from a point source and it can be calculated by

$$A_{div} = 20\log d_r + 11 \tag{5.7}$$

where d_r is the source–receiver distance.

5.1.4 Atmospheric absorption

Atmospheric absorption A_{atm} is calculated by

$$A_{atm} = \alpha d_r / 1000 \tag{5.8}$$

where α is the octave band atmospheric attenuation coefficient in dB/km (ISO 1993).

5.1.5 Ground effect

The ISO 9613 method for calculating the ground effect A_{gr} can only be used for ground that is approximately flat, either horizontally or has a constant gradient. Three separate regions are used when determining ground attenuation. For the source region, the minimum distance is 30 × height of source above ground, h_s, and the maximum distance is the source–receiver distance, d_p, as projected on the ground plane. For the receiver region, the minimum distance is 30 × height of receiver above ground, h_r, and the maximum distance is d_p. The middle region covers the distance between source and receiver regions, and there will be no middle region if $d_p < (30h_s + 30h_r)$. The attenuations of the above three regions are calculated for each octave band and then added to give the total ground attenuation per octave band.

The effect of the ground surface is accounted for by the ground factor G. With hard ground such as water, concrete and paving, $G = 0$; with porous ground such as ground surfaces covered by vegetation or farmland, $G = 1$; and with mixed ground $G = 0–1$. The total ground attenuation, in octave band, is determined by $A_{gr} = A_s + A_r + A_m$, where the component attentions for the three regions A_s, A_r, and A_m are calculated based on G and relevant dimensional factors (ISO 1993).

If the sound is not a pure tone and G is close to 1, an alternative method can be used to calculate A-weighted SPL

$$A_{gr} = 4.8 - \frac{2h_m}{d_r}\left(17 + \frac{300}{d_r}\right) \geq 0 \tag{5.9}$$

where h_m is the mean height of the propagation path above the ground. When using Equation (5.9), the increase in source power level due to ground reflections near the source should be accounted for by including the term D_Ω to the directivity correction D_C in Equation (5.3)

$$D_\Omega = 10\log\left[1 + \frac{d_p^2 + (h_s - h_r)^2}{d_p^2 + (h_s + h_r)^2}\right] \tag{5.10}$$

5.1.6 Screening

According to ISO 9613, screening effect A_{bar} is considered only if the surface density of the screen is >10kg/m², the screen has a closed surface without gaps, and the horizontal dimension of the screen normal to the line from source to receiver is larger than the wavelength of sound at the nominal midband frequency for the octave band of interest, λ. A_{bar} is given as an insertion loss, considering the diffraction over the top of the barrier ($A_{bar} = D_\delta - A_{gr} > 0$) and around the edge of the barrier ($A_{bar} = D_\delta > 0$), where D_δ is the octave band barrier attenuation, A_{gr} is the ground attenuation with no screening present, and δ is the path length difference between the diffracted and direct sound.

When one significant sound propagation path is assumed, D_δ can be calculated by

$$D_\delta = 10\log\left(3 + \frac{C_2}{\lambda}C_3\delta K_{met}\right) \tag{5.11}$$

where $C_2 = 20$ when the effect of ground reflections is included, and $C_2 = 40$ when ground reflections are taken into account separately by image sources; $C_3 = 1$ for single diffraction and $C_3 = [1 + (5\lambda/e)^2]/[1/3 + (5\lambda/e)^2]$ for double diffraction, with e as the distance between the two diffraction edges; and K_{met} is the meteorological effect correction factor where $K_{met} = \exp\{-(1/2000)[d_{ss}d_{sr}d_r/(2\delta)]^{0.5}\}$ for $\delta > 0$ and $K_{met} = 1$ for $\delta \leq 0$. For single diffraction, further to Equation (1.15),

$$\delta = \sqrt{(d_{ss} + d_{sr})^2 + a^2} - d_r \tag{5.12}$$

For double diffraction

$$\delta = \sqrt{(d_{ss} + d_{sr} + e)^2 + a^2} - d_r \tag{5.13}$$

where d_{ss} is the distance between the source and the first diffraction edge, d_{sr} is the distance from the receiver to the diffraction edge in case of single diffraction and to the second diffraction edge in case of double diffraction, and a is the component distance parallel to the barrier edge between source and receiver.

Guidelines have also been given for other barrier configurations (UK DfT 1988).

5.1.7 Reflections

In ISO 9613 reflections from outdoor ceilings or building façades are considered using image sources. Ground reflections are not included here since these are already included in A_{gr}. Reflections are only considered when a specular (geometrical) reflection occurs, the surface reflection coefficient is greater than 0.2, and the surface is sufficiently large, namely

$$\frac{1}{\lambda} > \frac{2}{(l_{min}\cos\beta)^2}\frac{d_{so}d_{or}}{d_{so} + d_{or}} \tag{5.14}$$

where d_{so} is the distance from the source to the point of reflection, d_{or} is the distance from the point of reflection to the receiver, β is the angle of incidence in radians, and l_{min} is the minimum dimension of the reflecting surface.

The sound power level of the image source is calculated by

$$L_{W,im} = L_W + 10\log\rho + D_{Ir} \tag{5.15}$$

where ρ is the reflection coefficient at the angle of incidence β, and D_{Ir} is the source directivity index in the direction of the receiver image.

5.1.8 Meteorological correction

In ISO 9613, downwind standard meteorological conditions for a wind speed of 1–5m/s at a height of 3–11m above the ground are defined, with wind blowing from source to receiver at an angle of ±45°. These conditions are regarded to represent reasonable meteorological conditions and are consistent with ISO 1996 (2003a) regarding description and measurement of environmental noise.

Meteorological correction C_{met} is used in the prediction of long-term average A-weighted SPL to include the effect of varying weather conditions that occur over a time period of several months or a year:

$$C_{met} = 0 \qquad\qquad\qquad \text{if} \quad d_p \leq 10(h_s + h_r)$$

$$C_0 \left[1 - 10\frac{h_s + h_r}{d_p} \right] \qquad\qquad \text{if} \quad d_p > 10(h_s + h_r) \qquad\qquad (5.16)$$

where C_0 is the meteorological factor that is dependent on local statistics for wind speed, wind direction and temperature gradients.

In Harmonoise, in order to assess the effects of meteorological refraction the radius of curvature from source to receiver is determined for each propagation path based on wind speed, wind direction and atmospheric stability estimated from cloud cover and period of day. A combined linear/logarithmic sound speed profile is assumed

$$c\,(h) = c_0 + Ah + B \log \left(\frac{h}{h_0} \right) \qquad\qquad\qquad\qquad (5.17)$$

where $c(h)$ is the speed of sound at a height h, and A, B, c_0 and h_0 are constants. These profiles can be converted to equivalent linear sound speed gradients (Watts 2005).

5.1.9   Miscellaneous attenuation

In ISO 9613, miscellaneous attenuation A_{misc} includes: A_{fol} due to sound propagation through foliage, depending on the density of the foliage close to the source and/or receiver; A_{site} due to sound propagation through an industrial site, namely scattering from installations unless they are covered in A_{bar}; and A_{house} due to combined effects of screening and reflections.

It is suggested in ISO 9613 that A_{house} should be calculated, at least in principle, using the methods described in Sections 5.1.6 and 5.1.7. It is also indicated, however, that because A_{house} is very situation-dependent, a more useful alternative, particularly for the case of multiple reflections where the accuracy of calculation suffers, may be to measure the effect, either in the field or by modelling.

A major difference between micro- and macroscale acoustic modelling is the consideration of built-up areas. Further examinations relating to A_{house} are included in Section 5.4, Chapter 4, and Chapter 7.

5.1.10   Noise mapping software

With the development of computer resources, a number of noise mapping software packages have been developed, such as Cadna, ENM, fluidyn, GIpSynoise, IMMI, LIMA, Mithra, Noisemap, Predictor, and SoundPLAN, to implement various algorithms such as those previously described for the calculation of sound propagation in large areas. This section briefly describes basic steps of noise mapping software, and discusses various features different software packages may have (Kang *et al.* 2001a).

A common feature of all noise mapping software is the combination of noise propagation calculations with a mapping and scheme editing facility, consisting of georeferenced input data, often associated with geographical information systems (GIS). Building a three-dimensional model is the next important step. Depending on the scale of the terrain in terms of vertical differences, ground elevation data is normally required at 5m intervals. Positional and height information is also required for buildings and any major structures. This information can be obtained from maps, aerial photographs, or by survey. Positions and characteristics of various types of sound sources are also needed. Then appropriate standards/algorithms should be chosen, as well as a series of calculation parameters, such as reflection order and the radius within which sources should be searched, evaluation parameters and reference time periods, and grid factors for dividing line or area sources. After the calculation process, noise maps can be produced, either horizontally above the ground or vertically in front of building façades, as well as other outputs such as exposure levels of a population for risk estimation purposes.

Whilst the results of various noise mapping software packages should be the same/similar if they use the same standards, different software may have different features, including computer requirements; input capabilities for data and graphic files; libraries of various sound sources and materials/ elements; calculation limits in terms of the number of sources, receivers and barriers; calculation results in octave or one-third octave bands; standards and algorithms implemented for various source types including road, rail, air and industry; output capabilities for two-dimensional, three-dimensional and digital maps; possibility for acoustic and visual animation; facilities for noise mitigation strategies such as designing the optimal shape of environmental barriers; and various aspects of acoustic environment assessment such as consideration of time profile and connection with GIS.

5.2   Noise mapping: accuracy and strategic application

While noise mapping has become an essential requirement, especially in Europe (EU 2002; UK DEFRA 1999, 2001, 2003; Abbott and Nelson 2002), and corresponding software/techniques have been widely used in practice, there are still debates about their usefulness (Shield 2002; Jopson 2002; Manning 2002; Tompsett 2002; Turner and Hinton 2002). This section first briefly discusses the usefulness of noise mapping. It then focuses on accuracy and efficiency in noise mapping by considering a series of typical configurations, with a commonly used software package, subsequently referred to as NMS. By choosing ISO 9613 (ISO 1993) and CRTN (UK DfT 1988) in NMS, the study examines the valid range of the algorithms, such as the suitability for a square or a street; possible errors caused by simplifications in three-dimensional models, such as pitched roofs and gaps between buildings; and effects of calculation parameters such as the reflection order (Huang 2003; Kang and Huang 2005). Broadband noise is considered in the calculation except where indicated. For the convenience of relative comparison, meteorological conditions are not considered. Issues relating to data acquisition for source power and three-dimensional maps are then discussed. Finally, a case study is presented based on an actual urban area.

5.2.1   Usefulness of noise mapping

An advantage of noise mapping techniques is that they can be used for larger urban areas compared to the various microscale simulation techniques. Noise mapping is a powerful and effective way to visualise and assess the acoustic environment. Ideally, it should give an accurate statement of noise levels in a specific location, provide noise trend data, establish exposure levels of a population for risk estimation purposes, identify pollution hotspots or quiet areas, yield information as to the effectiveness of noise management schemes, and indicate management/legislative/policy changes that may be required.

One of the main concerns is the accuracy of noise mapping techniques. For example, ISO 9613 provides some indication of the estimated accuracy of the principal elements of the procedure for a range of mean source and receiver heights up to 30m as a function of propagation distance up to 1km for long-term average downwind conditions. Whilst the estimated values of ± 3dB (± 1dB for source–receiver distance within 100m and source/receiver height 5–30m) are for A-weighted SPL of broadband noise, the accuracy of prediction for pure tone sources and individual octave band levels is likely to be lower. Moreover, these estimations do not include the additional and inherent uncertainty in urban areas associated with attenuation due to multiple diffractions and multiple reflections, especially those between ground and façades. Furthermore, most noise mapping algorithms are based on generic configurations and may not be applicable to many practical situations where the combined effects of various factors is important, for example, in the case of noise barriers combined with vegetation. In Harmonoise, the engineering model has been validated against measurements at two road locations where the roads ran on embankments above essentially flat terrain. The differences were never greater than 1.5dB L_{den} even at a distance greater than 1km from the road. However, further validation is required, for example, in the more complex situations typically found in urban areas (Watts 2005).

Cost and necessity are also concerns with regard to noise mapping. The acquisition and input of data with the production of a noise map would take a considerable amount of time, especially if database availability is poor, whereas normally only traffic is mapped on major urban roads and 'most experienced noise consultants could produce an accurate noise map if given a street map and a red pen' (Shield 2002). Some noise mapping projects are criticised for wasting money in showing what is already known, namely that roads, railways and airports are main noise sources.

Since noise mapping will have a major influence on policymaking (UK DEFRA 2001), there is a concern that if noise mapping is accepted without question, the corresponding policies could be misleading, although accurate noise mapping could well contribute to the implementation of the strategy (Shield 2002). It has also been argued that noise policies should not be too reliant on the development of detailed noise maps, which is unnecessary for many of the national measures currently available (Pease 2002). Another concern is that if noise contours have been produced using a particular noise model/database, and new and inconsistent contours are produced using a different modelling system, it might be difficult to determine which set of conditions or policies should take precedence (Jopson 2002).

On the other hand, Tompsett (2002) argued that noise levels were pretty much as predicted, often with an accuracy of a standard error of ±2dBA on façades exposed to traffic noise, for a properly constructed noise model used with its design limits. Moreover, a clear definition of accuracy is not obvious and the calculation is usually compared with a measurement under certain conditions rather than the 'real' answer. While noise maps can make politicians and the public better understand and treat noise as a serious issue, Turner and Hinton (2002) argued that strategic noise maps would also provide a platform for further refinement and development through which investigations into the noise impact in specific areas might be undertaken.

Noise mapping techniques are also useful for relative comparisons. Actual measurements can only be a small sample of the sound environment over much longer periods of time. Since the actual sound levels tend to vary considerably, due to the source variation, temperature and wind conditions, for example, short-term measurements could be rather unrepresentative. This variability could affect strategic comparisons. For instance, comparing sample measurements taken under different meteorological conditions between two different sites might be unreasonable, but a set of comparable measurements considering various conditions would often be impractical. In a calculated map, however, any differences in the outputs are solely due to the differences in the input data that are taken into account by the computation (Kang *et al.* 2001a).

Overall, whilst noise mapping is a useful tool for noise strategies and policies, attention must be paid to its accuracy and strategic application, including improving algorithms along with advancement in research and computation power, examining efficient and accurate ways of building models, consulting the people who live and work in the vicinity as well as local planning and environmental officers in the mapping processes, using noise mapping in various strategic ways, and introducing effective procedures for validation and quality control. For validation, it is important to compare calculated and measured SPL at various typical locations in octave or one-third octave bands, rather than only A-weighted values at locations where direct sounds are dominant.

5.2.2   Comparison with image source method

To examine the accuracy of noise mapping algorithms at the microscale, especially in terms of the effect of multiple reflections, a comparison is made between NMS and the image source model (see Section 4.1) in an idealised urban square and an idealised street canyon, both 50 × 50m, as shown in Figure 5.1. The surrounding buildings have an identical height of 20m. A single point source is positioned at (10m, 10m), with a height of 1.5m and a sound power level of 100dB, considering middle frequencies (from 500Hz to 1kHz). Ten receivers are considered along a diagonal of the square/street. The calculation using the image source model with a reflection order $R = 20$ is regarded as the accurate result.

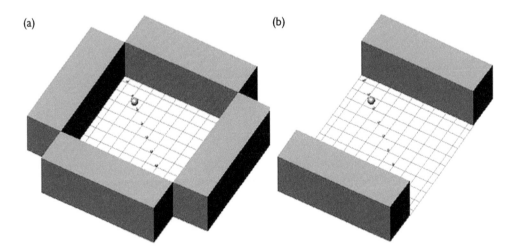

Figure 5.1 (a) The idealised square and (b) street, both 50 × 50m, showing the source (large dot) and receiver (small dots) positions. The origin of coordinates is at a corner of the square/street.

Figure 5.2 shows the difference in SPL between the image source model result with $R = 20$ and the NMS calculation with reflection orders $R = 1, 3, 5, 10$ and 20, where the absorption coefficient of the ground and building façades is assigned as $\alpha = 0.1$. It can be seen that the NMS result increases towards that of the image source model with increasing R until about $R = 10$, then it becomes rather stable/unchanged. The results with the two methods are similar in the near field, whereas with the increase of source–receiver distance, NMS underestimates the SPL by about 4dB in the square and 3dB in the street.

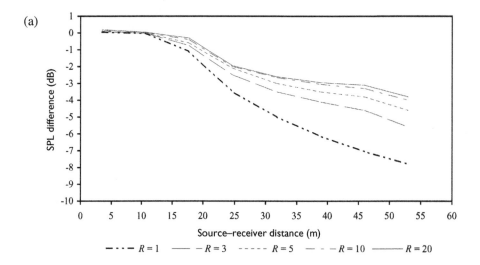

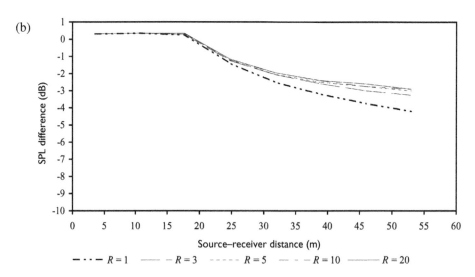

Figure 5.2 SPL of the NMS calculation with reflection orders $R = 1-20$, with reference to the image source model result with $R = 20$; (a) square; (b) street.

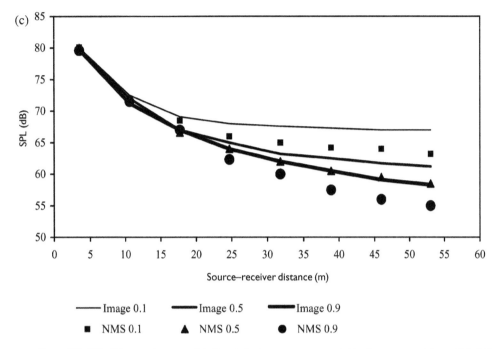

Figure 5.3 SPL difference between NMS and the image source model in the square, with α = 0.1, 0.5 and 0.9.

Comparisons between the two methods are also made with different boundary absorption coefficients. Figure 5.3 shows the results in the square with α of all the boundaries being 0.1, 0.5 and 0.9, where $R = 20$ for both models. As expected, the SPL attenuation along the receiver line becomes greater with increasing α, but it is interesting to note that the difference between the two models is almost consistent with various α. This suggests that there is a systematic error in the noise mapping software, probably caused by the simplified way of dealing with multiple reflections between façades and ground.

Overall, it seems that the noise mapping algorithms may not be applicable for microscale urban areas such as squares and streets, especially when reflections play an important role. There is a need to integrate more accurate modelling techniques into noise mapping software yet still retaining simplicity and speed.

5.2.3   Simplification of pitched roofs

In noise mapping practice, pitched roofs are often simplified to flat roofs in order to reduce the time for model construction and calculation. However, the assumed height of a flat roof can affect the results considerably. To examine this simplification, a typical residential building with a 45° pitched roof, eaves height of 5m and ridge height of 10m is used to compare with three simplified building blocks with a height of h = 5m (that is, eaves height), 7.5m (that is, eaves height plus half of the roof height) and 10m (that is, ridge height), respectively, as shown in Figure 5.4.

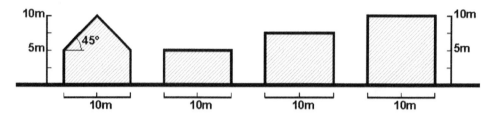

Figure 5.4 Cross section of the pitched roof and the three simplified building blocks.

Two typical building arrangements are considered, as illustrated in Figure 5.5, where the width and depth of each building is 20m and 10m, respectively. The road has a width of 10m and a line source along the road is assumed. In the arrangement shown in Figure 5.5a, 14 receivers are evenly positioned with a spacing of 10m and a height of 4m, along two lines, namely the middle line of a building gap and the middle line of a building. In Figure 5.5b, ten evenly distributed receivers are considered.

Corresponding to the configuration illustrated in Figure 5.5a, the broadband SPL differences between pitched and flat roofs are shown in Figure 5.6, where $\alpha = 0.1$ and $R = 20$. It can be seen that with a block height $h = 5$m the SPL is overestimated by 0.9dB on average over the 14 receivers, whereas with $h = 7.5$ and 10m the SPL is underestimated by 1.8dB and 3.7dB on average respectively. Consequently, $h = 5.7$m is tested and the difference is found to be much less, only 0.04dB in average, with an STD of 0.12dB. This suggests that an additional height of $\Delta = 0.7$m above the eaves height might be a good approximation for simplifying pitched roofs.

In order to examine the applicability of this rule further, three different eaves heights are considered, 3, 10 and 30m, representing bungalows, multistorey flats and high-rise buildings, respectively. From Figure 5.6 it can be seen that with those eaves heights the differences between pitched roofs and flat roofs of $\Delta = 0.7$m are once again negligible, at 0.14dB (STD = 0.16dB), 0.02dB (STD = 0.08dB) and 0.04dB (STD = 0.11dB) on average, respectively.

For the more complex configuration as illustrated in Figure 5.6b, with $\Delta = 0.7$m the difference between pitched and flat roofs is also very small, with average = 0.16dB, maximum = 0.5dB, and STD = 0.27dB, where the eaves height is 5m.

The calculation time is typically reduced by 50 per cent by simplifying pitched roofs into flat roofs for the above configurations.

5.2.4   Gaps between buildings

Ignoring gaps between buildings is another common way of simplifying noise mapping, even though a previous study based on the radiosity model has shown that a gap between buildings can provide about 2–3dB extra sound attenuation along a street, and the effect is more significant in the vicinity of the gap (Kang 2002d). To further examine the effects of building gaps, a comparison is made among building gap widths $w = 1, 2, 3, 4$ and 5m. As illustrated in Figure 5.7, three building configurations are considered, simulating typical UK house types, namely building width 5m as a detached house, 10m as semi-detached houses and 20m as terraced houses. The building depth is 10m for all three house types.

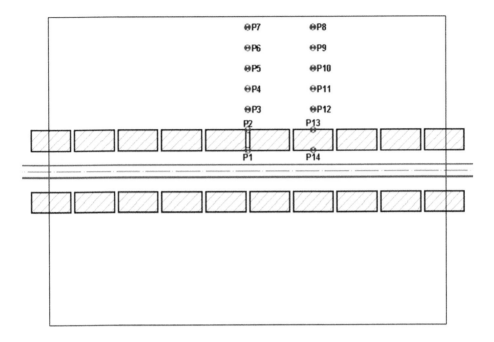

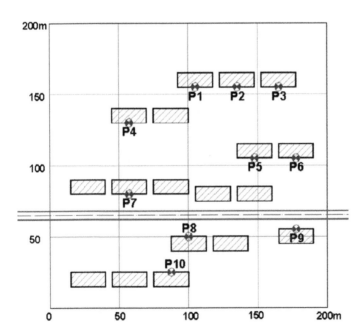

Figure 5.5 Site plan of the two calculation configurations, showing the positions of road, buildings and receivers.

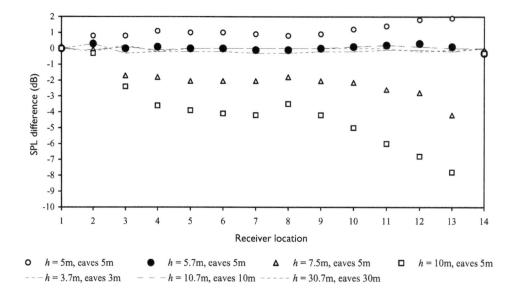

○ $h = 5$m, eaves 5m   ● $h = 5.7$m, eaves 5m   ▲ $h = 7.5$m, eaves 5m   ☐ $h = 10$m, eaves 5m
--- $h = 3.7$m, eaves 3m   – – – $h = 10.7$m, eaves 10m   ----- $h = 30.7$m, eaves 30m

Figure 5.6 SPL with various flat roof heights, with reference to the SPL of pitched roof, based on the configuration illustrated in Figure 5.5a.

Figure 5.8a shows the broadband SPL with various building gap widths with reference to the situation without any gap between buildings, where $\alpha = 0.1$, $R = 20$, and the receiver positions correspond to those shown in Figure 5.5a. As expected, the SPL difference increases with increasing w, and it is interesting to note that this increase is approximately proportional to w, as can be seen in the situation of detached house in Figure 5.8a. Where the direct sound plays an important role, such as at receivers 1 and 14, the SPL difference is negligible, whereas immediately behind the buildings, such as at receiver 2, the difference is the greatest, over 10dB with $w = 5$m. With the increase of source–receiver distance, the effect of gaps generally becomes less.

Compared to detached houses, the effect of building gaps is less, typically by 0.5–2dB and 2–4dB respectively, with semi-detached and terraced houses. This can be seen in the colour maps in Figure 5.8b. The difference between various house types becomes less with increasing source–receiver distance, decreasing building gaps and increasing importance of the direct sound.

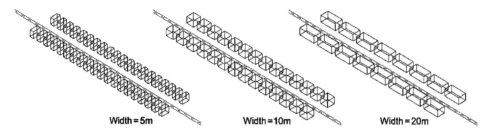

Width = 5m   Width = 10m   Width = 20m

Figure 5.7 Three configurations with different building widths used to examine the effects of building gaps.

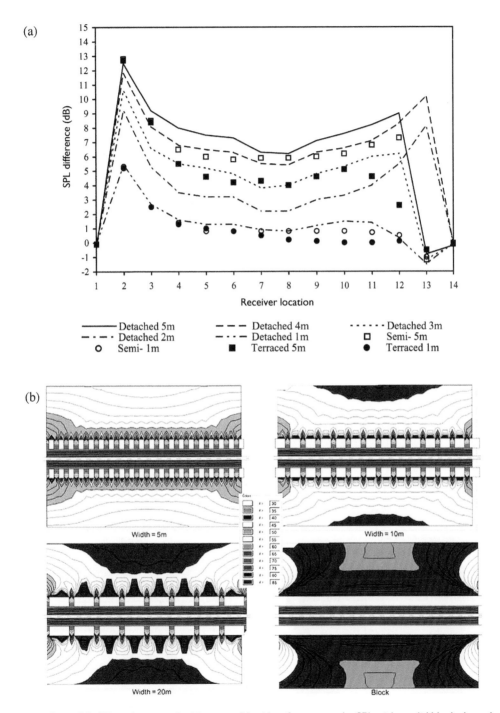

Figure 5.8 SPL with various building gaps (a) with reference to the SPL with a solid block along the street, and (b) the colour map with four building arrangements where the building gap is 5m. A colour representation of this figure can be found in the plate section.

Overall, it is suggested that the gaps between buildings should not be ignored when receivers behind buildings are considered, unless the building gap is less than 1–2m and the receivers are far from the source and buildings, say more than 30–50m.

5.2.5   Calculation parameters

Whilst it can be seen in Figure 5.2 that with increasing reflection order the NMS result becomes more accurate, it is useful to determine appropriate reflection order for more complex configurations, considering both accuracy and calculation time. Figure 5.9 shows the broadband SPL with reflection orders $R = 0, 1, 3$ and 8 for the configuration shown in Figure 5.5a, where $\alpha = 0.1$. It can be seen that although the difference is negligible between $R = 3$ and 8, it is significant between $R = 0$ and 1. In the vicinity of building gaps the differences between various reflection orders are the greatest, up to 4dB, whereas when the direct sound is dominant, such as at receivers 1 and 14, the SPL is almost not affected by the reflection order.

In Figure 5.10 a comparison between $R = 0$ and 1 is made for an actual urban area in Sheffield with changing ground levels, various building types and dimensions, and traffic noise from a number of roads. The average broadband SPL difference is 2.4dB, which is again significant.

Corresponding to Figure 5.9, Figure 5.11 compares the calculation time with various reflection orders. It can be seen that the increase in calculation time with increasing reflection order is approximately exponential. Considering both accuracy and efficiency, it is suggested that a reflection order of 1–3 should be used.

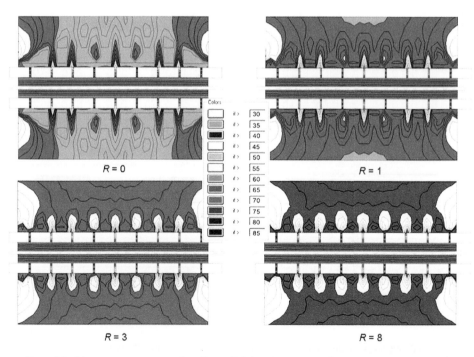

Figure 5.9   Noise maps in a street (see Figure 5.5a) with various reflection orders. A colour representation of this figure can be found in the plate section.

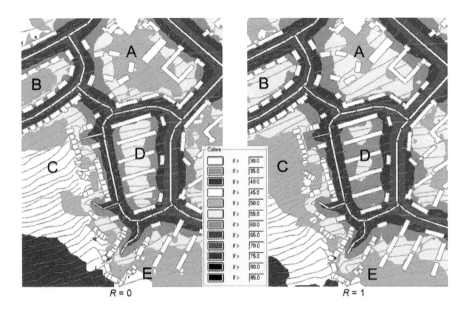

Figure 5.10 Comparison between noise maps with reflection order 0 and 1 in an urban area in Sheffield. A colour representation of this figure can be found in the plate section.

The maximum error margin (MEM) defines which sound sources can be ignored when their contribution is negligible, assuming that the neglected emitters might propagate freely towards the receptor point. For the configuration in Figure 5.10, compared to MEM = 0, with MEM = 0.5dB and 1dB the average difference is 1dB (STD = 0.76dB) and 1.7dB (STD = 1.16dB) and the calculation time is reduced from 2,950s to 700s and 600s, respectively,

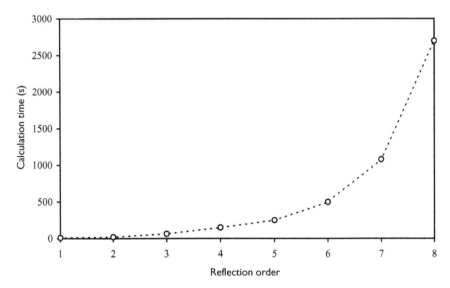

Figure 5.11 Increase in calculation time with increasing reflection order, based on the configuration shown in Figure 5.9.

whereas for the configuration in Figure 5.9, there is no difference in SPL and calculation time between MEM = 0, 0.5 and 1dB. Some other calculation parameters such as grid interpolations and emission point grids have also been examined and the results are rather different with different configurations (Huang 2003).

Overall, it is suggested that for large-scale noise mapping, a pilot study would be necessary with sample areas to determine appropriate calculation parameters, as well as suitable simplifications such as discussed in Sections 5.2.3 and 5.2.4.

5.2.6   Data acquisition

In addition to the effects of model simplifications and calculation parameters, the accuracy in noise mapping depends heavily on the availability and quality of input data, as well as their potential incompatibility with the software tools. For industrial noise mapping, David and Hessler (2003) suggested that the accuracy of any noise model is generally much more dependent on the quality of the input power levels rather than on the specific modelling program used or on the details of the propagation calculations.

Whilst much effort has been made to build three-dimensional maps incorporating road, rail and industrial sources, in connection with GIS data (Cogger 2003), models/data from other sources, such as for air-quality assessment (Stocker and Carruthers 2003), are also used in noise mapping exercises.

Although some simplifications on three-dimensional models could increase the noise mapping efficiency, as examined in Sections 5.2.3 and 5.2.4, it is recommended to avoid such simplifications wherever possible. This will make it easy to update the original data and also, a complete model could be applied to more detailed investigations for other purposes (DataKustik 2005).

5.2.7   A case study

Based on the above analysis, a case study was carried out in an area of Sheffield city centre, approximately 700 × 400m, to examine the accuracy and strategic application of noise mapping (Huang 2003; Kang and Huang 2005). The sound sources considered included road traffic as well as the central fountains. In order to obtain the traffic data and sound power levels of certain sources such as the fountains, measurements were made at 38 positions, as illustrated in Figure 5.12a. For the purpose of validation, measurements were also made at four representative locations, as illustrated in Figure 5.12b, where locations A and D were at the centre of two squares surrounded by buildings, and locations B and C were in semi-open spaces. The site plan and contour data were based on the EDINA Digmap, and the building heights were based on site measurements. Building gaps were not omitted, and pitched roofs were generally simplified following the rules discussed in Section 5.2.3, except for some main buildings, where the actual building forms were modelled. The three-dimensional model for noise mapping is shown in Figure 5.13.

The maximum search radiuses for source and emission point were greater than the site dimension, whereas for reflections, the source search radius was 100m. All the building surfaces had a reflectivity factor defined. The grid interpolation was 17 × 17m, the grid resolution was 10 × 10m, and the maximum error margin was 0.5dB.

In Figure 5.14 two noise maps are compared, one with a reflection order $R = 1$ and the other with $R = 3$. There are some differences between the two maps, particularly in the enclosed

(a)

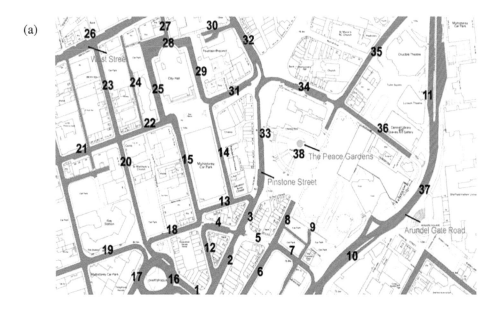

(b)

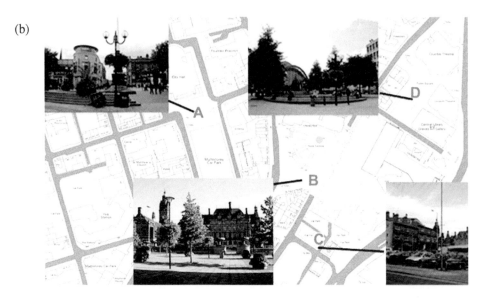

Figure 5.12 Site plan of an area of Sheffield city centre showing the measurement points for (a) sound source data and (b) validation.

spaces, but they are generally insignificant. At location D, for example, with $R = 3$ the SPL is 1dBA higher than that with $R = 1$.

In Table 5.1 a comparison is made between calculation and measurement. It can be seen that the agreement is rather good, with the calculated levels generally slightly higher than the measured values by up to about 2dBA. It is noted, however, that in this case study the site is a

Table 5.1 Comparison between calculated and measured results (dBA).

Receiver	A	B	C	D
Calculated, reflection order 1	64.2	69.5	62.9	60.2
Calculated, reflection order 3	64.8	69.7	63.2	61.2
Measured	63.9	68.7	62.4	59.3

relatively busy urban area, and the direct sound is generally dominant. Further studies are still needed for situations where reflection and diffraction play a more important role.

5.3   Noise mapping application

5.3.1   Europe

In Europe noise mapping has been given attention for about 20 years, and some countries have been particularly active (UK DEFRA 1999). For example, all towns with populations over 50,000 in the Netherlands were mapped before 1998. Noise mapping techniques are now in use in Europe at city or town levels as well as at the national level for road and rail networks, although approaches in different countries vary in terms of input data generation, mapped area, and information presented. The EU made provisions to allow suitable interim computation methods such as existing national methods to be used prior to the development of a common EU method.

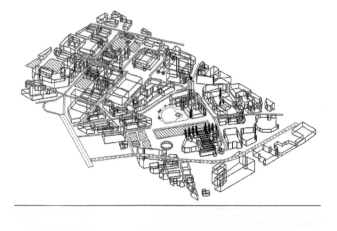

Figure 5.13 Three-dimensional model of an area of Sheffield city centre for noise mapping. A colour representation of this figure can be found in the plate section.

(a)

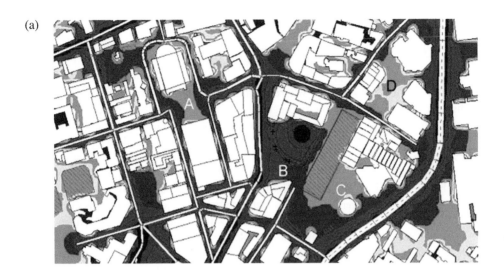

(b)

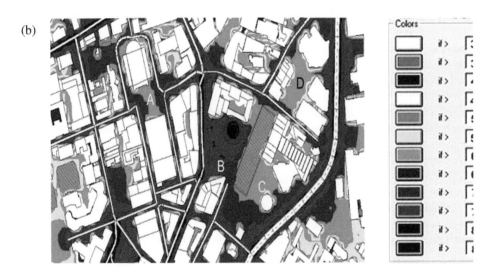

Figure 5.14 Noise map of an area of Sheffield city centre with a reflection order of (a) 1 and (b) 3. A colour representation of this figure can be found in the plate section.

In the EU Directive on environmental noise (EU 2002) it is stated that Member States shall ensure that no later than 30 June 2007 strategic noise maps showing the situation in the preceding calendar year must be made and, where relevant, approved by the competent authorities, for all agglomerations with more than 250,000 inhabitants and for all major roads that have more than 6 million vehicle passages a year, major railways that have more than 60,000 train passages per year and major airports within their territories. Moreover, Member States must adopt the measures necessary to ensure that no later than 30 June 2012, and thereafter every 5 years, strategic noise maps showing the situation in the preceding calendar year must be made for all agglomerations and for all major roads and major railways within their territories.

5.3.2   United Kingdom

Early noise mapping in the United Kingdom was restricted to areas in the vicinity of airports. The first attempt at providing a noise map of England was published by the Council for Protection of Rural England (CRPE) and the Countryside Commission in 1995. In 1999, a pilot noise mapping scheme was undertaken in Birmingham. This was followed by a large-scale noise mapping project for England, where England was divided into 30 zones of varying size. The first stage including London began in 2002 and the following stages are being carried out. An altered version of the CRTN procedure was developed (Abbott and Nelson 2002) for calculating road traffic noise so as to obtain L_{Aeq} levels rather than L_{A10} levels, and other procedures adopted include CRN for railway noise and ISO 9613 for other sources.

The Birmingham noise mapping project (Hinton 2000; Hinton and Bloomfield 2000) was carried out based on the German Standard DIN 18005 (DIN 1987) and ISO 9613. Building heights were set to 8m, except for tower blocks whose heights were estimated by noise consultants from aerial photographs. Noise levels were calculated at a 10 × 10m grid work, detailed at a 4m height to the nearest 5dB. No attempt was made to use GIS data and population data to calculate the numbers of people exposed to noise. The results showed day and nighttime noise emission levels based on busy roads, railways, Birmingham airport, 21 industrial sites and a combination of these. Based on PPG24 (see Section 2.3.4), it was suggested that 2 per cent of Birmingham fell into category D, 11 per cent into category C, 23 per cent into category B, and 64 per cent into category A.

Noise mapping of Greater London was launched in 2002, making the city the second to be mapped (UK DEFRA 2003) in England. DEFRA worked in partnership with the Greater London Authority (GLA) and 33 London borough authorities to map road traffic noise in the Greater London area which was the most comprehensive survey of traffic noise ever undertaken in the capital. For the prototype noise map all buildings were modelled with a height of 8m above local ground, although the model could be updated to reflect any adjustments to the building heights (Atkins 2003). Further projects were to cover other sources of transport and industrial noise in London.

Noise mapping has also been made/explored in a number of other cities at various scales. A typical example was the noise mapping of Cambridge city centre, where only road traffic sources were taken into account, the building heights were taken as 8m, the reflection order was 1, and no error margin or grid interpolation was allowed. A comparison between calculated and measured results showed good agreement, with the model predicting an average value of 1.3dBA higher than the measurement (Stocker and Caruthers 2003).

5.3.3   Noise mapping of industrial sites

For noise mapping of industrial sites, determination of sound power of noise sources is often a main concern. Many plants have point sources, line sources and area (plane) sources. Point sources typically include motors, pumps, gearboxes, fans and exhaust stacks/chimneys/vents. Line sources typically include pipes, conveyors and structural steel elements. The base information required for sound power level is usually generated either by direct measurement or a combination of measurement and subsequent calculation. For pipes there are certain measurement tools using a small portable anechoic chamber (Kang *et al.* 2001a), with which measurement can be made in a high noise environment. These devices have a limited low-frequency range, depending on the physical size of the box. Area sources typically include external walls of buildings, furnaces, inlets/outlets of cooling towers and aerial coolers. Some noise mapping software packages can calculate noise emitted from a building based on the internal SPL and the transmission loss of the building envelope. In addition to source type, for industrial sources particular attention needs to be paid to the tonal components, as well as source directivity.

As an example, the noise distribution around a plant is shown in Figure 5.15 (Kang *et al.* 2001a). For those receivers where the noise level is above the limit, it is possible to use noise mapping software to rank various noise sources in terms of their SPL contribution. For example, at receiver J the overall SPL is 46.4dBA, and it can be identified that the fan inlet of the flare blower is the most dominant noise source contributing 45.8dBA. If this source were turned off, or subject to significant noise control, then the overall SPL at receiver J would fall significantly by 8.6dBA to 37.9dBA.

The emission limits could be in the form of fence line limits, on-plant limits, or receiver based limit. If the limits are receiver based, combined effects from the plant and other sources should be considered. As an example, in Figure 5.16 a comparison is made between road only, plant only, and the combination (Kang and Huang 2002).

5.4 Other models

Along with the development of the noise mapping techniques described above, a number of other models for sound propagation in urban areas at mesoscale or macroscale have also been developed. Some such models are reviewed below.

5.4.1 Flat city model

This simplified model (Thorsson *et al.* 2004) considers an urban area with homogeneous buildings (in terms of building height and absorption characteristics) as an almost flat plane with canyons containing streets and backyards crisscrossing the landscape. In other words,

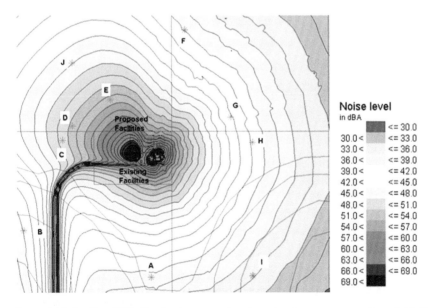

Figure 5.15 Predicted SPL distribution around a plant. A colour representation of this figure can be found in the plate section.

(a)

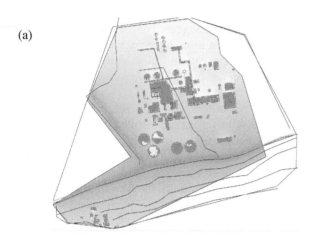

(b)

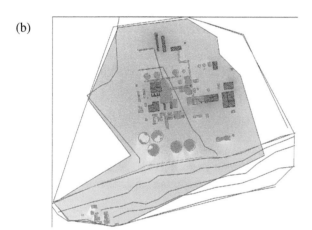

(c)

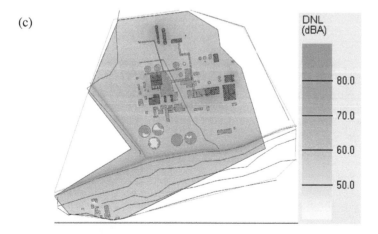

Figure 5.16 Comparison of SPL distribution between (a) road only (b) plant only, and (c) the combination. A colour representation of this figure can be found in the plate section.

this corresponds to raising the streets up to the rooftop level or removing all buildings. It is also possible to consider an inhomogeneous case if the transmission loss between all source and receiver canyons is known (Thorsson and Ögren 2005).

Simple point source propagation over a hard surface, namely, only with spherical spreading, is assumed. With the distribution of sources the SPL at any receiver position within the modelled area can then be calculated. All other effects, such as ground effect, screening and atmospheric absorption are not considered at this stage.

A number of point sources are assumed to be evenly distributed over the road network. To calculate the source strengths, data on traffic flows and road geometry are input into the model. A prediction of the equivalent level for certain time periods can then be obtained. The simplicity of this approach makes it easy to include a large number of roads as sources.

This approach overestimates the SPL, since many factors are neglected, including screening effects around the edges, and multiple reflections inside the canyons. To compensate for this, a correction factor is introduced, by assuming that the same correction applies independently of the distance between canyons. Note that the correction factor contains many mechanisms, including screening and ground effect. The attenuation during propagation from different source canyons into a single receiver canyon can be grouped into a single correction factor.

The correction factor is different for various urban structures, and an approximate correction factor can be deduced from measurements. A study in Sweden showed that the correction factor is between 6 and 10dBA. The correction factor can also be derived using a more detailed model such as described in Section 4.6.2 (Thorsson *et al.* 2004). The correction factor could be based on 24-hour average, or on other time periods.

It is important to note that the predictions from the flat city model are only valid in an area close to the calibration points, that is, the points from where data have been used to estimate the correction factor. Moreover, when predicting the noise level at positions not directly exposed to road traffic noise, it is important to include sources from a relatively large area. Using only the closest road as sound source will underestimate the sound levels substantially.

A similar approach was made by Yeow *et al.* (1977). The vehicles were treated as point sources and sound was propagated over a flat plane. Another similar model was developed by Shaw and Olson (1972), where the city was divided into patches with one point source with equivalent sound power located at the centre of each patch. A good agreement was obtained with long-term measurements. It was found that the correction factor was relatively constant with frequency.

5.4.2 Linear transport model

Most of the microscale models described in Chapter 4 assume continuous building façades, and the propagation around buildings is not considered. Thorsson (2006) developed a statistical model dealing with this type of propagation, so that a larger urban area can be considered. In the model the sound propagation including multiple reflections and diffractions is treated as the flow of small packages of sound energy, namely phonons, as also described in Section 4.5.

The model is derived using linear transport of phonons. Linear transport models are commonly used for the calculation of molecular or particle dynamics (Sommerfeld 1971; Davison 1957). An important difference between molecular dynamics applications and acoustic applications is that molecules can bounce off each other while phonons pass through each other and hence only collide with and scatter from objects in the propagation domain such as buildings, cars and trees. It is assumed that the scatterers are in the far field of each other. Only if the distances between scatterers are larger than their size is it possible to describe the sound propagation through the scatterers with linear transport theory, at least in a mean value sense.

Since the general equations are difficult to solve, further simplifications are made. Buildings are considered as cylinders of constant cross section, although various cylinders can have different cross sections. Isotropic scattering is assumed. In a single scattering process this assumption is unrealistic but it is probably fulfilled in an average sense over many collisions in various buildings, unless the building length is very long. Two-dimensional propagation environment is assumed with infinitely tall cylinders, namely propagation over rooftops is neglected, although it is also possible to apply the theory to three-dimensional motion.

The sound energy density $U(\mathbf{x})$ in a two-dimensional environment with isotropic scattering can be calculated by (Thorsson and Ögren 2005)

$$U(\mathbf{x}) = \int_{\Omega} \frac{\exp\left(-\left|\int_{\mathbf{y}}^{\mathbf{x}} \gamma(\mathbf{r})d\mathbf{r}\right|\right)}{|\mathbf{x}-\mathbf{y}|} [\eta(\mathbf{y})U(\mathbf{y}) + f(\mathbf{y})]d\mathbf{y} \tag{5.18}$$

where Ω is the geometrical propagation domain, $\mathbf{y}$ is the source point, $\mathbf{x}$ is the receiving point, $\mathbf{r}$ is a point on the straight line in between, f is the distribution function, $\gamma(\mathbf{y})$ is a function proportional to the absorption and scattering strengths, and $\eta(\mathbf{y})$ is proportional to the scattering strength:

$$\gamma(\mathbf{y}) = \alpha(\mathbf{y}) + \iota^{-1}(\mathbf{y}) \tag{5.19}$$

$$\eta(\mathbf{y}) = [2\pi\iota(\mathbf{y})]^{-1} \tag{5.20}$$

where ι is the mean free path length (Kuttruff 1982) and α is the absorption factor considering all mechanisms that decrease the number of phonons available for further propagation. The integral $\Lambda = \int_{\mathbf{y}}^{\mathbf{x}} \gamma(\mathbf{r})d\mathbf{r}$ is called the optical distance in electromagnetics (Ishimaru 1997).

The mean free path length can be understood as the mean length a sound particle is likely to travel before colliding with an obstacle. It depends on various factors including size, shape and density of the scatterers. It is difficult to accurately estimate the mean free path length since the effects of these factors are not entirely known, but an approximation can be made using the mean distance between the buildings in the area.

The transport equation (5.18) can also be derived from the wave equation using Twersky's multiple scattering theory, which assumes that scattering objects are in the far field of each other and the back-scattering is small. For more dense environments it is also possible to apply the diffusion approximation, namely using the diffusion equation (see Section 4.5). Equation (5.18) can be solved either with matrix inversion or through iteration. The former is often faster for moderate numbers of nodes, whereas the latter is advantageous for a very large number of nodes.

Thorsson (2006) has applied an iterative solution method based on finite element triangulation of the calculation area, making it possible to consider actual city geometries including the street grid and areas with special properties like recreational areas or parks. For a point source in a homogeneous scattering environment, the result agrees well with a previous model by Kuttruff (1982). For an actual urban area, the calculated SPL is too low compared with measurements, mainly due to the omission of sound propagation over rooftops.

The model has then been extended to include the transmission over rooftops (Thorsson and Ögren 2005). The domain over the rooftops is considered as a half-space with flat ground and homogeneous properties. Sound energy is transmitted to the half-space above the buildings from the sound field in between the buildings. Since the sound field is considered to be diffuse, and the building sizes are generally larger than the wavelengths, this transmission can be approximated by multiplication of the sound field between buildings with a transmission factor ξ_{up}. Due to reciprocity, a similar factor ξ_{down} can be used for the transmission from the half-space down to receivers. The contribution from the path over the rooftops can then be taken into account in Equation (5.18) as an item of additional source strength, although it seems to be rather difficult to estimate the transmission factors accurately.

The concept of sound particles was also used by other researchers for urban areas (Bullen 1979; Kuttruff 1982). Kuttruff obtained approximate solutions describing the gross dependence of sound energy density on the distance from the noise sources, the average building height, scattering cross section and absorption of the buildings. Particular consideration was given to freely flowing traffic, for which not only average values of the energy density but also expressions describing the range and frequency of fluctuations were presented.

5.4.3 Dynamic traffic noise

Whilst most prediction models are for steady-state SPL or based on the average over a relatively long time period, De Coensel *et al.* (2005) developed a tool for dynamic traffic noise prediction. The model includes a GIS-based traffic microsimulation part coupled with an emission model, and a beam tracing based 2.5D propagation part considering multiple reflections and diffractions (De Muer 2005).

The traffic is modelled using the commercial package Paramics (Quadstone 2005). It is based on the discrete cellular automata models of vehicle traffic (Nagel and Schreckenberg 1992), namely micromodels, where each vehicle, represented by a particle, is simulated individually. Road networks are simulated using nodes which correspond to junctions, and links which can be subdivided into several lanes. Traffic is simulated as a system of interacting particles. During the simulation process, vehicles are created at random with an origin and destination zone pair, and then loaded onto the network on a link inside its origin zone. The nature of the interactions between these particles is determined by the way vehicles influence each other's movement. The demands between different zones are described by an origin–destination (OD) matrix, where properties of various vehicle types can be taken into account. A vehicle is cleared from the network after reaching the destination.

The acoustic property of vehicles is included as a plugin to Paramics. A view-port is set, and at each time step, positional data of each vehicle inside the view-port are gathered, along with other information about the vehicle and the link the vehicle is travelling on, including vehicle type, age, velocity, acceleration, travelling direction, road surface type, and the link gradient. For each vehicle, one or more source spectra are associated, and are then mapped on a set of emission points.

There are two methods to map the sources on emission points. The first is to use a grid of emission points, and during the simulation process the vehicle sources are mapped to the nearest emission point. On the nodes a rectangular grid of emission points is placed, which allows to account for wide and complex crossings and for small roundabouts. The links are segmented per lane with a user-defined segment length, with one emission point placed at the centre of each segment. In the second method, a set of emission points is constructed on the fly for each vehicle in the network at their exact locations. This is relatively accurate, but the propagation calculation would be time-consuming since the positions of the emission points are different at each time step in the simulation. In this model, the calculations at receivers are based on a mixture of the above two methods. For the emission points in the vicinity of receivers the exact source positions are used since the source direction is important. Sources at a large distance from the receivers are mapped to an emission point grid, given that it is unlikely that there is a direct path and the source direction can thus be neglected.

In the propagation model an object precise polygonal beam tracing model is used to generate paths between the emission points and the receivers (Heckbert and Hanrahan 1984; Funkhouser *et al.* 1998). The model consists of a terrain model with superpositioned blocks representing the buildings. It is assumed that all façades are upright and roofs are flat. A set of beams is first two-dimensionally traced through the geometric network, where a beam consists of a group of rays, bounded by the buildings as objects. A convex cell subdivision of the environment is used (de Berg *et al.* 1997), so that each beam can have a local view on the environment, and the tracing can be performed efficiently. A cell boundary could be a portal, namely a virtual boundary. The constrained Delaunay triangulation scheme is used for the cell subdivision. With a triangulation over a convex polygonalization, operations on the beams are easy to formalise and implement, although more beams need to be traced. Through beam tracing, paths are constructed between each emission point and receiver within a reasonable distance. Meanwhile, various diffraction and reflection points in a vertical section (0.5D) are computed. Considerations of geometric divergence, atmospheric attenuation, ground effects and meteorological effects are based on ISO 9613 (ISO 1993) and Nord 2000 (Plovsing and Kragh 2000). In the model diffuse reflections from boundaries are not considered, so are the multiple reflections between facades amd ground.

In addition to commonly used indices such as L_{Aeq} and various statistical sound levels, the model can also calculate the spectrum of level fluctuations based on time series of a certain time period using FFT. Such a spectrum has many interesting characteristics, including so-called $1\!/f$ behaviour (Voss and Clarke 1975, 1978; De Coensel *et al.* 2003). For example, periodic events will show up as peaks in the spectrum. When a vehicle passes at each 10s, there will be a peak at 0.1Hz.

The model has been compared with measurements of $L_{Aeq,\,1s}$ over 15min and generally a good agreement has been found for all the statistical properties.

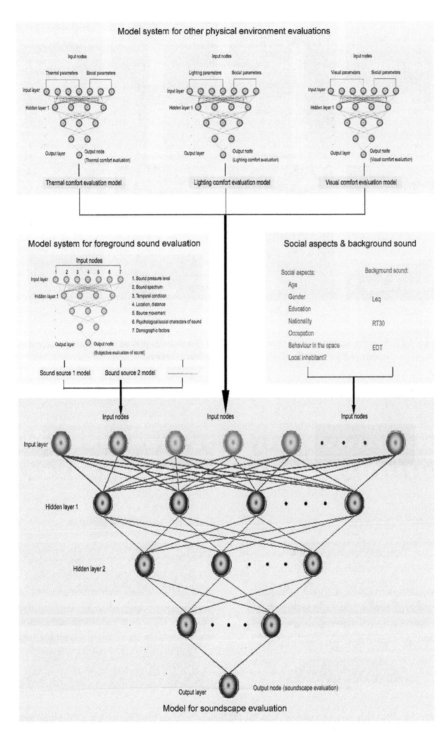

Figure 3.20 Framework for using ANN for soundscape evaluation.

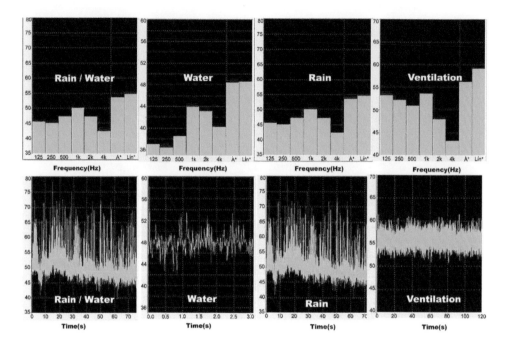

Figure 3.23 Spectra and temporal characteristics of the sounds played in the architectural reading room (AR) at the Sheffield University Main Library.

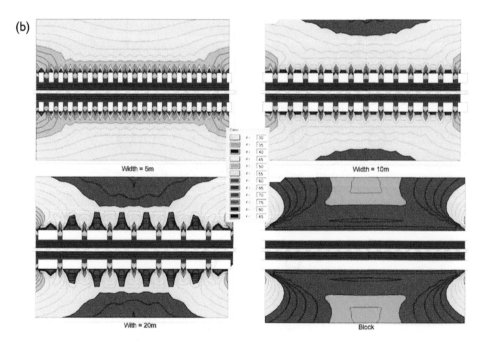

Figure 5.8 SPL with various building gaps (a) with reference to the SPL with a solid block along the street, and (b) the colour map with four building arrangements where the building gap is 5m. See p. 161 for (a).

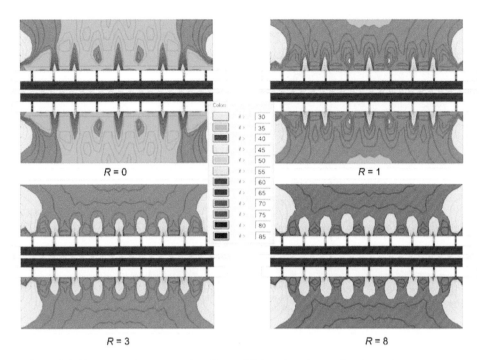

Figure 5.9 Noise maps in a street (see Figure 5.5a) with various reflection orders.

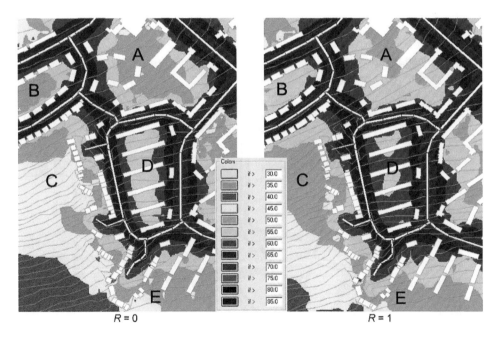

Figure 5.10 Comparison between noise maps with reflection order 0 and 1 in an urban area in Sheffield.

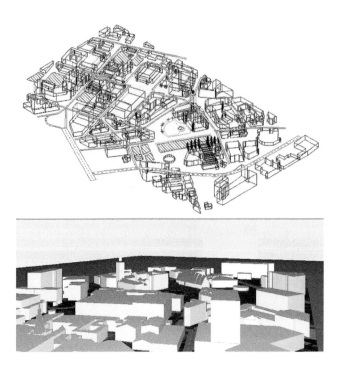

Figure 5.13 Three-dimensional model of an area of Sheffield city centre for noise mapping.

(a)

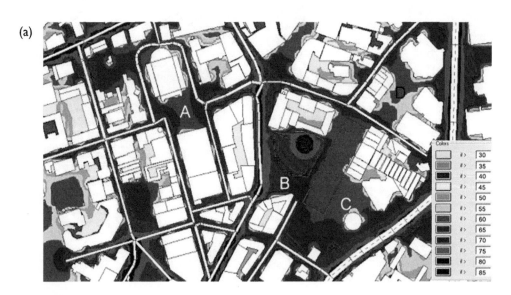

Figure 5.14 Noise map of an area of Sheffield city centre with a reflection order of (a) 1 and (b) 3.

(b)

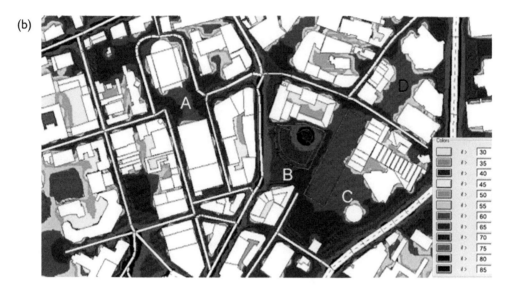

Figure 5.14b

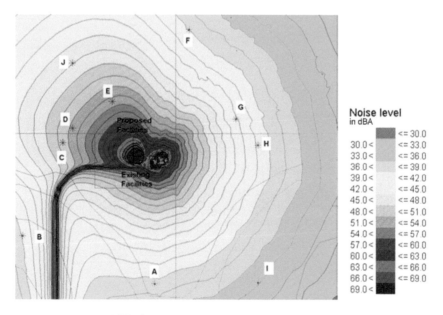

Figure 5.15 Predicted SPL distribution around a plant.

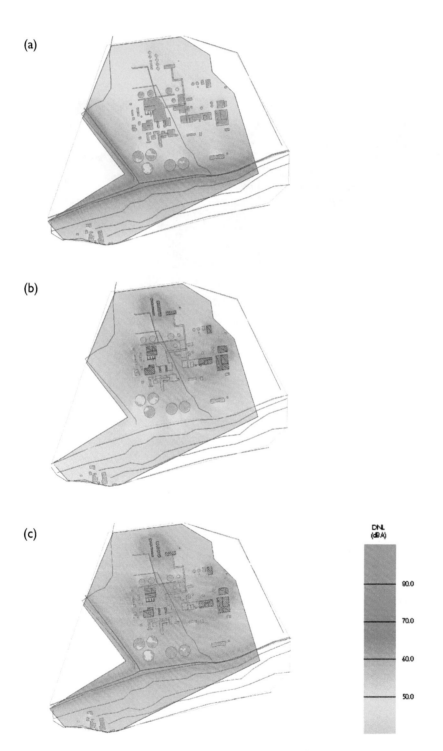

Figure 5.16 Comparison of SPL distribution between (a) road only (b) plant only, and (c) the

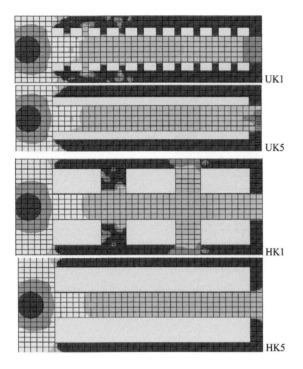

Figure 7.18 SPL on a horizontal plane at 1.5m above the ground. All the façades have a diffusion coefficient of 0.3 whereas the ground is geometrically reflective. Point source. Each colour represents 5dBA.

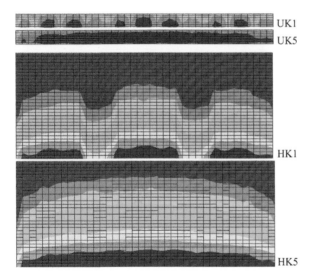

Figure 7.19 SPL distribution on a vertical plane at 1m from a façade. All the façades have a diffusion coefficient of 0.3 whereas the ground is geometrically reflective. Line source. Each colour represents 1dBA.

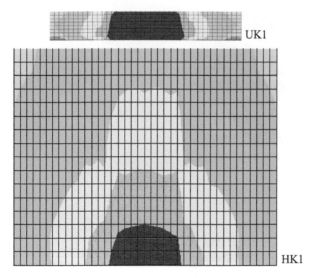

UK1

HK1

Figure 7.20 SPL distribution on a receiver plane perpendicular to the street length. All the façades have a diffusion coefficient of 0.3 whereas the ground is geometrically reflective. Line source. Each colour represents 3dBA.

Chapter 6

Urban noise mitigation

Urban noise mitigation includes many aspects from designing quieter vehicles to improving road surfaces (Sandberg and Ejsmont 2002) to planning traffic systems. The focus of this chapter (and this book) is on the mitigation measures relating to urban and building design.

This chapter discusses planning considerations (Section 6.1), building envelope design (Section 6.2), principles and applications of various environmental noise barriers (Section 6.3), and nonacoustic issues in designing environmental noise barriers (Section 6.4).

6.1 Planning considerations

6.1.1 Building planning

Compared to other pollutions, an important feature of noise pollution is that the spatial variance is rather significant even at a small scale due to the effect of urban texture and transportation infrastructure (Yu and Kang 2005). It is therefore important to strategically plan buildings.

Since large and hard building façades can effectively reflect sound energy, it is possible to arrange buildings so that the reflections can be directed to less sensitive areas. Alternatively, reflective façades can be made absorbent or diffuse. Curved façades should be given particular attention because they could focus sound energy to a small region in the receiving area and, thus, significantly increase the noise level. In Chapter 7 the effects of building arrangements in urban streets and squares are examined.

Although it would be useful to separate a building as far as possible from external noise sources, according to the inverse square law (see Section 1.5.1), this is only effective when the source and receiver are originally in close proximity. For example, moving from 10m to 20m apart and from 100m and 200m apart each gives a 6dB reduction for point source and 3dB for line source.

6.1.2 Self-protective buildings

Building forms can be designed to be self-protective from external noise to a certain extent. Figure 6.1 illustrates some examples and principles. In Figure 6.1a, the podium, usually for commercial use, acts as a noise barrier for the main building which is typically residential. In Figure 6.1b the higher floors, typically bedrooms, are farther from the noise source and they are also protected by the lower building blocks due to the screening effect. In Figure 6.1c balconies can effectively stop the direct sound from the source to the

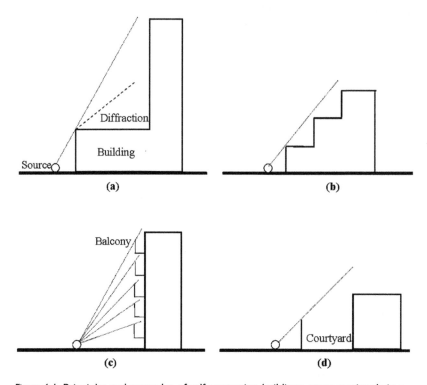

Figure 6.1 Principles and examples of self-protection buildings, cross-sectional view.

windows/doors. A number of numerical and experimental studies have been carried out on the effectiveness of balcony treatments (Mohsen and Oldham 1977; May 1979; Hothersall *et al*. 1996; Hossam El-dien and Woloszyn 2004). In Figure 6.1d, the courtyard wall acts as a noise barrier.

6.1.3   Vegetation

Research on the acoustic effects of vegetation has generally been limited. Whilst most studies have predominantly centred on the reduction of sound transmission through trees in open fields, recently it has been demonstrated that vegetation would be more effective in urban areas.

Open field

It has been demonstrated experimentally that tall vegetation can cause significant sound reduction compared to open grassland (Heisler *et al*. 1987; Parry *et al*. 1993). Mature evergreen vegetation (say, >7m wide) may provide a modest attenuation of 2–4dBA if the belt is sufficiently high and long, has dense foliage extending to the ground and can be well maintained (Egan 1988). Wide belts of tall, dense trees of a depth of 15–40m appear to offer an extra noise attenuation of 6–8dB at low (around 250Hz) and high frequencies (>1kHz) (Kotzen and English 1999). Aylor (1972) made a series of measurements that demonstrated significant attenuation of sound above 1kHz through tall vegetation including corn, hemlock, brush and pine. For dense homogeneous forests of either coniferous or deciduous trees with associated shrubbery it was found that the attenuation was of the order of 7dB per 30m of forest from 200 to 3kHz. However, within the first 30m from the source, amplification was found at about 1kHz, probably due to the resonance of trunks and branches, and this was of the order of about 4dB (Embleton 1963).

Price *et al.* (1988) used the multiple scattering theory of Twersky (1983) for an idealised random infinite array of identical parallel impedance-covered cylinders. In this model, the foliage was represented by arrays of much smaller cylindrical scatterers than the trunks and the attenuation effects were assumed to be arithmetically additive. Huisman and Attenborough (1991) used a stochastic particle bounce method for predicting propagation and reverberation. In conjunction with a two parameter impedance model and an adjustable parameter for the dependence of incoherent scattering on distance and frequency, good agreement was obtained with measurements in a monoculture of Austrian pines.

Tree and shrub arrangements are important. In random arrangements, the scattering contribution of trunks and branches is relatively minor and good sound attenuation requires high densities of foliage extending to ground level. Regular tree planting arrangements have been shown to offer useful 'sonic crystal' effects including 'stop-bands', giving rise to more than 15dB reduction in transmitted sound in a particular frequency range, as long as the filling ratio is sufficiently high (Umnova *et al.* 2006). The control of noise by trees arranged like sonic crystals has also been studied based on outdoor experiments (Martínez-Sala *et al.* 2006).

Because attenuation from trees is mainly due to branches and leaves, sound energy near the ground will not be significantly reduced, and deciduous trees will provide almost no attenuation during the months when their leaves have fallen. It is also noted that the attenuation from dense plantings, say more than 30m deep, will be limited by the flanking of sound energy over the top of the canopy of trees.

Some research was also carried out based on individual plants, especially in terms of the absorption by leaves. It was shown that factors affecting leaf absorption included biomass, size and orientation of leaves (Burns 1979; Martens 1980; Martens and Michelsen 1981; Watanabe and Yamada 1996). Tests were made in anechoic chambers and reverberation boxes, and a laser Doppler vibrometer system was used to measure the leaf vibration.

Urban context

Vegetation could be more effective in urban areas such as in a street canyon or in a square (Kang and Oldham 2003), where the problem to be dealt with is multiple reflections which cause sound levels to rise. In an urban context, the acoustic effects of vegetation arise through three mechanisms: sound absorption and sound diffusion, which occur when a sound wave impinges on the vegetation and is then reflected back; and sound level reduction, when a sound wave is transmitting through the vegetation. It is shown in Chapter 7 that increasing boundary absorption can achieve a substantial extra SPL attenuation, and compared to geometrically reflecting boundaries, there is a significant SPL reduction with diffusely reflecting boundaries. Consequently, when vegetation is used on building façades and ground, the effectiveness of absorption can be greatly enhanced since there are multiple reflections. Similarly, due to multiple reflections, the diffusion effect of vegetation will be significant even when the diffusion coefficient is relatively low. The absorption and diffusion effects are also useful for reducing negative ground effects that often occur in outdoor sound propagation. While the transmission effect in an open field may not be significant unless the density and depth are considerable, the effectiveness could again be significant if multiple reflections are considered. In sound propagation simulation models, such effects can be treated in a similar way as air absorption.

6.2   Building envelope

6.2.1   Acoustic enclosures

If the noise level of a single source or a group of sources is considerably higher than that of other noise sources, it is often effective to use an acoustic enclosure to cover the high level source(s). On the other hand, acoustic enclosures can also be used to protect certain sensitive receivers. An acoustic enclosure could range from a small box to a whole building. It is a common misunderstanding that an acoustic enclosure should be made using porous materials. This is because porous materials absorb sound but do not effectively prevent its transmission. A more appropriate acoustic enclosure should be solid, with sufficient mass, and sealed airtight around the edges. The sound insulation performance of some typical solid materials can be seen in Figure 1.5, whereas the transmission loss of a sheet of porous material is only a couple of decibels typically (Egan 1988). Nevertheless, if the two treatments are combined, namely, when the noise source is surrounded by a solid enclosure lined with porous absorbers, the noise reduction is greater than that without lining, because the sound absorbers reduce the build-up of reflected sound energy within the enclosure and, consequently, reduce the noise level outside the enclosure. The SPL reduction caused by absorbent linings can be estimated using Equation (1.21). It is noted that the efficiency of a given amount of absorber generally becomes less with the increase of absolute amount of absorber.

6.2.2   Combined walls

In many cases a wall or roof does not consist of one material but contains several elements such as ventilation openings, windows and doors. It is possible to theoretically estimate the sound transmission loss of a system with N elements if the performance of each element is known:

$$R = 10\log\left(\frac{\sum_{n=1}^{N} S_n}{\sum_{n=1}^{N} \tau_n S_n}\right) \qquad (6.1)$$

where τ_n is the sound transmission coefficient of element n, which can be calculated from R_n, namely the sound transmission loss of element n, using

$$\tau_n = \frac{1}{10^{R_n/10}} \qquad (6.2)$$

Using Equation (6.1) it can be demonstrated that even a small opening on a solid wall can noticeably increase the transmitted sound. For example, if an unglazed opening occupies an area of 10 per cent of a wall or roof the overall sound transmission loss will not exceed 10dB, whatever the construction of the wall or roof itself might be.

Similarly, small cracks around the perimeter of a partition and gaps around doors can seriously reduce the efficiency of sound insulation.

6.2.3   Silencers for ventilation openings

Ventilation openings are required for many acoustic enclosures and factory buildings. The pressure differences, due to either the wind or the stack effect, that can be used to drive a natural ventilation system are usually rather small and thus large apertures are required, which are potential acoustic weak points. An effective way of reducing the noise through ventilation openings is to use silencers, which are also widely used for noisy machines and duct systems. Silencers are normally classified into two basic types – absorptive and reactive (Irwin and Graf 1979; Wilson 1989; Bies and Hansen 2003). The former often employs porous materials, which are effective for broadband noise, especially at higher frequencies (see Section 1.4.2). Typical configurations of this type of silencers include lined ducts, lined bands, and parallel and blocked-line-of-sight baffles, as shown in Figure 6.2a. The latter does not primarily depend upon absorptive materials for the effectiveness. Employing one or more chambers that serve to reflect and attenuate the incident sound energy, they are most useful when the noise contains discrete tones. Typical reactive silencers include expansion chambers and cavity resonators, as shown in Figure 6.2b.

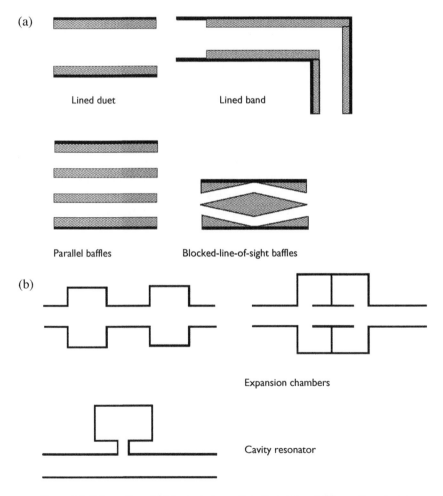

Figure 6.2 Schematics of (a) typical absorptive silencers and (b) reactive silencers.

Various strategic designs have been explored for ventilation openings in buildings. Field and Fricke (1998) used quarter-wavelength resonators to attenuate noise entering buildings through ventilation openings, and about 6–7dB extra attenuation was achieved over a relatively wide frequency range. The combination of passive and active controls was also employed (Maillard and Guigou-Carter 2000), where the opening walls were covered with porous materials to reduce noise at mid–high frequencies, and at low frequencies a single-channel active control system was used. With this approach the sound insulation was increased by 10dBA for traffic noise excitation. De Salis *et al.* (2002) proposed the use of hybrid treatments including an integral barrier placed in front of an aperture and the use of reactive linings around the aperture.

Similar to ventilation openings, doors could also be acoustic weak points when they are kept open due to operation requirements. An effective solution for this is to use an 'acoustic lock', as illustrated in Figure 6.3.

For an aperture of a given cross-sectional area, a lining is more effective with a high aspect ratio than for an aperture with a square cross section. However, the high aspect ratio cross section will result in greater flow resistance and impede airflow performance. Oldham *et al.* (2005a, 2005b, 2005c, 2005d) modelled simple apertures lined with porous absorbers in order to identify optimum configurations, in terms of both acoustic and ventilation performance, using the SYSNOISE (LMS 2005b) and FLUENT (Fluent 2005) software packages, respectively. Figure 6.4 shows the effect of reducing aperture width and increasing aperture length on effective free area (BSI 2004) and element normalised level difference, $D_{n,e}$, which is essentially the SPL difference between two reverberation chambers linked by the ventilator, normalised with respect to an area of absorption in the receiving room of 10m (BSI 1992). It can be seen that although reducing the aperture width to less than 4cm reduces the effective free area it also increases the element weighted normalised level difference, $D_{n,e,w}$ (ISO 1996). For example, for a short aperture, 0.3m, reducing the width from 4cm to 1cm reduces the effective free area to approximately 70 per cent of its apparent value but results in an increase of approximately 9dB in $D_{n,e,w}$. Results such as those shown in Figure 6.4 form the basis of a more systematic approach to the specification of lined ventilation apertures.

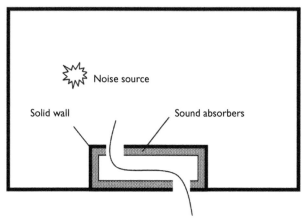

Figure 6.3 Schematics of an 'acoustic lock', plan view.

(a)

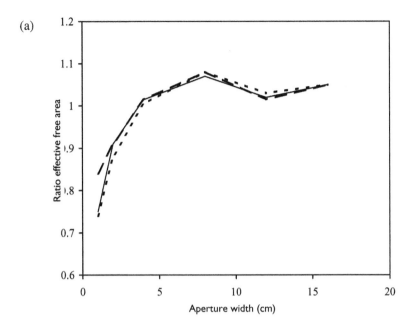

(b)

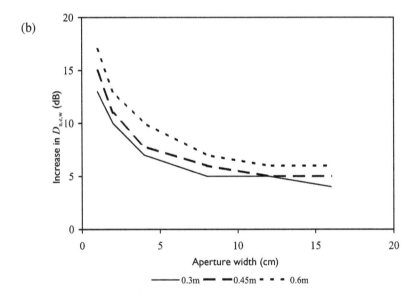

——— 0.3m ━━ ━━0.45m ■ ■ ■ 0.6m

Figure 6.4 Effect of aperture width on acoustic and airflow performance of a lined aperture for aperture lengths of 0.3, 0.45 and 0.6m; (a) variation of effective free area with aperture width; (b) variation of element weighted normalised level difference $D_{n,e,w}$ with aperture width. Data adopted from Oldham *et al.* (2005d).

6.2.4   Acoustic windows

Windows of buildings in noisy environments often need to be sealed, discouraging low-energy strategies based on natural ventilation. Various attempts have been made to produce suitable systems for solving the problem. Jakob and Möser (2003a, 2003b) carried out an experimental investigation of an actively controlled double-glazed window using loudspeakers and microphones inside the cavity, and the reduction in total SPL was typically 7dB with the feedforward controller and 3–6dB with the feedback controller. Mohajeri (1998) proposed a system with an 'intelligent' window which opens and closes depending on the type of sound being monitored. In an interactive window system, a glass baffle was positioned outside the window opening for noise reduction (Jones 1994; Evans 1994).

A relatively simple solution is to seal the windows and use a silencer-type element to allow natural ventilation, as discussed in Section 6.2.3. The element may be located within the window aperture, or be placed in the opaque part of a façade. It has been shown that a window system using a ventilation element filled with porous materials and a filtering system can typically give a weighted sound reduction index R_W of 30dB (Cotana 1999; Asdrubali and Cotana 2000).

A significant problem of such silencer-type elements is that they are usually made of nontransparent materials which impede daylighting if included in the original window aperture. Moreover, the use of fibrous materials may be a cause for concern on health grounds due to the potential hazard of fragments of fibre contaminating the air, and/or other contaminants being held in the fibre matrix, and released under certain conditions. Furthermore, conventional acoustic treatments with relatively rough surfaces tend to increase airflow resistance in the ventilator system.

A window system has been developed to overcome the above problems (Kang and Brocklesby 2003, 2004a, 2004b; Li 2004; Kang et al. 2005; Kang and Li 2006). The core idea is to create a ventilation path by staggering two layers of glass and using transparent microperforated absorbers (MPA) (see Section 1.4.2) along the path created to reduce noise. The system considers the ventilation performance by focusing on the need to achieve occupant comfort by means of air movement, rather than only the requirement for minimum air exchange. Figure 6.5 illustrates the generic/basic configurations of the window system.

The effectiveness of the window design has been investigated using FEM-based software FEMLAB (COMSOL 2004), considering the effects of opening size, air gap, louvers, hood and absorbers (Kang and Li 2006). It has been shown that with hard boundaries, the external and internal opening sizes as well as the air gap width affects the SPL difference considerably at different frequencies, whereas in terms of the average performance the difference is rather small, with about 2dB being typical. Louvers with hard boundaries barely bring acoustic benefit, but with absorbent surfaces the performance of the louvers can be significantly improved. A hood hung outside the opening is very effective in increasing the SPL difference and the effectiveness is improved by increased hood length. In Figure 6.6 some typical results are shown. Ventilation simulation has also been carried out using FEMLAB, confirming that the window systems, including the configurations with the external hood and louvers, will provide sufficient ventilation for comfort.

An integrated acoustics, ventilation and lighting test facility was developed between a semi-anechoic chamber and a reverberation chamber, to simulate as closely as possible the real world location of a window. The semi-anechoic chamber, simulating the external side, presented two surfaces with no absorption – the wall in which the window test section was

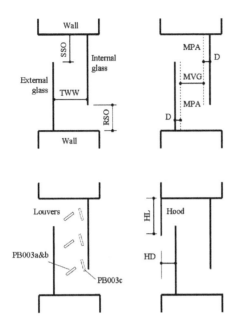

Figure 6.5 Generic/basic configurations of the window system. HL: hood length; HD: distance between hood and glass; SSO: source side opening; RSO: receiving side opening; MVG: minimum ventilation gap; TWW: typical window width; D: distance between glass and MPA.

located and the floor. In the reverberation chamber, foam absorbers were used to simulate the reverberation condition of a typical residential room and also, to reduce the unevenness in sound field caused by room resonances. Ventilation tests were carried out using a pair of fans located outside the semi-anechoic chamber and ducted through the wall. These fans were used to pressurise the room even when relatively large openings were being examined. At relatively low levels they were used to simulate the effect of low-pressure differentials across the test opening similar to those encountered in the real world. In addition to measuring the airflow (m/s) through the centre of the receiving side opening of the window system, smoke was used to visualise the ventilation performance. Lighting performance was tested using two large boxes – one to send a diffuse light field through the window and the other to collect the light.

Based on a large number of tests, it has been shown that considerable reduction in noise is possible with the window system, whilst providing sufficient ventilation for comfort (Kang and Brocklesby 2004b; Kang *et al.* 2005). In Figure 6.7 it is illustrated that such a window can perform better than closed single-glazed windows. The acoustic performance of the window systems can be adjusted using different MPA specifications/configurations. It has also been demonstrated that at normal flow speeds air movement has no adverse effect on the noise reduction achieved using MPA. The reduction in light levels produced with the use of a single sheet of MPA is similar to that from clear glass, but if multiple layers of MPA are applied the light reduction could be significant. Nevertheless, this effect may be useful in some scenarios – for instance, for solar control, especially where large areas of glazing are used.

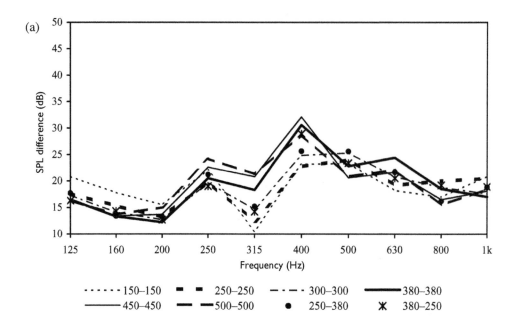

(a)

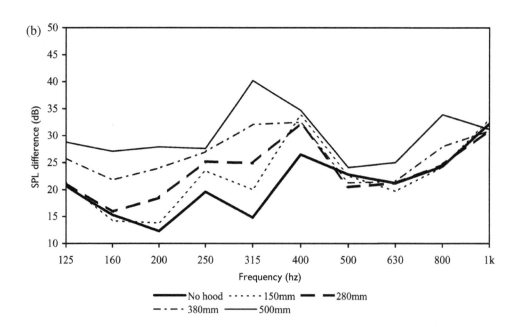

(b)

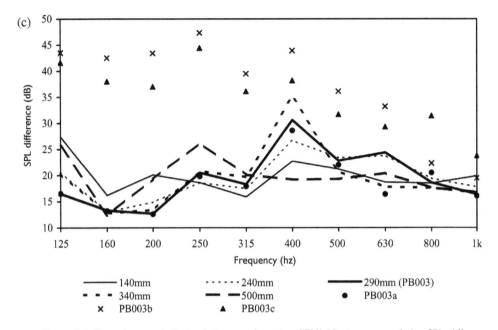

Figure 6.6 Typical numerical simulation results using FEMLAB, in terms of the SPL difference between source and receiving rooms; (a) with various opening size SSO-RSO (mm); (b) with various air gap TWW (mm) and the effect of louvers (see Figure 6.5 – PB003a: 45° louvers with hard surfaces; PB003b: 45° louvers with impedance 0.3 $\rho_0 c$; PB003c: 105° louvers with impedance 0.3 $\rho_0 c$); and (c) with various hood length HL (mm).

6.2.5 Vibration isolation

When a vibrating machine is rigidly mounted to a floor/ground/wall, structure-borne sound will be transmitted with little attenuation. A resilient element between the machine and the support can be effective for increasing the attenuation. Rubber, particularly in shear, has low natural frequencies and is useful for mounting small machines and motors. Steel springs are useful for many machine mountings, but usually require external damping. Cork and felt can also be used (Lawrence 1970).

6.3 Environmental noise barriers

Theories for basic barrier form, namely thin vertical reflective barrier, have been described in Section 1.5.4, and in Section 5.1.6 an engineering method is presented, for both single and double diffraction. In this section, some strategic barrier designs are reviewed (Ekici and Bougdah 2004; Ekici 2004).

Environmental noise barriers can be made with a range of materials including timber, sheet-metal, concrete, brick, plastic, PVC and fibreglass. Transparent barriers are also rather common, using laminated, toughened or reinforced glass, acrylic or polycarbonate sheet. In some cases, barriers are combined with solar panels.

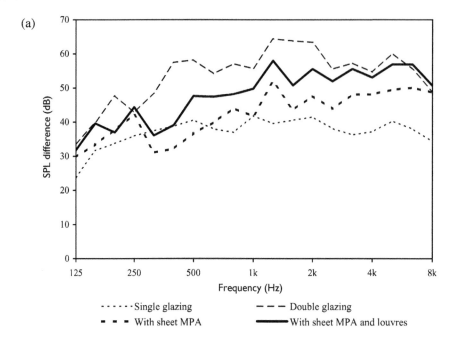

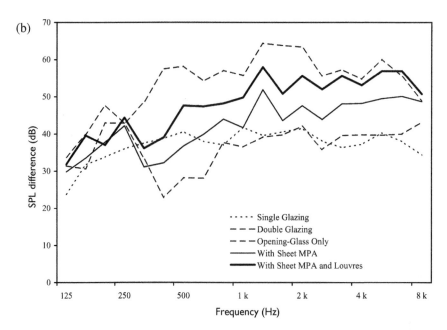

Figure 6.7 Measured acoustic performance of typical/strategic window configurations (Kang *et al.*, 2005), in terms of the SPL difference between source and receiving rooms. Acoustic performance of single- and double-glazing is also shown for comparison.

Helium-filled barriers have been developed as lightweight noise barriers for certain applications such as at construction sites or ballistic ranges (Ailman 1978). The sound waves being transmitted through such barriers are refracted away since they are travelling from a denser medium (air) into a less-dense medium (helium). This kind of barrier is capable of performing as efficiently as any similar sized barrier over the entire audible frequency spectrum.

6.3.1   Multiple-edged barriers

Considerable research has been carried out to refine the design of barrier top to maximise attenuation through diffraction. The beneficial effects of additional diffracting edges have been demonstrated with fir-tree profile (Alfredson and Du 1995), T-profile, Y-profile, arrow profile (May and Osman 1980a, 1980b), branched barriers (Shima *et al.* 1998), and U-sections which involve extra panels connected to the main screen by brackets (Watts *et al.* 1994). Some examples are illustrated in Figure 6.8a. It has been shown numerically (Alfredson and Du 1995) and experimentally (Watts 1996a) that the IL caused by extra diffracting edges could be about 3–5dBA.

6.3.2   Reactive barriers

Van der Heijden and Martens (1982) investigated the possibility of reducing traffic noise using a series of parallel grooves in the ground. The average insertion loss was around 4dBA, with much greater attenuation at low frequencies. An experimental study by Bougdah *et al.* (2006) demonstrated that with strategic designs, rib-like structures can be very effective in providing IL, typically 10–15dB over a rather wide frequency range. It was suggested that quarter-wavelength resonance and surface wave generation played a significant role in determining the performance at lower frequencies. It was noted that at frequencies lower than the limiting frequency, the attenuation could be negative at certain receiver locations due to surface wave generation.

Okubo and Fujiwara (1998) found that a barrier with a waterwheel top could provide an average improvement of 10dB in the frequency range it was intended for. Fujiwara *et al.* (1998) found that a T-profile with a reactive surface produced an improvement of 8.3dB in the mean IL, although a smaller gain might be expected since in practice the soft surface would not be expected to be equally effective over the whole range of frequencies. These designs consisted of a series of tubes, open on one side and rigid at the other. The depth of the tubes could be tuned to the quarter wavelength of the resonant frequency to be reduced. Some typical examples of reactive barriers are illustrated in Figure 6.8b.

6.3.3   Phase interference barriers

The concept of interfering type barriers was reported by Mizuno *et al.* (1984, 1985). As illustrated in Figure 6.8c, the basic configuration is a three-sided barrier consisting of hollow passages at an angle to the ground. The difference between any two adjacent hollow passages is constant, and thus sound waves are refracted when passing through this structural phase lag circuit. Top diffracted sound waves interfere with the refracted sound destructively in some areas resulting in noise reduction. Full-scale tests of a further developed configuration as illustrated in Figure 6.8c indicated that the maximum SPL reduction due to the device was 6dB (Iida *et al.* 1984). However, if the additional height of the barrier is taken into account the extra attenuation caused by the device becomes less than 1–2dB (Watts and Morgan 1996).

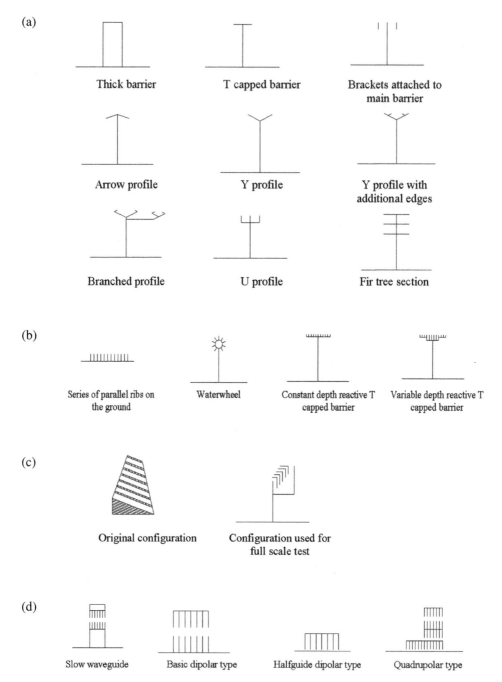

(a)

Thick barrier   T capped barrier   Brackets attached to main barrier

Arrow profile   Y profile   Y profile with additional edges

Branched profile   U profile   Fir tree section

(b)

Series of parallel ribs on the ground   Waterwheel   Constant depth reactive T capped barrier   Variable depth reactive T capped barrier

(c)

Original configuration   Configuration used for full scale test

(d)

Slow waveguide   Basic dipolar type   Halfguide dipolar type   Quadrupolar type

Figure 6.8 Schematics of strategic barrier designs, cross-sectional view: (a) multiple-edged barriers; (b) reactive barriers; (c) phase interference barriers; and (d) phase reversal barriers. Adopted from Ekici and Bougdah (2004).

6.3.4   Phase reversal barriers

Similar to the above, some examples of wave guide (Nicholas and Daigle 1986) and phase reversal barriers (Amram *et al.* 1987) are shown in Figure 6.8d, where the sound passing along the rigid stripes propagates more slowly than in free air which results in refraction. The noise reduction is achieved by the destructive interference of the waves going through the openings with the waves diffracted over the top. These devices would be better suited to unique and dominant pure tones at low frequencies, with an improvement of up to 5dB compared with a solid barrier, whereas at higher frequencies degradation of performance would be possible.

6.3.5   Longitudinal profiles

Along the length the barrier edge could be considered as an infinite number of point sources. Since the wave front from a stationary point source meets the barrier top at different locations, the phase and strength of each source varies monotonically along the edge. Coherent addition of the sound pressure from the effective sources results in destructive and constructive interference patterns. Barriers with randomly jagged edges along the longitudinal direction have been investigated, aiming at preventing the constructive interference pattern by interrupting the monotonic phase variation of the edge sources (Ho *et al.* 1997; Menounou and Busch-Vishniac 2000; Shao *et al.* 2001). It has been shown that such barriers could give enhanced performance at high frequencies, at about 3–7dB, and the improvement increases as the distance from the barrier to the receiver decreases. However, the low-frequency performance could be even poorer than that of a straight-edged barrier. The effectiveness of saw-tooth profiles has also been studied and an extra attenuation of 1.5–4.5dBA could be obtained (Wirt 1979).

6.3.6   Picket and louvered barriers

Examples of picket barriers and vertically louvered barriers are shown in Figure 6.9. With such a profile the dead weights wind loading on barrier foundations could be reduced. Thnadners create deeper shadow zones by varying amplitude or phase gradients (Wirt 1979). With a flat top picket of 25 per cent, open area improvements of 1–4dB have been reported compared to a reflective barrier. As the proportion of the open area increases, as in saw-tooth pickets (50 per cent) and splitter panels (85 per cent), the effectiveness diminishes.

Wassilieff (1988) investigated the performance of picket fences with regular perforations using diffraction theory, and it was found that an improvement can be achieved due to destructive interference of low-frequency sound between the sound transmitted through the gaps and that passing over the barrier top. At high frequencies the performance could be improved using sound absorbers in the gaps.

Compared with a solid 3m-high barrier, a louvered noise barrier with a louver angle of 9° gave a noise increase of 9.5dBA (Watts *et al.* 2001) behind the barrier. Fully absorptive louvers on both the source and receiver side reduced this increase to 3dBA.

6.3.7   Absorptive treatment around diffracting edges

The noise shielding efficiency of different shapes of absorbing obstacles on top of a barrier edge has been studied numerically and experimentally, and it has been shown that this is generally up to 3dB (Fujiwara and Furuta 1991; Fujiwara *et al.* 1995; Yamamoto

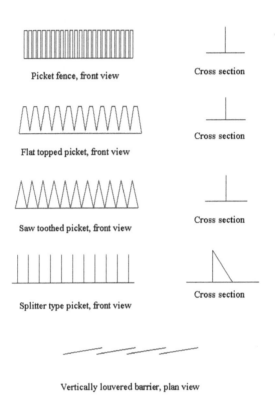

Figure 6.9 Schematics of various picket barriers and vertically louvered barriers. Adopted from Ekici and Bougdah (2004).

et al. 1995; Matsumoto *et al*. 2000). Tests on a 4.2m high barrier have indicated that, with an effective height of 0.46m of absorptive material at the top of a barrier, an equivalent noise reduction to that of an additional barrier height of up to 1.1m can be achieved (Gharabegian 1995). For T-profile barriers the effectiveness of adding absorptive material on top of the horizontal cap is over 2dB according to numerical modelling, but less improvement has been found in full-scale measurements (May and Osman 1980a, 1980b; Hothersall *et al* 1991b; Watts *et al*. 1994). Generally speaking, for the full improvement in attenuation to be realised, absorbent treatment is only needed within one wavelength of the edge of a rigid screen (Butler 1974; Rawlins 1976). Figure 6.10 shows some practical applications where absorptive materials could be used around the free edges of a barrier.

6.3.8   Dealing with reflections

Reflection is an important consideration when barriers are used in urban environment. This may occur, for instance, when there is another barrier or a building on the opposite side of the road, or when high-sided reflective vehicles run close to the barrier. The effectiveness of a barrier could be significantly reduced since after single or multiple reflections, the effective

Cylindrically capped   Mushroom capped   Horizontally louvered

Figure 6.10 Schematics of barriers with absorptive treatment, cross-sectional view. Adopted from Ekici (2004).

height of the barrier could become much less. Additional annoyance may be caused due to reverberation (Kang 1988). For a vehicle pass-by, the peak noise level will be more significantly affected since this usually occurs when the vehicle is in the same cross section as the receiver. Consequently, some noise indices will be more affected than others (Kotzen and English 1999). There are several ways to diminish the negative effects of reflection, including using sloped, dispersive and absorbent barriers.

Sloped barriers

Generally speaking, relatively small angles of tilt can restore almost all of the single barrier IL, counteracting the degradation due to multiple reflections. Slutsky and Bertoni (1988) suggested that a barrier tilt of 3° for wide roadways is enough but greater angles of 10–15° are required for narrow roadways. A study by Menge (1978) showed that the IL increases to a maximum when the angle of the tilt reaches 10° and then drops as the angle further increases. It is noted that the use of sloping barriers may not be appropriate where there are high-rise buildings near the road.

Dispersive barriers

Dispersive barriers provide an alternative solution to the reflection problem by scattering the sound waves that impinge onto the barrier surface. Some examples are shown in Figure 6.11. These types of barriers take up slightly more space on the ground but also have structural benefits. The pockets of free space could potentially be used for vegetation, which could further increase diffusion/scattering. It is noted that such treatments are less effective for a line source parallel to the barrier, compared to point sources (May and Osman 1980a, 1980b).

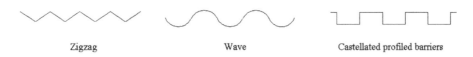

Zigzag   Wave   Castellated profiled barriers

Figure 6.11 Schematics of dispersive barriers, plan view.

Absorbent barriers

Hothersall and Tomlinson (1997) examined the reflections from high-sided vehicles and it was shown that absorbers on the source side of the barrier could restore the attenuation to the levels when no multiple reflections existed. A progressive improvement was observed with an increase in absorbent area. A study by Watts (1996b) on the reverberation build-up of traffic noise in parallel barrier configurations showed that the sound absorptive barriers were effective in counteracting the degradation in single barrier performance due to unwanted reflected paths, although later measurements suggested that the effect of absorptive barriers, on both one or two sides of a road, was generally less than 1dB in terms of L_{Aeq} and L_{A10} (Watts and Godfrey 1999).

A number of absorbent materials can be used with barriers. Absorptive concrete barriers have been used, including wood-fibre concrete and granular concrete, where wood fibres or small cementaceous balls are used as aggregate. Microperforated PVC (Kang and Fuchs 1999) can be used for transparent barriers. Thatch or vegetation on the barrier surface could also bring absorption. Other absorbers include perforated bricks and some recycled materials which are suitable for outdoor use. It is important to note that the absorption performance of porous materials may be significantly affected if they are wet or damp.

6.3.9   Strategic architectural/landscape designs

Earth mounds

Strategically designed earth mounds can be effective in reducing noise. A noise barrier may be erected on top of an earth mound to reduce the horizontal land take, although there is inconclusive evidence that this could in some cases diminish the acoustic performance of the earth mound (Ekici 2004). Factors affecting the performance include the effective barrier height, scattering and double diffraction losses on the barrier top, absorption effects of grass-covered slopes, and the slope angle of the wedge (Hutchins *et al.* 1984a, 1984b; Hothersall *et al.* 1991a).

Vertical alignment of road

Similar to the above, roads in cuttings, roads elevated above the surrounding ground on embankments or on other structures such as viaducts can create shadow zones, as shown in Figure 6.12a.

Cantilevered and galleried barriers

By angling the top section of a barrier towards the source, the diffracting edge of the barrier will come closer to the source, so that the barrier is more efficient and, thus, the barrier height could be reduced if needed (Jin *et al.* 2001). A galleried barrier is a substantial cantilevered barrier which covers the nearside traffic lane, forming a partial enclosure. Some examples are illustrated in Figure 6.12b, where it is noted that similar principles are also applicable in plan.

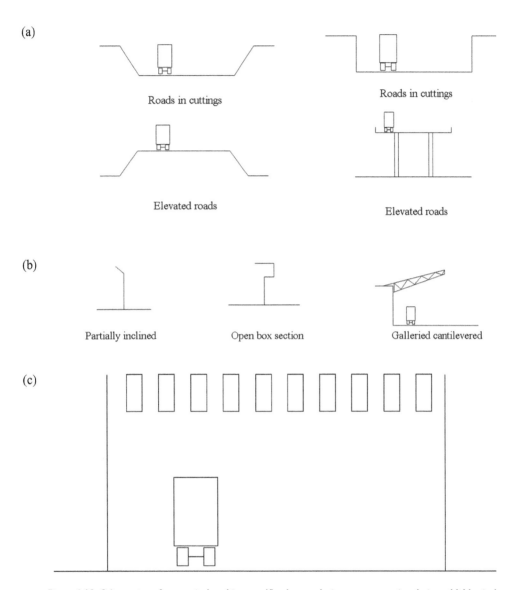

Figure 6.12 Schematics of strategical architectural/landscape designs, cross-sectional view. (a) Vertical alignment of road; (b) cantilevered and galleried barriers; and (c) suspended panels. Adopted from Ekici (2004).

Suspended panels

Arrays of sound absorbing panels can be hung above a road to prevent any direct line of sight to the source when viewed at an oblique angle from receivers, as shown in Figure 6.12c. In other words, the panels stop/absorb direct sounds and only allow the diffracted sounds (Kotzen and English 1999).

Vegetation and biobarriers

When trees or vegetation are used as noise barriers, a considerable depth is needed, as discussed in Section 6.1.3, which is not always practical. Alternatively, trees and nonarboreal vegetation can be used conjunctively with noise barriers. Measurements have shown that the presence of trees produces an SPL decrease of 2–4dBA downwind when the wind speed is 6–12m/s (Van Renterghem and Botteldooren 2002, 2003; Van Renterghem *et al.* 2002).

Biobarriers are also often used; these may be divided into four generic types, including A-frame and vertical corten steel, box wall, woven-willow, and stack and crib biobarriers (Kotzen and English 1999).

6.4   Nonacoustic factors in barrier design

In addition to reduction in noise level, the success of an environmental noise barrier depends on many other nonacoustic factors, for example, the consideration of structure, fixing, viewing at speed, pattern, texture, colour, light and shade, material and design, visual neutrality and compatibility, safety, environmental impact, and cost (Kotzen and English 1999; Hong Kong EPD 2003). The UK Highways Agency (1995) has published guidance on the use of barrier materials. Besides some basic acoustic rules, guidance is given in other aspects, including barrier appearance, design consideration for rural, semi-urban and urban areas, construction and operation factors, design process and assessment framework, maintenance, and costs.

In this section, three nonacoustic factors, including public participation in the design process, lifecycle assessment (LCA), and perception of various barrier materials, are discussed.

6.4.1   Public participation

Public participation has been commonly used in the design process of various sectors (Sanoff 2000). In order to demonstrate the importance and effective ways of public participation in the design of environmental barriers, a case study was carried out in a UK community that had recently received a noise barrier (Joynt 2005). In the first stage of the study, semi-structured interviews were carried out. Through the application of the grounded theory approach (Glaser and Strauss 1967; Strauss and Corbin 1990), analysis of the qualitative data was generated into a theory by achieving a close fit between the concepts raised by each respondent. In the second stage a quantitative investigation was made through a questionnaire survey.

It was found that it was not only a lack of public participation that could evoke negative perceptions of a noise barrier's effectiveness, but the adoption of an inappropriate method, which was ineffectual at accessing and involving all affected members of the public into the public participation process, was equally influential. Of those who attempted to become involved a strong sense of disenfranchisement was apparent, as their opinions were neither actively sought nor adopted. The use of the traditional patriarchal methods of public participation, associated with a 'tokenistic' approach, was responsible for feelings of exclusion and disenfranchisement that could lead to a negative perception of a noise barrier's benefits.

A positive finding was the overall willingness to become involved in the process. Whilst the 'public' is often branded as apathetic, the interviews illustrated a keenness and awareness of locally salient issues, and a desire to share responsibility in the processes of addressing them.

These feelings were not confined to the individuals that worked within and for the community as volunteers and professionals, but included people of all genders, ages and ethnicities.

A general lack of trust between the professionals and the public was also highlighted, with the public perceiving the professionals' role as largely ineffectual, which consequently led to a general feeling of hopelessness when attempting to integrate. The professionals, on the other hand, perceived the public as largely incapable of grasping the concept of realistic commercial constraints. This problem would be significantly diminished, should methods for effective dissemination of facts be adopted, which impart nontechnical and realistic expectations of a noise barrier's abilities.

The results of the quantitative investigation suggested that a lack of effective public participation can negatively impact the perception of the barrier's effectiveness of mitigating noise. For example, the high percentage of the unawareness of the local forum that was adopted as the focal point of the public participation process was significantly correlated to the high percentage of people who did not perceive noise level reduction by the barrier. There was also a disparity between the objective reduction in noise levels based on detailed noise mapping, and the subjective perception in noise levels of 'decreases', 'no change' or 'increases', which could again be related to the consequences of the ineffectual public participation (Joynt and Kang 2002, 2003).

As those most likely to have their opinions heard are not necessarily those most adversely affected, but those most likely to present themselves at public meetings, it is proposed that having the ability to attend a public meeting and being amongst the 'vociferous' minority, should not be the qualifying criteria for being integrated into the planning and design process of a noise barrier. The qualifying criteria should be the salience of the problem to the individuals, based on their properties, location and received noise level, regardless of ethnicity, language barriers, mobility restrictions, commitments and other potentially limiting factors. Those who would be directly affected by the barrier installation could be identified using noise mapping techniques. It is also noted that one method of participation does not necessarily suit all members of a community, but there are correlations in the preferred approach between subsections of the community. Thus, if these subsections were identified prior to participation, then a mixture of methods, including using the Internet, could be adopted to generate the best outcome (Joynt 2005).

6.4.2 Lifecycle assessment

The choice of preferred noise barriers often lies foremost on an ability to mitigate the noise problem at the best available cost. This cost is invariably accounted for primarily as an economic one, largely overlooking the potential environmental costs that are also attributed to it. An approach has been devised (Joynt 2005; Joynt and Kang 2006) by adopting the existing frameworks laid out by the Building Research Establishment (BRE) (Anderson *et al.* 2002), and other LCA programs, into a specific methodology/framework for the assessment of the lifecycle impacts of environmental noise barriers. The result of this approach is a comprehensive and systematic list of the expected use of nonrenewable resources, embodied energy and pollutants emitted for a number of noise barrier materials.

The first section of the LCA, known as the 'cradle-to-gate', gives a full account of the most to least environmentally sustainable materials, weighted in accordance with the impacts on wider environmental problems, such as global warming, atmospheric and water pollution. The findings of the 'cradle-to-gate' analysis is that timber is the most environmentally sustainable material to be used in a noise barrier structure, followed by recycled aluminium, recycled steel, precast concrete, living willow, woven willow, polymethyl methacrylate (PMMA), unrecycled steel, and the least environmentally sustainable being unrecycled aluminium.

The impacts of material transportation and maintenance beyond the factory gate up to the point of disposal are an equally important factor influencing a material's overall sustainability. An increased release of environmentally harmful pollutants and greenhouse gases, during the processes from 'gate-to-grave', can change the balance from a sustainable material to an unsustainable material dramatically.

The impacts of recyclability at the end of a barrier's useful life also create a large determining factor on how sustainable materials can be. By taking this into account, aluminium and steel change from one of the least to one of the most environmentally sustainable noise barrier choices. Equally, the disparity between the environmental claims of the willow barriers, and the reality of the environmental impact caused by the use of the mineral wool inner core, illustrates the importance of having a method of evaluating the overall sustainable assets of a barrier. It is also important to consider the impacts of transportation in the recycling process. In that, a barrier that is recycled in a vastly different location from that where it was used may reduce its overall sustainability considerably (Joynt 2005).

6.4.3   Perception

There is evidence that the visual shielding of the noise source by a noise barrier has a considerable psychological effect (Magrab 1975). In an experiment by Aylor and Marks (1976), a selection of four noise barriers were positioned around the circumference of a circle with a swivel chair in the centre for the respondent to sit in. The experiment was under free-field conditions, and used a sound source projected from speakers behind each of the barriers as a stimulus. Using the method of magnitude estimation (see Section 2.1.3), it was found that visual shielding by a barrier dramatically affected the perception of sound transmitted through the barrier. When the barrier only partially obscured the sight of the sound source, loudness was judged less than when there was no intervening barrier, whereas when the barrier totally obscured the sight of the source, loudness was judged greater than when the source could be seen, either in whole (no barrier) or in part.

Mulligan et al. (1987) found that the assessment of loudness increased as the percentage of vegetation between source and receiver increased, when the ambient noise level was held constant. In the experiment the sound was a single tone at 500Hz, varying between 50dB and 80dB and being played back through headphones.

The question of obscuring the traffic source by vegetation was also investigated experimentally by Watts et al. (1999). The first stage consisted of an in situ experiment with different densities of vegetation between listeners and noise source. The second stage was undertaken under controlled conditions. It was shown that there were differences in the sensitivity to noise depending on the degree of visual screening, which was largely independent of the noise exposure level. This was illustrated by the fact that the difference in the noise exposure level needed to incite the same subjective response was 4dBA, between the site with 30 per cent vegetation cover and 90 per cent vegetation cover, in the direction that the listeners were more sensitive to noise where the screening was highest.

It is also important to consider the effects of visual preference on the human response to sound (Anderson et al. 1983; Warren et al. 1983). In an experiment by Viollon et al. (2002; Viollon 2003), stereo sound tracks were used and large colour slides were projected to create the visual settings. It was found that perceptions of road traffic noise transmitted through barriers varied according to visual degrees of pleasantness and efficiency. The more pleasant the noise barrier was, the less stressful was the road traffic noise.

Joynt (2005) recently carried out an investigation on the impact of barrier materials on the perception of noise reduction. Different from other studies, a moving visual stimulus was used under laboratory test conditions in a virtual reality environment RAVE. Five sets of recordings were heard by the respondents at 71.6, 76.6, 81.6, 86.6 and 91.6dBA, respectively. Five barrier materials were studied, including concrete, metal, timber, transparent acrylic, and a hedgerow of deciduous vegetation.

Figure 6.13 shows the preconceptions of various types of barriers. It can be seen that the majority of the respondents, based purely on preconceptions, predicted the concrete barrier to be the most effective noise attenuator, followed by metal and timber ($p < 0.0005$). In other words, regardless of which noise barrier was presented to the respondents, preconceptions of a material's ability to attenuate noise were imbedded.

These preconceptions were reflected in the results of the further perception exercise as well, with the respondents largely perceiving the more solid looking and opaque barriers as more effective at attenuating noise, despite the noise level being held constant. Strong significant correlations were found between the preconceived ideas and the perceptions especially at the lower SPL representative of being adjacent to a motorway.

In terms of the link between aesthetics and perception of noise attenuation, the results showed that the transparent and deciduous vegetation barriers, judged most aesthetically pleasing, were inferior to those judged as most effective at attenuating noise, such as concrete barriers.

Whilst the results of the above studies seem to show some discrepancies and further research is needed, the findings have clearly demonstrated the complex nature of the psychological effect of barriers on perceived noise levels and the importance of expectory features of familiar environments.

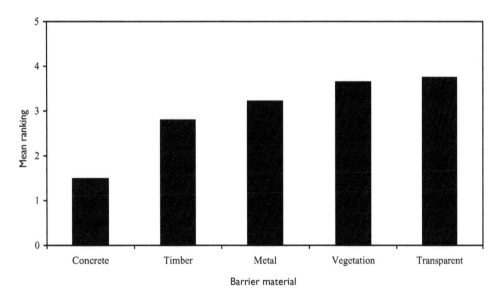

Figure 6.13 Preconceptions of various types of barriers on their potential to attenuate noise, with 1 as most effective and 5 as least effective. Data adopted from Joynt (2005).

6.4.4   Design process

An overall model/framework for the design process has been proposed by Joynt (2005), in order to ensure the longevity, success and sustainability of an environmental noise barrier with fully integrated approach considering various acoustic and nonacoustic factors.

Chapter 7

Sound environment in urban streets and squares

Streets and squares are important elements of urban environments, and their acoustic conditions are receiving great attention, especially with increased city centre regeneration. This chapter analyses the basic characteristics of sound fields in urban streets and squares surrounded by reflecting building façades, and the effects of architectural changes and urban design options, including boundary reflection pattern, street/square geometry, boundary absorption and building arrangements (Kang 2000a, 2000b, 2000d, 2001, 2002a, 2002b, 2002c, 2002d, 2004a, 2005a, 2005b; Kang *et al.* 2001b).

Two computer models, as described in Chapter 4, one based on the radiosity method for diffusely reflecting boundaries according to the Lambert cosine law (see Section 4.4), and the other based on the image source method for geometrically reflecting boundaries (see Section 4.1), are used to carry out a series of parametric studies with hypothetical urban streets and squares. While the models are accurate for simulating the two kinds of idealised sound fields, which is the main attention of this chapter, it is also useful to consider the conditions in between, namely with partially diffuse and partially geometrical boundaries, or in other words, with a diffusion coefficient between 0 and 1. For this purpose, Raynoise (LMS 2005a), an acoustic simulation software package based on beam tracing, is used (also see Section 4.9.1). Since random numbers are involved in the Raynoise computation when a diffusion coefficient is considered, causing variations in results, computation for each configuration is repeated a number of times to obtain an average.

For the sake of convenience, the boundary absorption is assumed to be independent of the angle of incidence. The façades and ground are assumed to have a uniform absorption coefficient of 0.1, except where indicated. With some strong absorption patches, such as open windows or gaps between buildings as sound energy sinks, it has been demonstrated that the trend of the comparison results will not change significantly. Except where indicated, absorption from air and vegetation is not included. Excess attenuation due to ground interference and temperature or wind-gradient induced refraction is generally not taken into account.

Acoustic indices include the steady-state SPL, RT based on RT30 and EDT. The SPL is relative to the source power level, which is typically set at 0 or 100dB. For urban streets, the source–receiver distance in this chapter refers to the horizontal distance along the length, except where indicated. The terms near field and far field in this chapter refer to close and remote receiver positions, rather than dimensions relating to the size and wavelengths of the sound source.

Most calculations are based on a single point source, which is useful to gain a basic understanding of sound propagation. The results are representative of certain types of urban noise, such as low-density traffic. They are also useful for considering noise propagation from a junction to a street. Generally speaking, if an architectural change or urban design option is effective with a single source, it is also effective with multiple sources such as a line source along a street canyon (Kang 2002a), although the effectiveness diminishes if a receiver is very close to the source, since architectural changes and urban design options are normally only useful for reducing reflections, except barriers, which reduce the direct sound.

7.1   Urban streets

A single street canyon as well as an urban element consisting of a major street and two side streets are both considered, as illustrated in Figures 7.1a and 7.1b respectively. For the single street, except where indicated, the street length is 120m, a point source is positioned at (x = 30m, y = 6m, z = 1m), and the source–receiver distance along the length is 1–60m (x = 31–90m). With this arrangement the calculation results should not be significantly affected when the street length is extended. In other words, the results can be generalised to represent longer streets.

The study of the sound field in the cross street urban element can lead to an improved understanding of noise control in a network of interconnecting streets typical of urban areas. The size of the urban element is 120 x 120m. Only diffusely reflecting boundaries are considered. The element is divided into five areas, which are called streets N, S, W, E and M, respectively. The diffraction over buildings is ignored because in the configurations considered, the energy transferring through street canyons is dominant.

Except where indicated, the buildings in the above two configurations are continuous along a street and of a constant height on both sides, and the sound attenuation along the length is generally based on the average of 5–10 receivers across the width. The source and receiver heights are both 1m.

7.1.1   Basic characteristics of sound field

Single street

The SPL distribution on a series of horizontal planes, cross sections and longitudinal sections are calculated in a single street canyon with diffusely reflecting boundaries, where the street length, width and height are 120, 20 and 18m, respectively. On a horizontal plane at 1m above ground, along y = 10m the SPL attenuation is 22.7dB at source–receiver distances of 5–90m. The SPL variation becomes less when the horizontal plane is farther from the source. On a plane of 18m above the ground, for example, along y = 10m the SPL attenuation at source–receiver distances of 5– 90m becomes 16.8dB. In a cross section, the SPL variation is considerable in the very near field, and then becomes rather even. At x = 35m the variation is 7.5dB, whereas beyond x = 55m, it is less than 1.5dB. This result corresponds to the measurements by Picaut *et al.* (2005). In terms of longitudinal sections, at a distance of 1m from façade A, with x = 30m, namely in the same cross section as the sound source, the SPL variation along the height is significant, at about 8dB. With increasing distance from the source in the length direction, the SPL variation along the height decreases rapidly. Beyond x = 50m, it becomes less than 1dB. With different source positions in a given cross section, the SPL difference at a receiver is only significant in the near field but becomes negligible beyond a certain source–receiver distance along the length, say 15m.

When the boundaries are geometrically reflective the SPL variations are generally similar to the above.

Figure 7.2 compares the decay curves at three receivers, (35m, 2m, 1m), (50m, 2m, 1m), (90m, 2m, 1m), corresponding to source–receiver distances of 5, 20 and 60m respectively,

(a)

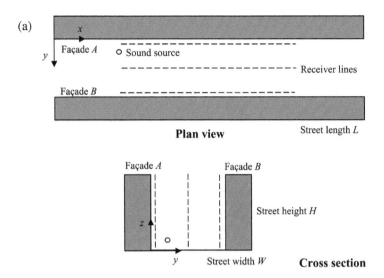

(b)

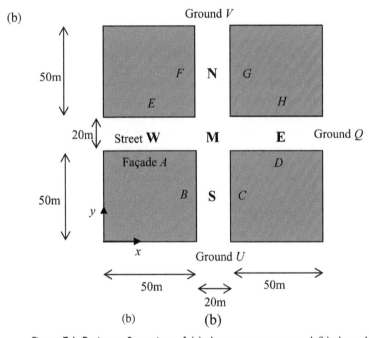

Figure 7.1 Basic configuration of (a) the street canyon and (b) the urban element used in the calculation.

where the boundaries are diffusely reflective. It is important to note that reverberation increases systematically with increasing distance from the source. This important phenomenon has also been observed from *in situ* measurements (Picaut *et al.* 2005), and in long enclosures (Kang 1996a, 1996b, 1996c, 1996d, 2000c). The RT is generally over 1s

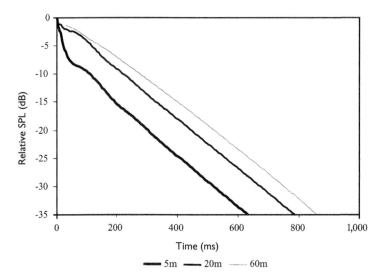

Figure 7.2 Decay curves with increasing source–receiver distance, where the boundaries are diffusely reflective, the street length, width and height are L =120, W =20 and H =18m, respectively, and a point source is positioned at (30m, 6m, 1m).

(see also Figure 7.6), suggesting that the reverberation effect is significant in such a street.

To examine the distribution of RT and EDT in a cross section, the ratio between two typical receiver sets, (31–90m, 18m, 18m) and (31–90m, 2m, 1m), is calculated, as shown in Table 7.1. It can be seen that for both diffusely and geometrically reflecting boundaries the RT is rather even throughout all the cross sections along the length, whereas the EDT is only even in the

Table 7.1 Ratio of reverberation time between two typical receiver sets (31–90m, 18m, 18m) and (31–90m, 2m, 1m).

Source–receiver distance (m)	Geometrical boundary		Diffuse boundary	
	$RT_{(18,18)}/RT_{(2,1)}$	$EDT_{(18,18)}/EDT_{(2,1)}$	$RT_{(18,18)}/RT_{(2,1)}$	$EDT_{(18,18)}/EDT_{(2,1)}$
1	1.2	1.9	1.1	4.0
5	1.2	1.9	1.0	1.5
10	1.1	1.8	1.0	1.1
15	1.1	1.2	1.0	1.0
20	1.0	1.1	1.0	0.8
25	1.0	1.1	1.0	0.9
30	1.0	1.0	1.0	0.9
35	1.0	0.9	1.0	0.9
40	1.0	1.1	1.0	0.9
45	1.0	1.1	1.0	0.9
50	1.0	1.1	1.0	0.9
55	1.0	1.1	1.0	0.9
60	1.0	1.0	1.0	0.9

cross sections beyond a certain distance from the source. This distance appears to be slightly shorter with diffusely reflecting boundaries than with geometrically reflecting boundaries.

Cross streets

Figure 7.3 shows the SPL distribution with a point source at five positions in streets S and M: (60m, 0, 1m), (60m, 15m, 1m), (60m, 30m, 1m), (60m, 45m, 1m) and (60m, 60m, 1m), as well

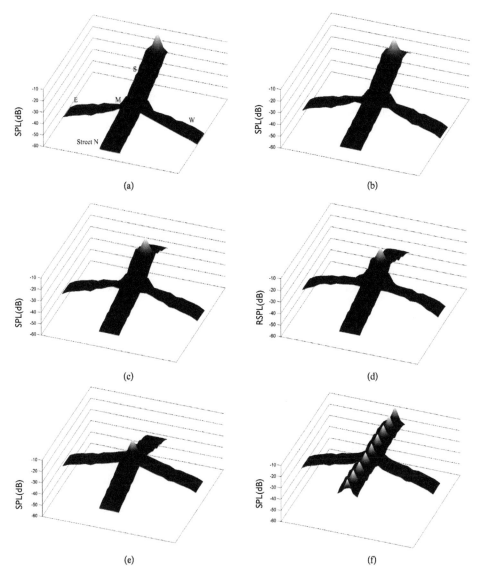

Figure 7.3 SPL distribution with a point source at five positions in streets S-M: (a) (60m, 0, 1m); (b) (60m, 15m, 1m); (c) (60m, 30m, 1m; (d) (60m, 45m, 1m); (e) (60m, 60m, 1m); and (f) with nine sources at a spacing of 15m along y =0–120m. The street height is 20m.

as with nine sources at a spacing of 15m along $y = 0$–120m. The street width and height are both 20m. It can be seen that when the source is closer to the middle of the street junction, the average SPL in the streets becomes higher because less energy from the source can be reflected out of the streets. In street W or E, for example, the average SPL difference between source positions (60m, 0, 1m) and (60m, 60m, 1m) is 18dB, which is significant. With multiple sources, the average SPL in the side streets, namely street W or E, is considerably lower than that in the major street, namely street S-M-N, at about 11dB on average. Also, the SPL attenuation along the side street W or E is significant, at about 15dB from $x = 50$ to 0m or from $x = 70$ to 120m. These results quantitatively demonstrate that if noise sources are along a major street like S-M-N, it is an effective way to reduce noise by arranging buildings in side streets such as W or E.

It is also interesting to investigate the effect of side streets on the sound field in a major street. A comparison is made between the configuration above and a configuration without streets W and E but the façades in street S-M-N are totally absorbent at the position of side streets, namely $y = 50$–70m. A point source is positioned at (60m, 15m, 1m) in street S. It is shown that despite the significant changes in the boundary condition in the side streets, the SPL in the major street only changes by about 0.5–1dB, and this is limited in the range of 50–80m. This suggests that with noise sources along a major street, the energy reflected from side streets to the major street is negligible.

Figure 7.4 shows the distribution of RT and EDT with three source positions from the end of the major street to the middle of the street junction, namely (60m, 0, 1m), (60m, 30m, 1m) and (60m, 60m, 1m). Similar to a single street, it is important to note that both the RT and EDT generally become longer with increasing source–receiver distance. When the source moves from the end of the major street to the middle of the street junction, the average RT in all the streets increases slightly, at about 10 per cent. In side streets W and E the reverberation, especially EDT, is systematically longer than that in the main street S-M-N, which is mainly due to the lack of direct sound. Overall, except in the near field, the RT is about 1–2s, and the EDT varies from 0.2 to 3s, suggesting that the reverberation effect is significant in such an urban element.

7.1.2   Boundary reflection pattern

A comparison of SPL attenuation along the length is made in Figure 7.5 between geometrically and diffusely reflecting boundaries. Two street configurations, $W = 20$m and $H = 6$m, and $W = 20$m and $H = 18$m, are considered. The receivers are along two lines in the length direction, namely (31–90m, 2m, 1m) and (31–90m, 18m, H), which represent relatively high and low SPL in a cross section, respectively. It is interesting to note that in comparison with geometrical boundaries, the SPL with diffuse boundaries decreases significantly with increasing source–receiver distance. The main reason is that, with diffuse boundaries, the total energy loss becomes greater because the average sound path length is longer, especially for the receivers in the far field. The difference between the two kinds of boundaries has also been found with other street dimensions as well as in long enclosures (Kang 1996e, 1997a). For a street 200m long, 20m wide and 30m high, for example, the difference is typically 10dB at the far end of the street (Kang 2000a, 2002a).

In the near field, with diffusely reflecting boundaries there is a slight increase in SPL, as can be seen in Figure 7.5. This is due to the energy reflected back from farther boundaries, which has also been experimentally observed (Picaut *et al.* 2005). In long enclosures, there is a similar phenomenon (Kang 1995, 1997b). From Figure 7.5 it can also be seen that with $H = 18$m, the SPL difference between the two kinds of boundaries is less than that with $H = 6$m, both along (31–90m, 2m, 1m) and (31–90m, 18m, H).

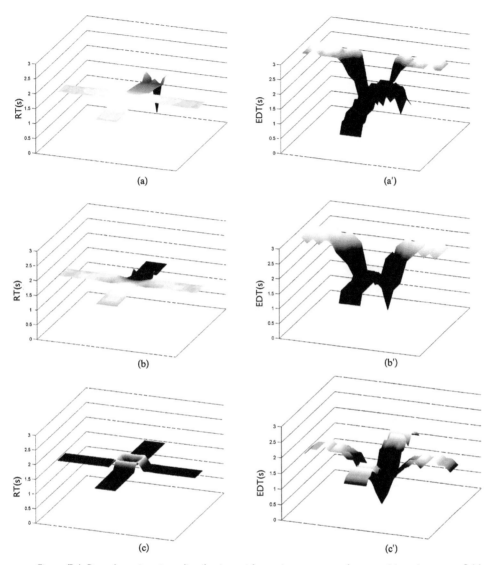

Figure 7.4 Reverberation time distribution with a point source at three positions in streets S-M. (60m, 0, 1m): (a) RT; (a') EDT. (60m, 30m, 1m): (b) RT; (b') EDT. (60m, 60m, 1m): (c) RT; (c') EDT. The street height is 20m.

A comparison of the RT and EDT between diffusely and geometrically reflecting boundaries is shown in Figures 7.6a and 7.6b, respectively. The calculation is carried out with a street width of 20m, three street heights of 6, 18 and 30m, and receivers along (31–90m, 2m, 1m). It can be seen that, in comparison with diffuse boundaries, the EDT with geometrical boundaries is shorter with a relatively low width/height ratio and is longer when this ratio is relatively high, whereas the RT is much longer with any aspect ratio. An important reason for the difference in RT is that, with geometrical boundaries, the image sources are well separated and, thus, the ratio of initial to later energy at a receiver is less than that with diffuse boundaries (Kang 2000d).

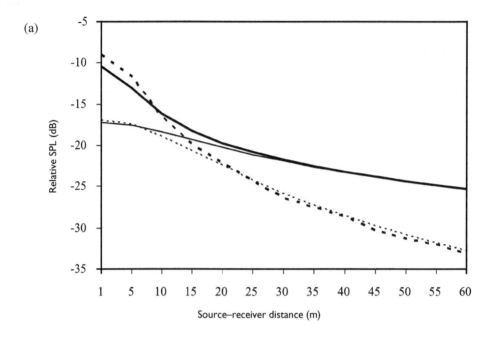

(a)

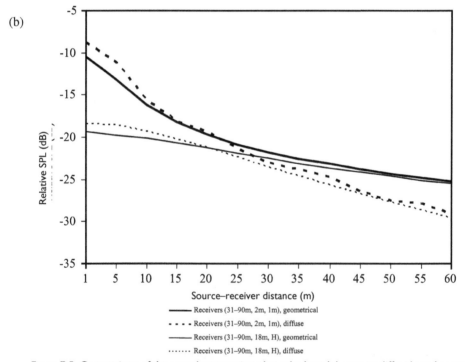

(b)

Receivers (31–90m, 2m, 1m), geometrical
Receivers (31–90m, 2m, 1m), diffuse
Receivers (31–90m, 18m, H), geometrical
Receivers (31–90m, 18m, H), diffuse

Figure 7.5 Comparison of the sound attenuation along the length between diffusely and geometrically reflecting boundaries in two streets. (a) $W = 20m$ and $H = 6m$; (b) $W = 20m$ and $H = 18m$.

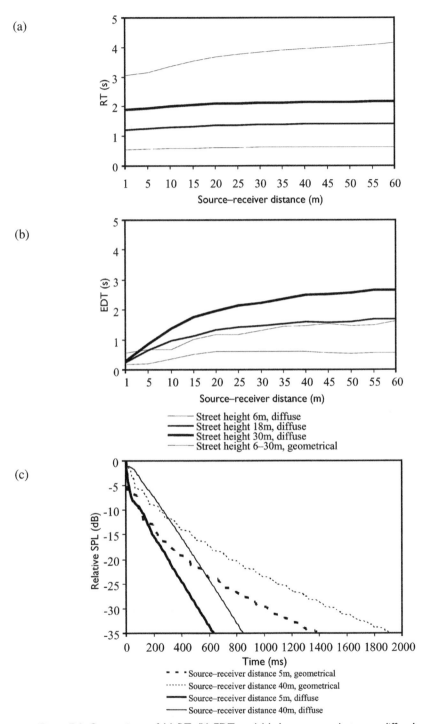

Figure 7.6 Comparison of (a) RT; (b) EDT; and (c) decay curves between diffusely and geometrically reflecting boundaries. The street width is 20m.

Figure 7.6c shows the decay curves at two typical receivers, 5 and 40m from the source, where the street height is 18m. From the decay curves the above-mentioned differences between diffuse and geometrical boundaries can be clearly seen.

It is noted that when a street canyon has a mixture of two kinds of boundaries, the sound field tends to be close to that formed by purely diffusely reflecting boundaries. For example, in Section 4.4.11, it has been shown that there is no significant difference in SPL and reverberation between a geometrically and a diffusely reflecting ground when the façades are diffusely reflective. Moreover, multiple reflections tend to make the diffuse reflection mechanism dominant at higher orders of reflection, even if the boundary diffusion coefficient is small (Ismail and Oldham 2005; also see Section 4.4.1).

7.1.3   Street geometry

Single street

While it is expected that when a street becomes wider, the SPL attenuation along the length will increase, it is interesting to examine the variation of the increase along the street length (Kang 2002a). With a street height of 18m, when moving façade B from $y = 10$ to 40m, corresponding to a change in width/height ratio of 0.56 to 2.2, with diffusely reflecting boundaries, the extra SPL attenuation, based on the average of two receiver lines of (31–90m, 2m, 1m) and (31–90m, 2m, 18m) is about 2.6dB in the near field, increases with source–receiver distance, reaching 5.2dB until 20m, and then decreases with further increase of source–receiver distance to about 3dB at 60m. A possible reason for the variation is that in the near field, the SPL is dominated by the early reflections from façade A such that façade B is relatively less effective. In the very far field, the average sound path length is already long and consequently the effect caused by moving façade B away is proportionally less. With geometrically reflecting boundaries the SPL changes are generally similar, although the decrease in extra SPL attenuation in far field is much less. A further comparison between street widths of 5 and 160m shows that the SPL difference is about 9dB at a source–receiver distance of 200m, where the boundaries are geometrically reflective, the street height is 20m, and the receiver plane is at 10m above the ground (Kang 2002a).

With geometrically reflecting boundaries, if the source height is lower than the street height, the sound field in the street remains unchanged with further increase of the street height because the increased street boundary will not cause any reflection down to the original street canyon. Whereas with diffusely reflecting boundaries, with increasing street height, less energy can be reflected out of the street canyons and, thus, the overall SPL in the streets should become higher. Figure 7.7 shows a comparison of sound attenuation between four street heights, 6, 18, 30 and 54m, which correspond to the width/height ratio of 3.3–0.37. In the figure the street width is 20m, the receivers are along (31–90m, 2m, 1m) and the sound levels are with reference to the SPL at (31m, 2m, 1m) when the street height is 54m. In the near field, say within 10m from the source, the difference between various street heights is insignificant, which indicates the strong influence of the direct sound. With the increase of source–receiver distance, the effect of boundaries becomes more important and, thus, the difference between different street heights becomes greater, at 8dB between street heights 6 and 54m with source–receiver distance of 60m.

With diffusely reflecting boundaries the RT and EDT become longer with the increase of street height, as can be seen in Figures 7.6a and 7.6b, and/or decreased street width, as shown in Figure 7.8, where the street height is 18m, and the receivers are along (31–90m, 2m, 1m). It can be seen that between street widths of 10 and 20m, the difference is typically 5–10 per cent in RT and 20

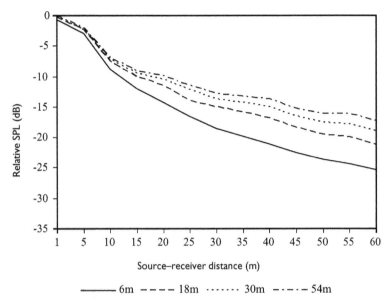

Figure 7.7 Sound attenuation along the length with different street heights, with diffusely reflecting boundaries. The street width is 20m.

per cent in EDT. In the case of geometrically reflecting boundaries, reverberation increases with increasing street width, but is unchanged with increasing street height if the source height is lower than the original street height (see Figures 7.6a and 7.6b), as discussed above.

Figure 7.9 compares the decay curves when the street height increases from 6 to 30m, where the boundaries are diffusely reflective, the street width is 20m, and the receiver is at

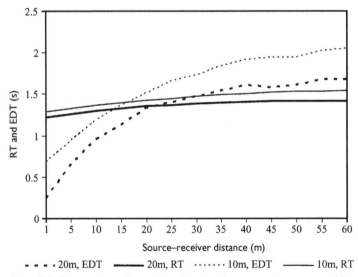

Figure 7.8 Comparison of reverberation time between two street widths, where the boundaries are diffusely reflective, and the street height is 18m.

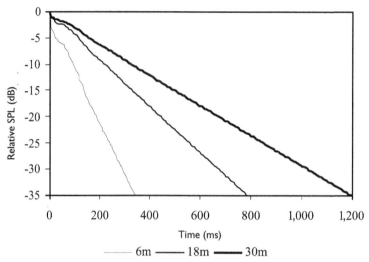

Figure 7.9 Decay curves with increasing street height, where the boundaries are diffusely reflective, the street width is 20m, and the source–receiver distance is 20m.

(50m, 2m, 1m), namely the source–receiver distance is 20m. The increase in reverberation with increasing street height can be clearly seen.

Cross streets

For the configuration shown in Figure 7.1b, a comparison of the SPL attenuation along the street centre (0–120m, 60m, 1m) is made between two street widths, 10 and 40m, which corresponds to a change in street width/height ratio from 0.5 to 2. The source is positioned at the middle of the cross street, (60m, 60m, 1m), and the street height is 20m. The results show that with a street width of 40m the SPL attenuation is about 3–7dB greater than that with a street width of 10m, except in the very near field where the direct sound is dominant. The maximum difference occurs at a source–receiver distance of about 25m, which is similar to the situation in a single street, as discussed above.

   Situations with varied street widths in the urban element are also examined. Figure 7.10 compares two cases: street S is 30m wide and street N is 10m wide; and street S is 10m wide and street N is 30m wide, where a source is positioned at (60m, 30m, 1m) in street S, and the street height is again 20m. It is interesting to note that with a wider street S, the average SPL is about 4dB less in this street, but in streets W-M-E the SPL is generally increased, also by about 4dB, due to increased energy from street S. In street N, the SPL difference between the two situations is insignificant, generally within 1dB. In terms of reverberation, the differences between the two configurations vary in different streets, but generally speaking, with a narrower street S, both RT and EDT are longer.

   The effect of street height is examined by comparing the sound fields with street height 20m and 60m, where a point source is near one end of street S-M-N, at (60m, 15m, 1m), and the street width is 20m. The street width/height ratios in the two cases are 1 and 0.33, respectively. Along street S-M-N the SPL is systematically increased by the increased street height,

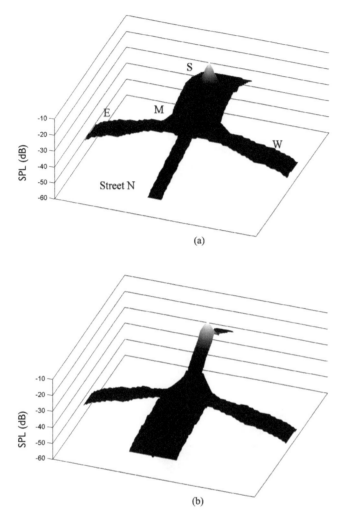

Figure 7.10 SPL distribution with varied street widths in the urban element: (a) street S 30m wide and street N 10m wide; (b) street S 10m wide and street N 30m wide. The street height is 20m.

typically at about 2–4dB except in the near field, and this increase generally becomes greater with increasing source–receiver distance. In street W-M-E the SPL difference between the two street heights ranges from 2 to 7dB, again increasing with increased distance from the source. A main reason for the difference between street S-M-N and W-M-E is that in the former direct sound plays an important role, whereas in the latter the sound field is dominated by reflected energy. In terms of reverberation, RT and EDT are approximately doubled when the street height is increased from 20 to 60m.

Figure 7.11 shows the SPL changes caused by staggering street S and N, where two configurations are considered: (1) street N is shifted to $x = 70$–90m; and (2) street S is shifted to $x = 30$–50m while street N is at $x = 70$–90m. The street height is 20m, and a point source is in

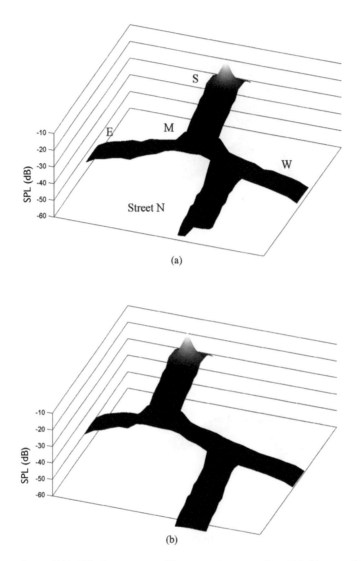

Figure 7.11 SPL changes caused by staggering street S and N; (a) street N shifted to x = 70–90m; (b) street S shifted to x = 30–50m and street N shifted to x = 70–90m. The street height is 20m.

street S, positioned in the middle of the street width, with $y = 15$m. From Figure 7.11 it can be seen that in street N, the SPL reduces by about 4–5dB when only street N is staggered, and 10–15dB when both street S and N are staggered. The SPL reduction is caused by the diminished direct sound, lengthened reflection path, and increased number of reflections. In street S, conversely, the SPL is almost unchanged when one or two streets are staggered. In street W-M-E, the SPL distribution pattern varies by the street staggering, but the average SPL is almost constant.

7.1.4   Boundary absorption and building arrangements

Boundary absorption is useful for diminishing reflection energy and, consequently, reducing the overall SPL in urban streets. Broadly speaking, absorbers on street boundaries include absorbent materials, open windows and gaps between buildings.

Single street: evenly distributed absorbers

In Figure 7.12 the SPL attenuation along the length with three boundary absorption coefficients is compared, where the street width is 20m and the street height is 18m. Two typical receiver lines are considered, along (31–90m, 2m, 1m), and (31–90m, 2m, 18m). The calculation is made for both kinds of boundaries (Kang 2000b).

Generally speaking, the extra attenuation caused by increasing boundary absorption is significant. From Figure 7.12a it can be seen that with diffusely reflecting boundaries the extra attenuation is relatively constant along the street length, especially at receivers (31–90m, 2m, 18m). This is because when boundaries are diffusely reflective, the SPL at any receiver is dependent on the contribution from all the patches and, thus, an even increase of absorption on all the boundaries should have a similar effect on all the receivers. For geometrically reflecting boundaries, the extra attenuation increases with the increase of source–receiver distance, as can be seen in Figure 7.12b. This is possibly because with a longer source–receiver distance the difference in sound path length between low and high orders of reflection becomes less, such that higher orders of reflection become relatively important. Given that increasing boundary absorption is more effective for higher orders of reflection, the extra attenuation becomes greater with a longer source–receiver distance.

Corresponding to Figure 7.12a, Figure 7.13 shows the effect of boundary absorption on decay curves at source–receiver distance of 20m, namely at receiver (50m, 2m, 1m), where the boundaries are diffusely reflective. It appears that the RT is approximately doubled when the absorption coefficient decreases from 0.9 to 0.5, or from 0.5 to 0.1.

Single street: strategically distributed absorbers

To study the effect of sound absorption distribution, a given amount of absorption is arranged with three distribution schemes in a street 20m wide and 6m high, where the SPL in a cross section is represented by the average of two receivers, namely (31–60m, 2m, 1m) and (31–60m, 18m, 6m). The three arrangements for façade A, façade B, and ground G are: $\alpha_A = 0.9$ and $\alpha_B = \alpha_G = 0.05$; $\alpha_A = \alpha_B = 0.475$ and $\alpha_G = 0.05$; and $\alpha_A = \alpha_B = \alpha_G = 0.209$. The results for both diffusely and geometrically reflecting boundaries show that the sound attenuation along the length is the highest if the absorbers are arranged on one façade and the lowest if they are evenly distributed on all boundaries. The SPL difference between different arrangements is typically 1–4dB with geometrically reflecting boundaries, and 1dB with diffusely reflecting boundaries. The difference between the two kinds of boundaries is probably because geometrical boundaries are more affected by the reflection pattern (Kang 2000d).

Further study is carried out for various absorber arrangements on the ground and façades in a street canyon of $W = 20$m and $H = 18$m with diffusely reflecting boundaries, where for the sake of convenience, absorbers are assumed to be totally absorbent (Kang 2002c). The calculation of the sound attenuation along the length is based on the average of four receiver lines,

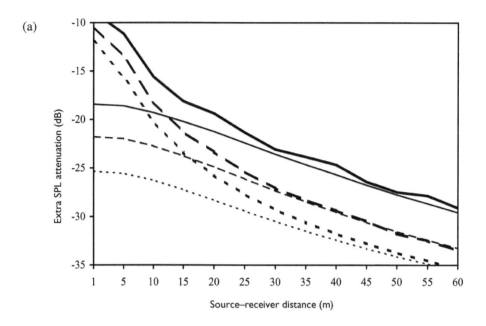

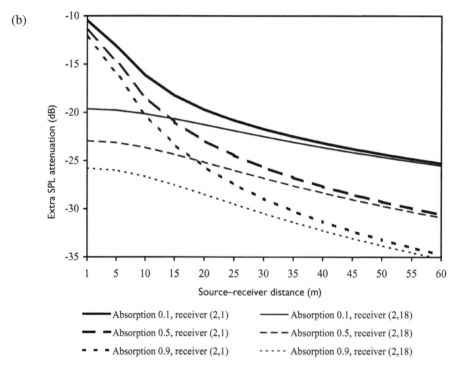

Absorption 0.1, receiver (2,1) ——— Absorption 0.1, receiver (2,18)

— — Absorption 0.5, receiver (2,1) – – – Absorption 0.5, receiver (2,18)

▪ ▪ ▪ Absorption 0.9, receiver (2,1) ····· Absorption 0.9, receiver (2,18)

Figure 7.12 SPL attenuation along the length with different boundary absorption coefficients: (a) diffusely reflecting boundaries; (b) geometrically reflecting boundaries. The street width is 20m and the street height is 18m.

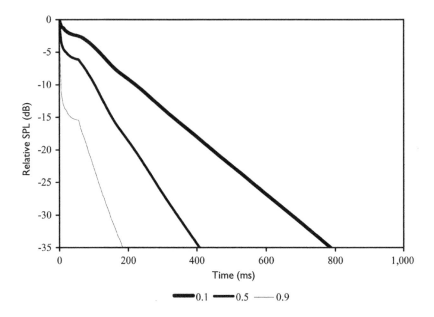

Figure 7.13 Decay curves with different boundary absorption coefficients at receiver (50m, 2m, 1m), with diffusely reflecting boundaries. The street width is 20m and the street height is 18m.

namely (31–90m, 2m, 1m), (31–90m, 2m, 18m), (31–90m, 18m, 1m), and (31–90m, 18m, 18m). First, four absorber arrangements along the ground are considered, with absorption coverage of 18.3, 41.5, 58.5 and 100 per cent, respectively. It appears that the extra attenuation is approximately proportional to the absorber area, reaching about 3–4dB with the ground totally absorbent. To examine the effect of façade absorption four cases are considered: absorbers vertically arranged on one or two façades, and horizontally arranged on one or two façades. For each treated façade, the ratio of absorber to façade area is 50 per cent. The results show that the extra attenuation is about 1–2dB with absorbers on one façade only, and around 2–4dB with absorbers on both façades. It is interesting to note that with a given absorber area there is almost no difference in sound attenuation between vertically and horizontally distributed absorption. This suggests that with diffusely reflecting boundaries, if a given amount of absorbers is evenly distributed on a boundary, the pattern of the absorber arrangement plays an insignificant role for the sound field.

Single street: building arrangements

In planning practice it is often useful to know, if there are buildings on one side of a street, what will happen when a new building is built on the opposite side. This effect is examined by calculating the sound field with various building heights on one side of street. It is shown that, when the boundaries are diffusely reflective and the street width and height are 20m and 18m, the extra SPL attenuation caused by reducing the height of façade B from 18m to 15.25m, 9m, 2.75m and 0m is 0.4–0.6, 0.8–1.8, 1.4–2.4 and 2–4.2dB respectively, where the extra attenuation increases with increasing source–receiver distance. In the calculation the sound attenuation along the length is again based on the average of four receiver lines, namely (31–90m, 2m, 1m), (31–90m, 2m, 18m), (31–90m, 18m, 1m), and (31–90m, 18m, 18m).

The effect of gaps between buildings is examined by calculating the extra attenuation with a gap on façade B between $x = 48$m and 72m, and then with a gap on façade A and a gap on façade B, both between $x = 48$m and 72m, where other configurations are the same as above. In the calculation, the absorption coefficient of the gap is assumed to be 1, given that reflection from side façades are insignificant, as discussed in Section 7.1.1. The results show that in the length range containing the gap(s) there is a notable extra SPL attenuation, which is about 2dB with one gap and 3dB with two gaps. Conversely, after the gap(s), say $x > 75$m, the extra attenuation becomes systematically less, and before the gap(s), say $x < 45$m, the extra attenuation is almost unnoticeable. This suggests that the sound field at a receiver is more affected by the nearby patches. Further analysis with more configurations also demonstrates that a gap between buildings can provide extra sound attenuation along the street and the effect is more significant in the vicinity of the gap.

Cross streets: evenly distributed absorbers

For the configuration illustrated in Figure 7.1b, with street height $H = 20$m, a calculation is carried out by evenly increasing the absorption coefficient of all the boundaries from 0.01 to 0.99. When a source is at (60m, 15m, 1m), the SPL decreases significantly in all the streets, and the decrease is approximately proportional to the linear increase of boundary absorption coefficient from $\alpha = 0.01$ to about 0.9, by 13dB, and then there is a sharp SPL decrease from $\alpha = 0.9$ to 0.99, by 4.8dB. When the source is at (60m, 60m, 1m), the average SPL decrease from $\alpha = 0.01$ to 0.99 is only about 5–6dB. An important reason for the SPL difference between the two source positions is that with the source at the middle of the street junction, (60m, 60m, 1m), the direct sound plays a dominant role in a considerable area and, thus, the boundary absorption is relatively less efficient (Kang 2001).

Cross streets: strategically distributed absorbers

Based on the street dimensions in Figure 7.1b, ten configurations are considered with absorbers at various positions, where the absorption coefficient of the absorbers is assumed to be 0.9, a point source is positioned in street S, at (60m, 15m, 1m), and the street height is 20m. The results show that absorbers are more effective when they are arranged on boundaries in the street with the source, due to the strong direct sound energy on those boundaries. Similarly, absorbers on boundaries which receive lower order of reflections are more effective. For example, absorbers on boundaries E and H are more effective than those on A and D, especially for streets W and E. In terms of the average SPL in the whole cross streets, with a given amount of absorption, the difference between different absorber arrangements is about 3–5dB.

It is interesting to note that the SPL in the source street S is almost not affected by the amount and arrangement of absorbers in other streets. This means that the energy reflected back to street S from other streets is negligible, within 1dB. The effects of absorbers in other streets are also rather 'local', mainly affecting the streets where the absorbers are located. As expected, the ground absorption is more effective when the façades are acoustically hard. The effectiveness of an absorbent ground is 4dB with façade absorption $\alpha = 0.1$, whereas this value is only 1dB with façade absorption $\alpha = 0.9$.

Absorption from air and vegetation

In addition to boundary absorption, air absorption may also reduce the sound level, especially at relatively high frequencies. Absorption when sound waves pass through vegetation may have similar effects (see Section 6.1.3). A calculation is made for a typical street 20m wide and 18m high, with $M = 0.005, 0.015$ and 0.025Np/m, corresponding approximately to the air absorption at 3, 6 and 8kHz at 20°C and 40–50 per cent relative humidity (ANSI 1999a). With diffusely reflecting boundaries, the average extra attenuation in the street caused by the air absorption is 0.6–2, 1.7–5.6, and 2.7–9dB with the three M values, respectively. The extra attenuation increases gradually with increasing source–receiver distance.

With geometrically reflecting boundaries in the street the extra SPL attenuation caused by air absorption is systematically less than that with diffusely reflecting boundaries, although the difference is only within 1dB. An important reason is that with diffusely reflecting boundaries the sound path is generally longer and thus air absorption is more effective.

The variation in extra attenuation caused by air absorption in a cross section is considerable in the near field and then becomes insignificant with increasing source–receiver distance. For example, between (31–90m, 2m, 1m) and (31–90m, 18m, 18m), beyond about 20m from the source cross section, the difference is less than 0.2dB with diffusely reflecting boundaries and less than 0.5–1dB with geometrically reflecting boundaries, whereas this difference is about 2dB at 1m from the source cross section for both kinds of boundaries.

As expected, calculation shows that the reverberation becomes shorter if air absorption is included. The difference in RT and EDT between $M = 0$ and 0.025Np/m is generally 30–60 per cent in the above street, suggesting that the reduction in reverberation caused by air absorption is more significant than that in SPL. This is because the SPL depends mainly on early reflections, whereas reverberation is dependent on multiple reflections, for which air absorption is more effective due to the longer sound path. Further analysis shows that with diffusely reflecting boundaries the reductions in RT and EDT caused by air absorption are similar, whereas with geometrically reflecting boundaries the reduction in RT is systematically greater than that in EDT.

7.2   Case study: comparison between UK and Hong Kong streets

To further examine the effect of urban texture, the sound fields in two very different kinds of street configurations, typical of the UK and Hong Kong (HK), are compared (Kang *et al.* 2001b). Main factors for comparison include street height, width, building types, and boundary absorption and diffuse conditions.

7.2.1   Configurations

For the UK situation, two types of streets are considered, based on the street configurations in the High Storrs area and the Hunters Bar area in Sheffield, respectively. The former contains two-storey semi-detached houses on both sides. Each building block is 10m wide and the gap between the buildings is 5m. The building height is 8.5m, the building depth is 7m, and the street width is 20m. The latter contains terraced houses on both sides. The buildings are continuous along the street. For comparison, two street heights, 8.5m and 15m, and two street widths, 12m and 20m, are considered. The building depth is also 7m. For the sake of convenience, the roofs are assumed to be flat except where indicated. In both types of streets, the

Table 7.2 Configurations used in the comparison between UK and HK streets.

Configuration	Length (m)	Width (m)	Height (m)	Roof/boundary	Building type
UK1	160	20	8.5	Flat roof	Semi-detached
UK2	160	12	8.5	Flat roof	Terraced
UK3	160	12	8.5	Sloped roof	Terraced
UK4	160	12	15	Flat roof	Terraced
UK5	160	20	8.5	Flat roof	Terraced
HK1	160	20	65	Concrete façade	Discrete lots
HK2	160	30	65	Concrete façade	Continuous
HK3	160	30	65	Glass façade	Continuous
HK4	160	30	30	Concrete façade	Continuous
HK5	160	20	65	Concrete façade	Continuous

façades are brick or stone, the roofs are covered with tiles, and the ground is concrete. The street length is considered to be 160m for both street types. Overall, five street configurations are used in the calculation, as listed in Table 7.2. In Figure 7.14 two typical configurations, UK1 and UK2, are illustrated.

For the Hong Kong situation, two street types are considered based on typical urban texture in the Mongkok area. One type is a street with discrete building lots on both sides. The lots are 20m apart and each lot is 40m long. The buildings are 65m high, 20m deep, and the street width is 20m. The other type is a street with continuous buildings on both sides. The building depth is again 20m. For comparison, two street heights, 65m and 30m, and two street widths, 20m and 30m, are considered. For both street types, the street length is 160m

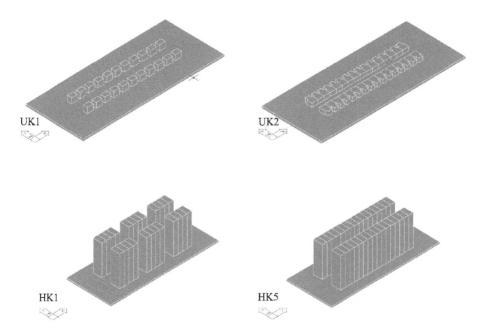

UK1

UK2

HK1

HK5

Figure 7.14 Three-dimensional representation of typical configurations used in the comparison between UK and HK streets.

and the building roofs are flat. Two kinds of façade materials, concrete and glass, are considered. The ground is concrete for both street types. Overall, five street configurations are included in the calculation, as listed in Table 7.2. Two typical configurations, HK1 and HK5, are illustrated in Figure 7.14.

Calculation is mainly made with a single point source, which is on the central line of the street and at 20m outside the street (see Figure 7.18). The sound power level is assumed to be 100dB. For selected street configurations, calculation is also made with a line source along the street centre. The source height is 0.5m in both cases.

Receivers are on three planes: a horizontal plane at 1.5m above the ground level, a horizontal plane at the street height, and a vertical plane at 1m from a façade. For the streets with semi-detached houses or discrete lots, a vertical plane at 1m from a side wall is also considered. This plane is approximately halfway between the two street ends.

7.2.2 Results

Geometrically reflecting boundaries

Figure 7.15 shows the sound attenuation along the centre line of the street with geometrically reflecting boundaries, where in Figure 7.15a the receivers are at 1.5m above the ground and in Figure 7.15b the receivers are at the street height, for example, 8.5m for UK2 and 65m for HK2. The absorption coefficient of all the boundaries is assumed to be 0.05, except for HK3, where $\alpha = 0.02$. The air absorption is 0.00042 Np/m. These data correspond to middle frequency values.

In Figure 7.15a it can be seen that in comparison with the UK streets, the SPL is generally lower in HK streets. This is mainly because the street width is greater in the HK cases. When the street width is increased from 12m to 20m and 30m, the SPL becomes about 1.5dB and 3dB less, respectively. The results in Figure 7.15b suggest that at the top floor level, the SPL in the UK streets is higher than that in HK streets, typically at about 3–7dB, mainly caused by the difference in source–receiver distance. As expected, the difference becomes less with increasing source–receiver distance.

By comparing HK2 and HK3, it is seen that the SPL difference between the absorption coefficient of 0.05 and 0.02 is about 1dB. This approximately indicates the difference between concrete and glass façades.

Corresponding to Figure 7.15, the decay curves at two typical receivers are shown in Figure 7.16. The results suggest that in the UK cases the reverberation time is systematically shorter than that in the HK cases, typically by 50–100 per cent. The difference is mainly caused by the difference in street width. The effect of absorption coefficient on reverberation can be seen in Figures 7.16a and 7.16b by comparing HK2 and HK3. As expected, the difference is much greater than that of the SPL.

From Figure 7.16 it can also be demonstrated that the reverberation time in these streets could reach about 5–10s if boundaries are geometrically reflective, which is significant. By comparing Figures 7.16a and 7.16b it can be seen that with increasing source–receiver distance there is an increase in RT, which corresponds to the results in Section 7.1.

Diffusely reflecting boundaries

Assuming the boundaries are diffusely reflective, a comparison is made between UK5 and HK5. The two streets have the same width, but the heights are 8.5 and 65m, respectively. The absorption of the boundaries is again considered to be 0.05, with air absorption as 0.00042Np/m.

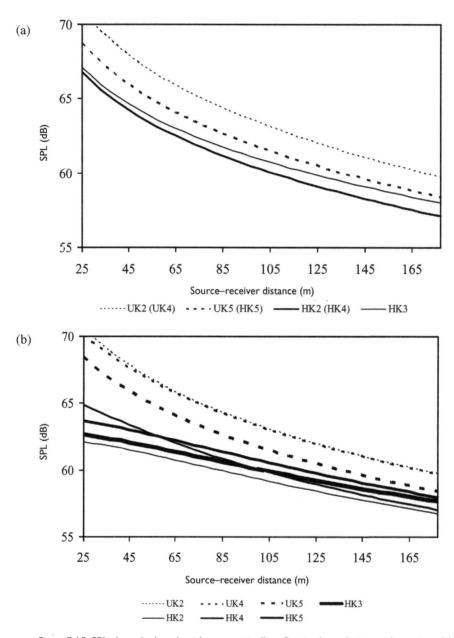

Figure 7.15 SPL along the length with geometrically reflecting boundaries, with receiver (a) at 1.5m and (b) at street height.

Figure 7.17a compares the SPL distribution between UK5 and HK5 along two horizontal receiver lines, a line along the street centre and at 1.5m above the ground, and a line at the street height and at 1m from a façade. It is seen for the first receiver line, in comparison with UK5, in HK5 the SPL is generally 3–5dB higher, indicating the effect of street height. For the second receiver line,

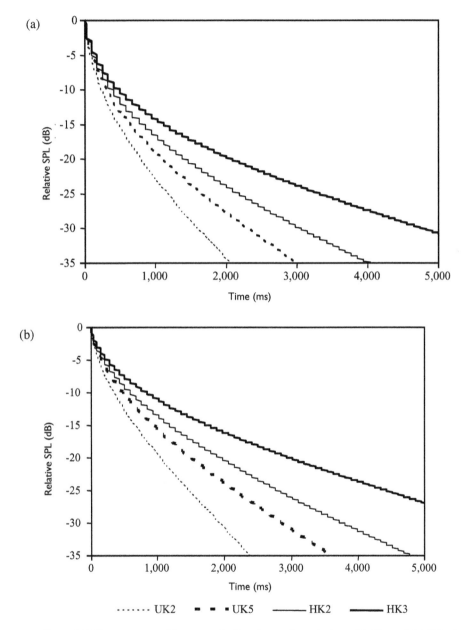

Figure 7.16 Decay curves with geometrically reflecting boundaries at (a) 40m and (b) 120m from the source. The receivers are at 1.5m above the ground.

compared with UK5, the SPL in HK5 is systematically higher beyond about 40m although the actual source–receiver distance is considerably greater. Corresponding to Figure 7.17a, the comparison in RT and EDT between HK5 and UK5 is shown in Figures 7.17b and 7.17c, respectively. It can be seen that the reverberation time in HK5 is substantially longer than that in UK5.

Mixed boundaries

Calculation is made using Raynoise for UK1, UK5, HK1 and HK5, assuming a diffusion coefficient of 0.3 for all the façades whereas the ground is considered as geometrically reflective. The surface absorption is based on the database in Raynoise, considering the materials listed in Table 7.2. The air absorption corresponds to a temperature of 10°C and relative humidity of 50 per cent for the UK cases and a temperature of 25°C and relative humidity of 85 per cent for the HK cases. A full frequency range of 63–8kHz is considered, with the source spectrum corresponding to typical traffic noise. Figure 7.18 shows the sound distribution on a horizontal plane at 1.5m above the ground in the case of a single source. By comparing UK1 and HK1 or UK5 and HK5, it can be demonstrated that the SPL is systematically increased by the increased building height, and this SPL increase becomes greater with increasing source–receiver distance. As expected, in comparison with continuous buildings, the SPL is lower with discrete lots, typically by 3–5dBA.

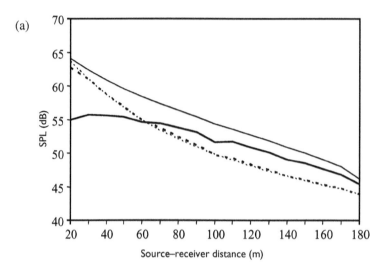

(a)

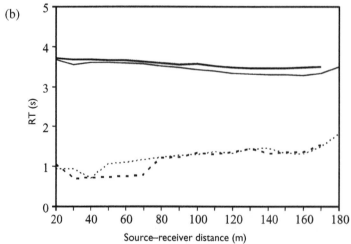

(b)

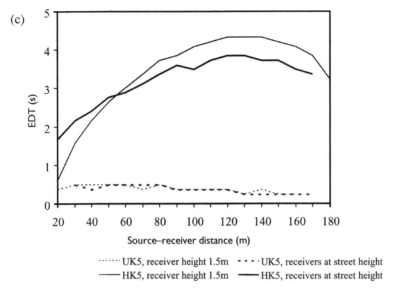

(c)

EDT (s)

Source–receiver distance (m)

········ UK5, receiver height 1.5m   · · · · UK5, receivers at street height
——— HK5, receiver height 1.5m   ——— HK5, receivers at street height

Figure 7.17 Comparison in (a) SPL; (b) RT; and (c) EDT between HK5 and UK5 with diffusely reflecting boundaries.

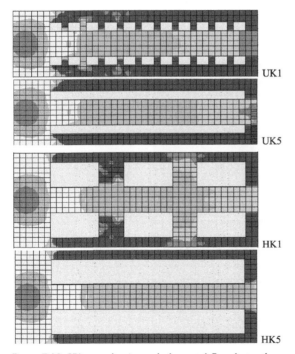

UK1

UK5

HK1

HK5

Figure 7.18 SPL on a horizontal plane at 1.5m above the ground. All the façades have a diffusion coefficient of 0.3 whereas the ground is geometrically reflective. Point source. Each colour represents 5dBA. A colour representation of this figure can be found in the plate section.

With a line source along the street centre, Figure 7.19 shows the SPL distribution on a vertical receiver plane at 1m from a façade. Contrary to the situation with a single source, the SPL in the UK streets is almost the same as, or only slightly lower than that in the HK streets. Clearly this is due to the effect of direct sound. From the figure it is also seen that the SPL in UK1 and HK1 is about 2dBA lower than that in UK5 and HK5, respectively, indicating the effect of the gaps between buildings.

With the line source, the sound distribution on a receiver plane perpendicular to the street length and at 1m from a side wall is shown in Figure 7.20. By comparing Figures 7.19 and 7.20, it can be seen that for HK1, the sound attenuation along the height is about 10dBA on the side wall, whereas this value is only 6dBA on the front façade (note, the two figures have different dBA scale).

Calculation using Raynoise also suggests that the RT is generally shorter with a line source than with a single point source located outside the street.

Summary

In the case of geometrically reflecting boundaries, the street width plays an important role and thus, the SPL in the UK streets is 1–3dB higher than that in the HK streets when a single source is considered. With boundary diffusion, conversely, the SPL in the UK streets is about 3–5dB lower. Since there are always some irregularities on building or ground surfaces, the latter case is closer to the actual situation. With a line source, the SPL difference between the UK and HK streets becomes less. The RT in the HK streets is substantially longer than that in the UK streets. For both the UK and HK streets, the gaps between buildings can typically bring 3–5dB noise reduction.

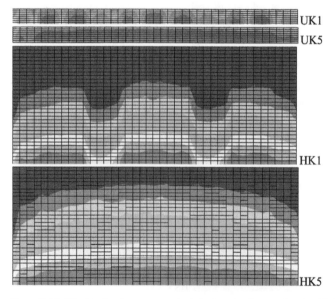

Figure 7.19 SPL distribution on a vertical plane at 1m from a façade. All the façades have a diffuse co-efficient of 0.3 whereas the ground is geometrically reflective. Line source. Each colour represents 1dBA. A colour representation of this figure can be found in the plate section.

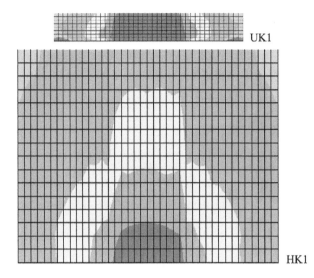

UK1

HK1

Figure 7.20 SPL distribution on a receiver plane perpendicular to the street length. All the façades
have a diffusion coefficient of 0.3 whereas the ground is geometrically reflective. Line
source. Each colour represents 3dBA. A colour representation of this figure can be
found in the plate section.

7.3   Urban squares

A series of hypothetical urban squares, as illustrated in Figure 4.2, are used for a parametric study
(Kang 2002b, 2005a; Kang and Zhang 2003). Except where indicated, a square is surrounded by
buildings, the absorption coefficient of all the boundaries is 0.1, and a point source with a sound
power level of 0dB is positioned at $(x = L/5, y = W/5, z = 1.5\text{m})$. Analyses are based on either 100
evenly distributed receivers or a receiver line along a diagonal. The receiver height is 1.2m.

7.3.1   Basic characteristics of sound field

A square of 50 × 50m, surrounded by buildings with a height of 20m, is used to analyse the basic
characteristics of sound field. Figure 7.21a shows the SPL distribution in the square, with both
diffusely and geometrically reflecting boundaries. It can be seen that with both kinds of boundaries, the SPL initially decreases significantly with increasing source–receiver distance, about
5–8dB from source–receiver distance 8m to 25m (corresponding to receiver 24 to 56 as shown
in Figure 4.2), for example, and then becomes approximately stable, with a variation of less than
2dB beyond 25m. This feature is similar to that in regularly shaped enclosures, although the SPL
values are different from those calculated using the classic theory.

The RT and EDT are shown in Figures 7.21b and 7.21c respectively. It can be seen that the
RT is very even over the entire square. The STD to average ratio is 0.9 per cent for diffusely
reflecting boundaries and 3.3 per cent for geometrically reflecting boundaries. The EDT is
very low in the near field, say 10–15m from the source, and then becomes relatively even after
a rapid increase. The STD to average ratio is 28.8 per cent for diffusely reflecting boundaries
and 32.5 per cent for geometrically reflecting boundaries. When the source–receiver distance
is increased from 8m to 25m, changes in RT and EDT are plus 10–20 per cent and plus 70–110

(a)

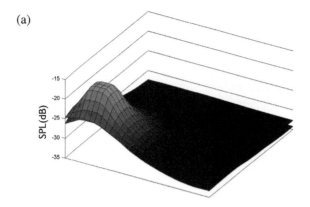

(b)

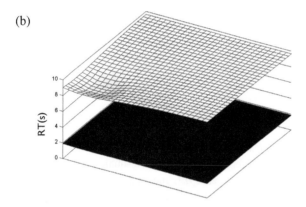

(c)

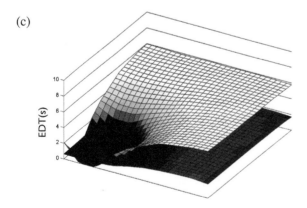

Figure 7.21 Distribution of the (a) SPL; (b) RT; and (c) EDT in a square of 50 × 50m, with diffusely (dark mesh) and geometrically (light mesh) reflecting boundaries. Square height 20m. Source at (10m, 10m, 1.5m). Boundary absorption coefficient 0.1.

per cent respectively, with both kinds of boundaries considered. Beyond 25m the variations are much less, <3 per cent and 2–8 per cent for RT and EDT. An important reason for the short reverberation in the near field, especially the EDT, is that direct sound plays an important role, and there is a lack of early reflections. Overall, the RT and EDT are rather long in such a square, about 2s with diffusely reflecting boundaries, and around 8–10s with geometrically reflecting boundaries.

7.3.2   Boundary reflection pattern

In Figure 7.21a it can be seen that the SPL with diffusely reflecting boundaries is generally lower than that with geometrically reflecting boundaries, which is similar to the situation in street canyons. This difference generally increases with increasing source–receiver distance, to a maximum of about 2dB, which is considerable, given that for both kinds of boundaries the variation in SPL across the receivers, except in the very near field, is only about 5–8dB. In the near field, with diffusely reflecting boundaries there is a slight SPL increase compared to that with geometrically reflecting boundaries, due to backscattering. It is noted that the SPL differences between the two kinds of boundaries depend on the square size, as further discussed in Section 7.3.3.

As can be seen in Figures 7.21b and 7.21c, the reverberation resulting from geometrically reflecting boundaries is significantly longer at about 400 per cent for the RT and 200 per cent for the EDT, than that from diffusely reflecting boundaries, again similar to the situation in street canyons. An important reason for the differences is that for a given order of reflection, with geometrically reflecting boundaries the sound path is generally much longer, mainly due to the flutter echo effect. In Figure 7.22 the decay curves with diffusely and geometrically reflecting boundaries are compared at receiver 100.

The results in SPL, RT and EDT with increasing diffusion coefficient ranging from 0 to 1 are shown in Figure 7.23, where three typical receivers, 24, 56 and 89, are considered (see

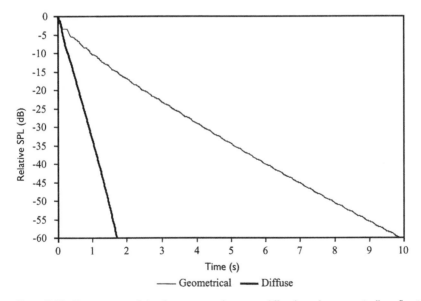

Figure 7.22 Comparison of the decay curves between diffusely and geometrically reflecting boundaries at receiver 100, where the source–receiver distance is 53m.

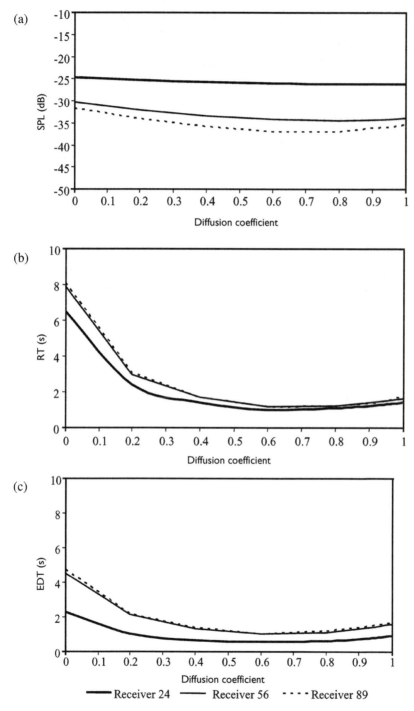

Figure 7.23 Variation in (a) SPL, (b) RT and (c) EDT with increasing diffusion coefficients at receiver 24, 56 and 89, where the source–receiver distances are 8, 25 and 46m respectively.

Figure 4.2), and the calculation is made using Raynoise. It can be seen that with partially diffuse and partially geometrical boundaries, the sound field is generally between the two idealised situations. It is important to note that when the diffusion coefficient is increased from 0 to about 0.2, the decrease in RT and EDT is significant, whereas when the diffusion coefficient is further increased, the changes become much less. The variation in SPL is similar to that in RT and EDT, although the diffusion coefficient value where the SPL becomes approximately steady is slightly greater than 0.2. Those results suggest that with only about 20 per cent of the energy incident upon the boundaries diffusely reflective, the sound field in an urban square is close to that resulting from purely diffusely reflecting boundaries. This is in agreement with the discussion in Section 4.4.1. With increasing source–receiver distance, the variation in SPL, RT and EDT with various diffusion coefficients generally becomes slightly greater, as can be seen by comparing the three receivers in Figure 7.23. This is possibly because reflections play a more important role at longer distances to the source. It is also noted in Figure 7.23 that when the diffusion coefficient is increased from about 0.7 to 1, there is a slight increase in RT and EDT, which is probably caused by the decrease in initial energy in the sound decay process.

Based on the above discussions for urban streets and squares, it is evident that by replacing geometrically reflecting boundaries with diffusely reflecting boundaries, the sound attenuation along the length becomes considerably more, the reverberation is significantly shorter, and the extra SPL attenuation caused by air or vegetation absorption is increased. As a result, from the viewpoint of urban noise reduction, it is better to design the building façades and the ground of a street canyon or a square as diffusely reflective rather than acoustically smooth. Although it might be unrealistic to design all the boundaries as purely diffusely reflective, some diffuse patches on a boundary, or boundaries with a high diffusion coefficient, are helpful in making the sound field closer to that resulting from diffuse boundaries, especially when multiple reflections are considered. Similar to diffuse boundaries, street/square furniture, such as trees, lampposts, fences, barriers, benches, telephone boxes, bus shelters, can act as diffusers and thus be effective in reducing noise. With the similar principle, diffuse boundaries and street furniture are also useful for reducing overall background noise of a city, which is produced by the general distribution of sources throughout the city.

7.3.3 Square geometry

Square height

The effect of square height on the SPL distribution is shown in Figure 7.24a, by considering three square heights, 50, 20 and 6m, where the square size is 50 × 50m, a point source is at (10m, 10m, 1. 5m), and the receivers are along the diagonal as illustrated in Figure 4.2. Only diffusely reflecting boundaries are considered since with geometrically reflecting boundaries increasing square height from 6m to 50m will have no effect on the receivers considered. It can be seen that in the near field the SPL is almost unchanged with various square heights, whereas with the increase of source–receiver distance, the SPL becomes higher with a greater square height. This is expected, because with more façade areas there are more reflections. By increasing the source–receiver distance the SPL difference between different square heights initially increases, and then becomes approximately stable. In the far field, between square heights 50m and 6m, the difference is about 8dB.

The effect of square height on the RT and EDT is shown in Figure 7.24b. The simulated RT values, averaged over the entire square, are 4.19, 1.84 and 0.83s with square heights of 50, 20 and 6m, which are close to the calculated values using the classic Eyring formula for room

(a)

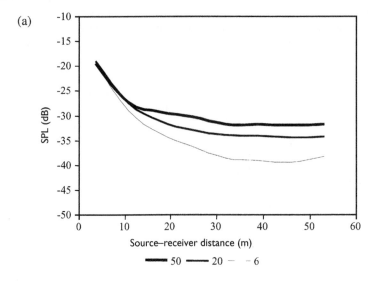

(b)

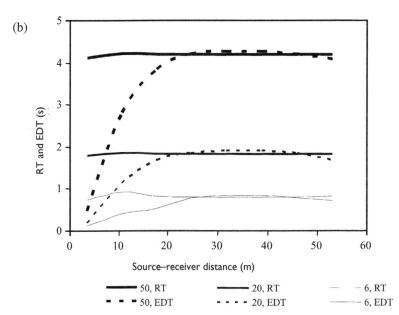

Figure 7.24 Comparison of (a) the SPL and (b) reverberation times between three square heights: 50, 20 and 6m, with diffusely reflecting boundaries.

acoustics (see Section 1.6.2), 4.31, 2, and 0.63s, where the ceiling absorption coefficient is assumed to be 1. In all the three squares the EDT becomes stable beyond about 25m from the source, although for a greater square height the increase in EDT with increasing source–receiver distance is more rapid, because of the increased reflections by the increased boundary area. In the relatively far field, the EDT values are close to RT with various square heights, indicating that the decay curves are close to linear.

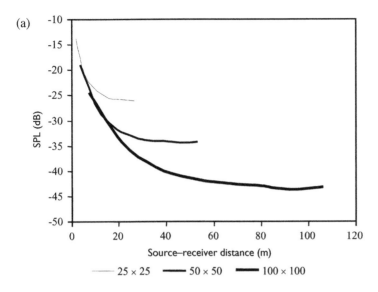

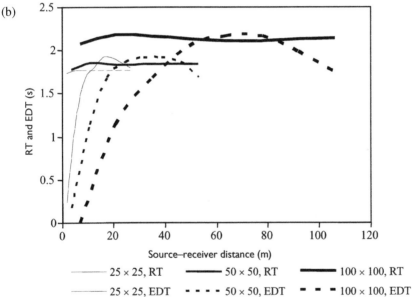

Figure 7.25 Comparison of (a) the SPL and (b) reverberation times between three square sizes: 25 × 25m, 50 × 50m and 100 × 100m, with diffusely reflecting boundaries.

Square size

Comparisons are made between three squares of different size, 25 × 25m, 50 × 50m and 100 × 100m, in Figures 7.25 and 7.26, for diffusely and geometrically reflecting boundaries, respectively, where the square height is 20m, the receivers are along a diagonal of each square, and a point source is positioned at (5m, 5m, 1.5m), (10m, 10m, 1.5m) and (20m, 20m, 1.5m) in the three squares respectively.

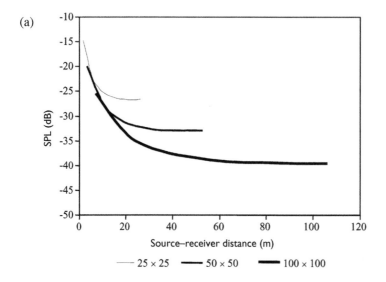

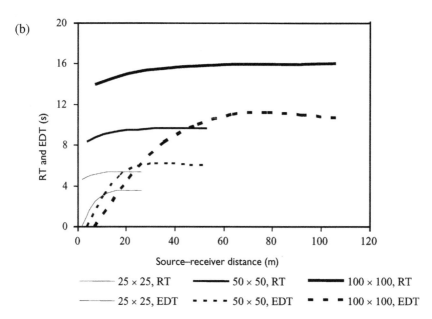

Figure 7.26 Comparison of (a) the SPL and (b) reverberation times between three square sizes: 25 × 25m, 50 × 50m and 100 × 100m, with geometrically reflecting boundaries.

For both kinds of boundaries, in the near field, say within a source–receiver distance of 5–10m, there is no significant difference in SPL between the three square sizes. Clearly this is due to the dominant role of the direct sound. With the increase of source–receiver distance, the SPL becomes systematically less with increased square size. In the far field of each square, the SPL is approximately 6–9dB lower when the square side is doubled.

Compared to geometrically reflecting boundaries, with diffusely reflecting boundaries there are more energy losses resulting from the scattering of sound out of the square. Moreover, the SPL attenuation with distance is greater since some energy is redistributed towards the source due to backscattering. However, with diffusely reflecting boundaries, reflections from all boundary patches will contribute to the SPL at a receiver, whereas with geometrically reflecting boundaries only the surfaces below the source and the receiver heights are effective. The importance of these three effects varies with different square configurations. With the square sizes of 50 × 50m and 100 × 100m, the first two effects are dominant, so that with diffusely reflecting boundaries the SPL attenuation is systematically greater than that with geometrically reflecting boundaries, at about 2–5dB, as can be seen by comparing Figures 7.25a and 7.26a. With the increase of height/side ratio, the third effect plays a major role and thus with diffusely reflecting boundaries, the SPL attenuation is systematically smaller than that with geometrically reflecting boundaries. This is the case of the 25 × 25m square, although the difference is within 1dB. Overall, the results suggest that compared to diffusely reflecting boundaries, with geometrically reflecting boundaries the SPL attenuation along a square is generally smaller unless the height/side ratio is high, say 1:1.

The reverberation results with diffusely reflecting boundaries are shown in Figure 7.25b. It can be seen that the RT increases with increasing square size, as expected. It is interesting to note that the simulated RT values in the three squares, 1.77, 1.84 and 2.14s, are close to the calculations using the Eyring formula, 1.88, 2 and 2.03s. However, using the two methods the RT ratios between various square sizes, namely 1.77:1.84:2.14 and 1.88:2:2.03, are different, suggesting that the direct use of the Eyring formula may not be appropriate (also see Section 4.7). In terms of the distribution of RT and EDT, the three squares are similar: the RT is even across a square, whereas the EDT increases rapidly with increasing source–receiver distance until about the square centre, and then becomes relatively stable, where the EDT is close to RT, indicating that the decay curves are close to linear. It is noted that the increase in EDT with increasing source–receiver distance is slower in a larger square. This is mainly because the sound paths become longer with increased square size, so that there is a lack of early reflections.

With geometrically reflecting boundaries, the RT and EDT variations are generally similar to the above, as shown in Figure 7.26b, although the absolute values as well as the RT and EDT ratios between various square sizes are rather different. Unlike the situation with diffusely reflecting boundaries, with geometrically reflecting boundaries the RT values are considerably higher than the EDT. This is probably due to the large number of late reflections corresponding to flutter echoes.

Square shape

The effect of square shape is examined by comparing two configurations, 50 × 50m and 100 × 25m, namely with the same ground area but different aspect ratios, 1 and 4 respectively. The square height is again 20m, the receivers are along a square diagonal, and a point source is at (10m, 10m, 1.5m) and (20m, 5m, 1.5m) in the two squares respectively. Figures 7.27 and 7.28 compare the two squares for diffusely and geometrically reflecting boundaries, respectively. It is interesting to note that for both kinds of boundaries, in the 100 × 25m square, the SPL in the near field is almost the same as, or only slightly higher than, that with the 50 × 50m square, whereas in the far field, say over 30m, the SPL in the 100 × 25m square is systematically lower, at 5dB with diffusely reflecting boundaries and 2dB with geometrically reflecting

(a)

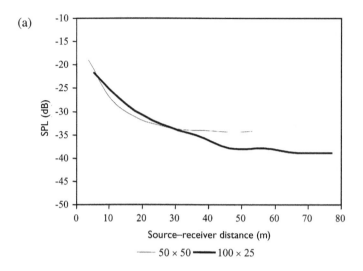

(b)

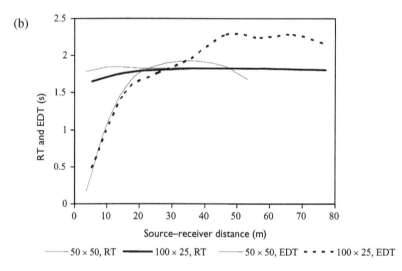

Figure 7.27 Comparison of (a) the SPL and (b) reverberation times between two square shapes: 50 × 50m and 100 × 25m, with diffusely reflecting boundaries.

boundaries. A main reason for the extra SPL attenuation is that the sound path is generally greater when a square becomes longer. The results suggest that from the viewpoint of noise reduction, it is better to design a square with a greater aspect ratio.

By comparing Figures 7.27a and 7.28a, it is noted that the SPL difference between diffusely and geometrically reflecting boundaries becomes greater when the aspect ratio of a square increases. For the 50 × 50m square, the difference between the two kinds of boundaries is about 2dB in the far field, whereas this value is about 5dB for the 100 × 25m square. A possible reason is that with diffusely reflecting boundaries, the relative increase in sound path when a square becomes longer is generally greater than that with geometrically reflecting

(a)

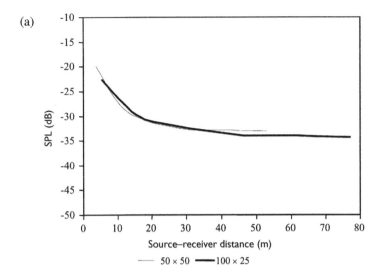

(b)

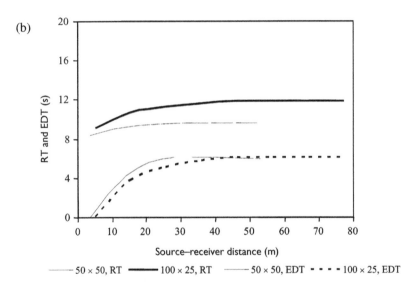

Figure 7.28 Comparison of (a) the SPL and (b) reverberation times between two square shapes: 50 × 50m and 100 × 25m, with geometrically reflecting boundaries.

boundaries. Also, with diffusely reflecting boundaries, a longer square will backscatter more energy to the near field.

The differences in reverberation between the two square shapes are generally less than those in SPL, both for diffusely and geometrically reflecting boundaries, as shown in Figures 7.27b and 7.28b, respectively. This is probably because the SPL is more affected by early reflections and reflection patterns, whereas the reverberation is more related to multiple reflections and the square volume, which is the same for the two square shapes. In Figure 7.27b it is noted that with diffusely reflecting boundaries, in the far field of the 100 × 25m square the

EDT is considerably longer than RT. Clearly this is caused by the strong early reflections relative to the direct sound in this area. With geometrically reflecting boundaries the RT in the 100×25m square is longer than that in the 50×50m square, by about 18 per cent on average, and this difference increases with increasing source–receiver distance, as can be seen in Figure 7.28b. This is probably because in the 100×25m square the sound paths are generally longer and thus in the decay process the late reflections are relatively stronger compared to the 50×50m square. For a similar reason, the near field EDT in the 100×25m square is slightly shorter than that in the 50×50m square, due to the lack of early reflections.

7.3.4 Boundary absorption and building arrangements

Evenly distributed absorption

The SPL attenuation with a range of boundary absorption coefficients is shown in Figures 7.29a and 7.30a for diffusely and geometrically reflecting boundaries, respectively, where the square is 50×50m, the square height is 20m, and the receivers are again along the diagonal. It can be seen that the maximum SPL difference between absorption coefficients 0.1 and 0.9 is 10.4dB with diffusely reflecting boundaries and 11.5dB with geometrically reflecting boundaries. These values are considerably greater than those in street canyons (see Section 7.1.4). An important reason is that in urban squares the SPL depends on reflections from all four façades and thus the effect of increasing boundary absorption is more significant. By comparing Figures 7.29a and 7.30a it is seen that the difference between the two kinds of boundary becomes less with increasing absorption coefficient. This is because with a higher absorption coefficient, boundary reflections become less important compared to the direct sound.

When the absorption coefficient is increased from 0.1 to 0.3, 0.3 to 0.5, 0.5 to 0.7, and 0.7 to 0.9, the decrease in SPL averaged over the entire square is 2.0, 1.7, 1.5 and 1.3dB respectively for diffusely reflecting boundaries, and 2.4, 1.6, 1.4 and 1.3dB respectively for geometrically reflecting boundaries. This indicates that a given increase of sound absorption is less efficient for noise reduction if there are already considerable absorbers in the square, which is expected. It is also noted that the decrease in SPL caused by increasing absorption coefficient becomes systematically greater with increasing source–receiver distance. Clearly this is because with a longer source–receiver distance the SPL depends more on reflections, for which boundary absorption is effective.

The effects of boundary absorption on reverberation are shown in Figures 7.29b and 7.30b, for diffusely and geometrically reflecting boundaries, respectively. The RT decreases with increasing boundary absorption, as expected, but it is interesting to note that, for both kinds of boundaries, the rate of decrease is considerably slower than that predicted using the Eyring formula, suggesting the nondiffuse feature of the sound fields in urban squares, especially when the boundary absorption is high. For both diffusely and geometrically reflecting boundaries, the RT/EDT ratio generally becomes greater with increasing boundary absorption, suggesting that, in this square, a given SPL reduction in reflections can cause a greater decrease in EDT than that in RT. It is also noted that when the boundary absorption coefficient is high, say 0.7–0.9, the energy in early reflections becomes very weak, especially in the near field where the number of reflections is small. The shapes of the decay curves thus become extremely nonlinear, especially for geometrically reflecting boundaries. As a result, the RT and EDT vary considerably at various receivers, as can be seen in Figures 7.29b and 7.30b.

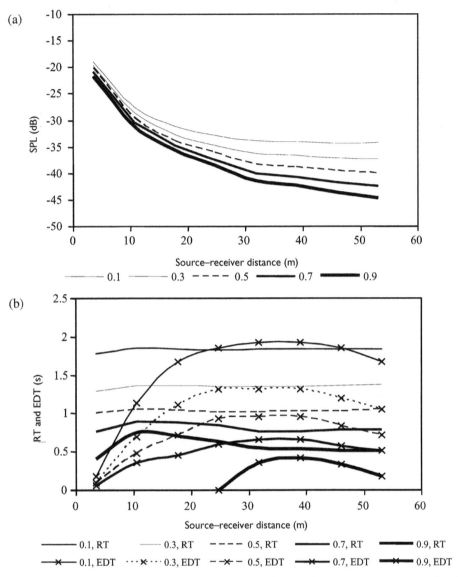

Figure 7.29 Effects of boundary absorption on (a) SPL and (b) reverberation times, with diffusely reflecting boundaries.

Strategic building and absorption arrangements

Urban squares are not always surrounded by buildings, and sound absorption is often unevenly distributed. To study such effects, simulation is carried out using the 50 × 50m square, with one, two and three façades only, as well as with absorbers of $\alpha = 0.5$ on one to four façades. The rest of the boundaries have an absorption coefficient of 0.1. In Figure 7.31 those configurations are illustrated, where the square height is again 20m. Both diffusely and geometrically reflecting boundaries are considered.

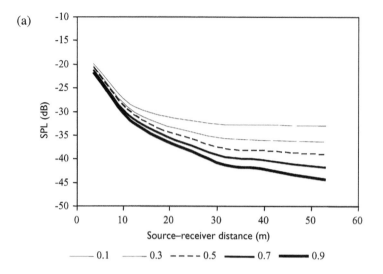

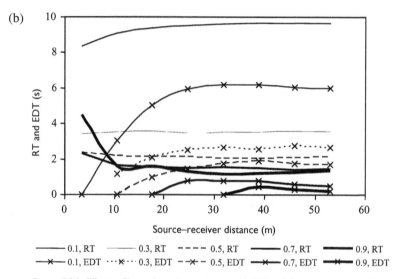

Figure 7.30 Effects of boundary absorption on (a) SPL and (b) reverberation times, with geometrically reflecting boundaries.

The effects of building and absorption arrangements on the SPL distribution are shown in Figures 7.32 and 7.33 respectively, where both diffusely and geometrically reflecting boundaries are considered. The tendencies with both kinds of boundaries are generally similar. When façade V is removed or made absorbent, the SPL near this façade becomes lower, showing a rather 'local' effect. When two opposite façades, U and V, are removed or made absorbent, the direct sound plays a much more important role, although the effect of multiple reflections between façades A and B can still be seen. When three façades, U, V and B are removed, the sound field is dominated by the direct sound. With the three façades

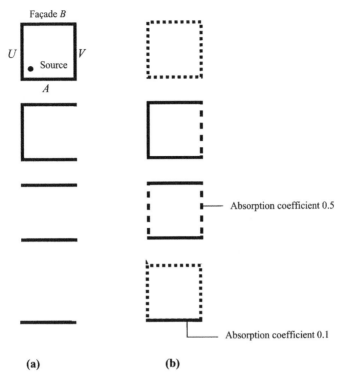

Façade *B*

U   *V*

Source

A

Absorption coefficient 0.5

Absorption coefficient 0.1

(a)   **(b)**

Figure 7.31 (a) Four building arrangements and (b) four absorber arrangements in the 50 × 50m square.

absorbent the SPL distribution shows a similar pattern, although the SPL is generally higher.

Figures 7.34 and 7.35 show the effects of building and absorption arrangements on the RT distribution respectively. With diffusely reflecting boundaries, the RT distribution is generally even across the square with various building and absorption arrangements, although the RT value reduces with decreased number of façades or increased number of absorbent façades, as expected. The STD to average ratios are 0.9, 1.4, 3.2 and 5.9 per cent with four, three, two and one façade, and 1.5, 1.6, 1.9 and 2.2 per cent with one, two, three and four façades absorbent. With geometrically reflecting boundaries the reflection pattern in the square is considerably changed with different building and absorption arrangements and, consequently, the RT distribution becomes more uneven. The STD to average ratios are 3.3, 11.4 and 12.7 per cent with four, three and two façades, and 4.7, 7.4, 9.2 and 1.6 per cent with one, two, three and four façades absorbent. It is noted that with only one geometrically reflecting façade, multiple reflections will not occur and, thus, the RT is not included in Figure 7.35.

The effects of building and absorption arrangements on the EDT distribution are shown in Figures 7.36 and 7.37 respectively. The tendencies of changes are generally similar to those of RT, but the spatial variations are much greater. For various configurations the ratio of the STD to the average EDT ranges 25–34 per cent with diffusely reflecting boundaries and 32–110 per cent with geometrically reflecting boundaries (Kang 2005a).

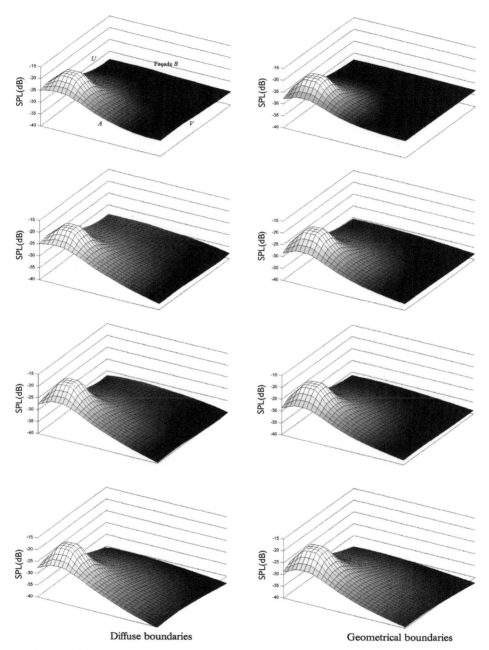

Figure 7.32 SPL distribution with four building arrangements (see Figure 7.31).

A further calculation is made using Raynoise, based on the 50 × 50m square, where three types of openings are considered: in the middle of one side, in the middle of each side, and in the four corners of the square (Yang 2005). For SPL, with one side opening of 2.5, 10 and 25m, the average decrease in the square is 0.8, 1.1 and 2.0dB respectively, compared to the

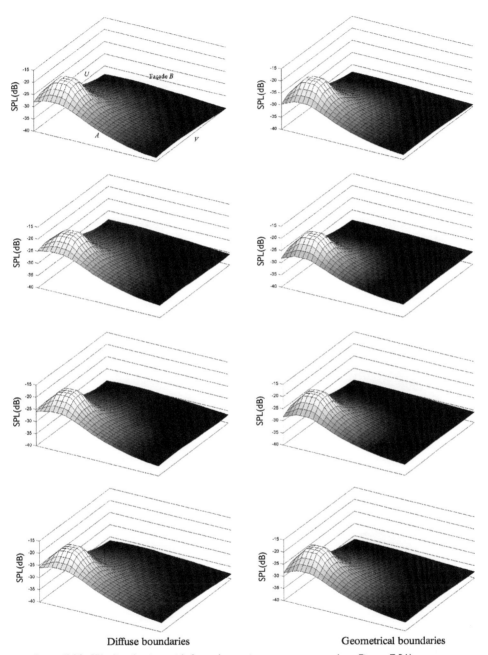

Diffuse boundaries   Geometrical boundaries

Figure 7.33 SPL distribution with four absorption arrangements (see Figure 7.31).

enclosed situation. With a 25m opening on each side, the maximum SPL decrease in the square is 7dB compared to the enclosed situation. In comparison with openings on four sides, when four corners are open, the SPL attenuation with increasing source–receiver distance becomes less, with a maximum difference of 3.6dB. In terms of RT, compared to the enclosed

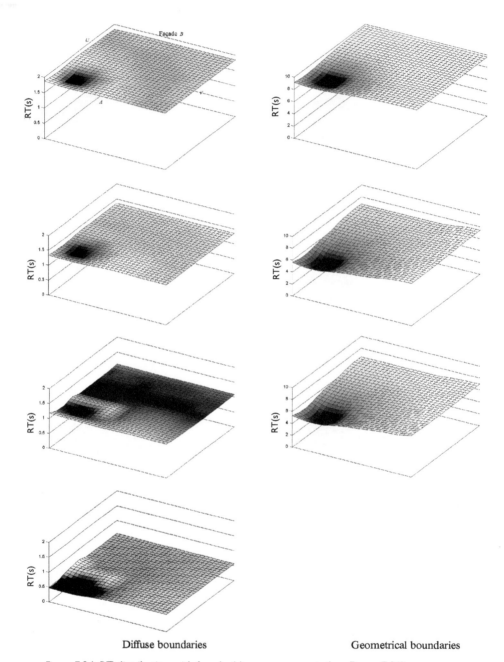

Diffuse boundaries   Geometrical boundaries

Figure 7.34 RT distribution with four building arrangements (see Figure 7.31).

situation, with a 25m opening on each side the decrease is 57 per cent. It is noted that the opening position has a rather significant influence on RT. Compared to a 2.5m opening on each side, the RT is systematically higher with a 10m opening on one side only. Corresponding to the results in Figures 7.34 and 7.35, the pattern of RT distribution is similar

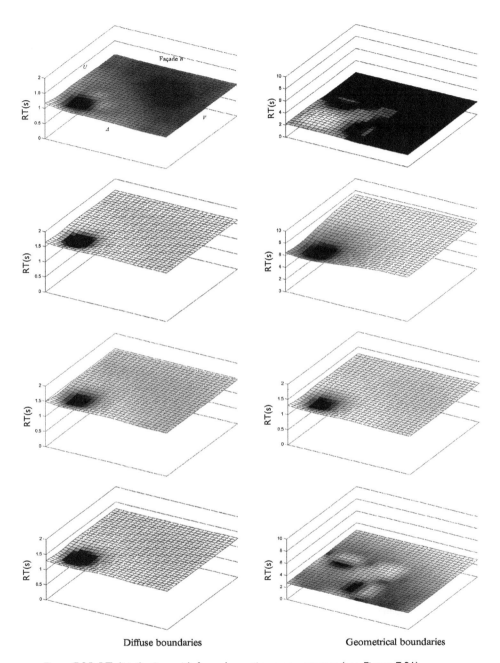

Diffuse boundaries   Geometrical boundaries

Figure 7.35 RT distribution with four absorption arrangements (see Figure 7.31).

between the enclosed square and with an opening on one side, but becomes rather different when openings are on four sides. For EDT, the variation in the square is greater than that in RT, which corresponds to the situation shown in Figures 7.36 and 7.37, suggesting that the EDT values strongly depend on the position and size of the openings near the receiver.

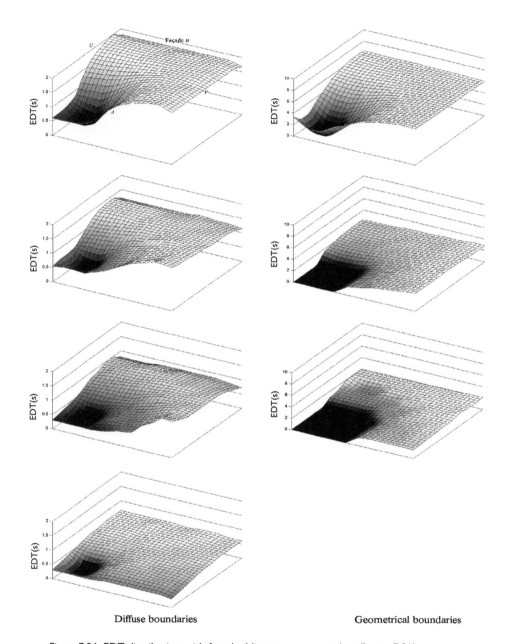

Diffuse boundaries   Geometrical boundaries

Figure 7.36 EDT distribution with four building arrangements (see Figure 7.31).

7.4   Case study: classic squares

The typological soundscape characteristics are analysed in three classic urban squares from Renaissance urbanism (Yang 2005), representing typical spaces of enclosure, continuity, and contrast (Bacon 1975; Trancik 1986; Moughtin 2003). The analysis below refers to middle frequencies.

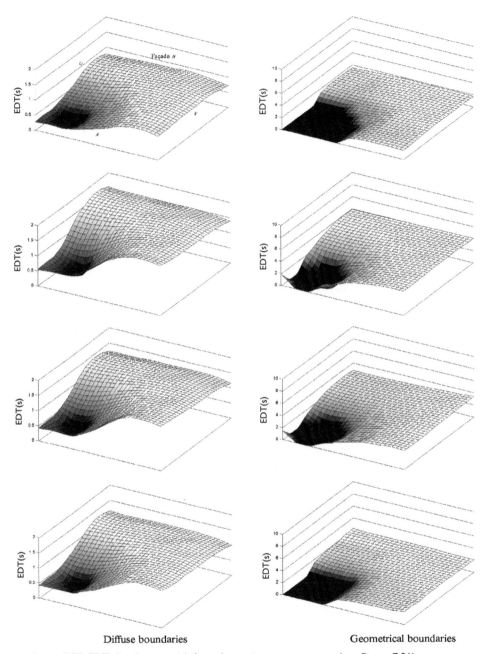

Figure 7.37 EDT distribution with four absorption arrangements (see Figure 7.31).

Campidoglio in Rome is a typical example of space enclosure and order. Three simplified models are considered, simulating the Campidoglio after Michelangelo's work; before Michelangelo's work, that is, without the Capitoline Museum; and an imaginary configuration where the two front buildings, the Palazzo del Conservatori and Capitoline Museum, are rotated to become parallel. It is shown that in the three configurations the variation in average SPL is within 1dB, the average RT are 1.87, 1.70, and 1.94s, and the average EDT are 1.18, 0.82, and 1.21s, respectively compared to the rectangular space, the angled façades in Michelangelo's design create a more even sound field but shorter reverberation.

Piazza Navona in Rome is a famous example of continuity, with a long plan at a ratio of approximately 1:5. The square is dominated by fountains that give soul and life to the place, both visually and aurally. Two configurations are simulated, an abstract model of the square without any opening, and the actual plan. In the latter, although with no more than 10 per cent of boundary opening, the sound field is rather different. The SPL is 2.1dB less on average, and the average RT and EDT values drop from 2.83 to 2.05s and from 2.28 to 1.23s, respectively. Corresponding to the calculations in Section 7.3, this result again suggests that relatively small changes in spatial form could have considerable effects on reverberation whilst the change in SPL is relatively small.

Piazza della Signoria in Florence forms two distinct but interpenetrating spaces. This contrast in space provides a great opportunity to create a dramatic urban soundscape sequence. The acoustic indices are rather different in various parts of the square. For example, the EDT increases by more than 60 per cent when the narrow shaft space joins the central square, suggesting that in different positions of the square people may have rather different perceptions for a given sound source such as the fountain. Moreover, when people walk through, different parts of the square could give very different responses to their footsteps or voices, reflecting/indicating spatial changes.

7.5   Sound propagation between two parallel streets

In the above sections main attention has been paid to the sound propagation within street canyons or squares, rather than the diffraction over building roofs. Using the coupled FDTD-PE model, Van Renterghem *et al.* (2006) carried out a parametric study of sound propagation between parallel street canyons. A two-dimensional idealised configuration is used with a coherent line source.

The results show that the shielding is rather insensitive to the width/height ratio of the canyons, except for very narrow canyons. For ratios larger than 1, relative SPL at the receiver canyon becomes more or less constant.

The degree of absorption on the façades is very important. Rigid walls result in very poor shielding towards the receiver canyon, and the effectiveness of boundary absorption could be more than10dB.

The effect of introducing diffusers such as recesses by windows and protrusions by windowsills, together with a roughened wall, increases with increasing frequency when comparing to flat façades. At 1kHz an extra shielding of about 10dB is gained with profiled façades. The presence of balconies results in an important increase in shielding. Inclining the parapet of the balconies also results in an extra increase at some frequency bands.

In the case of downwind sound propagation, shielding considerably decreases compared to a nonmoving atmosphere. With increasing incident wind speed and increasing frequency, shielding decreases. In the case of upwind sound propagation, turbulent scattering plays an important role and the shielding does not increase compared to a nonmoving atmosphere.

The effect of an incoherent line source, which is more appropriate for traffic noise, was estimated by performing a number of calculations in two-dimensional cross sections through source and receiver. The result suggests that with an incoherent line source the shielding decreases compared to a coherent line source.

References

01dB, 2005, *dBFA User's Manual*, 01dB-Metravib, Limonest, France.

ABBOTT, P.G. and NELSON P.M., 2002, *Converting the UK Traffic Noise Index L. to EU Noise Indices for Noise Mapping*. Project Report EPG1/2/37 – Traffic Noise Calculation Method for Noise Mapping, AEQ Division of UK Department for Environment, Food and Rural Affairs (DEFRA).

ABE, K., OZAWA, K., SUZUKI, Y., and SONE, T., 1999, The effects of visual information on the impression of environmental sounds. *Proceedings of Inter-Noise*, Fort Lauderdale, USA.

——, 2006, Comparison of the effects of verbal versus visual information about sound sources on the perception of environmental sounds. *Acustica united with Acta Acustica*, 92, 51–60.

AEAT, 2004, *IMAGINE – State of the Art*. Deliverable 2 of the IMAGINE project – Improved Methods for the Assessment of the Generic Impact of Noise in the Environment. Document identity: IMA10TR–040423-AEATNL32.

AES, 2001, AES–4id–2001: AES information document for room acoustics and sound reinforcement systems – characterisation and measurement of surface scattering uniformity. *Journal of the Audio Engineering Society*, 49, 149–65.

AILMAN, C.M., 1978, A light-weight helium-filled noise barrier. *Proceedings of Inter-Noise*, San Francisco, USA.

ALFREDSON, R.J. and DU, X., 1995, Special shapes and treatment for noise barriers. *Proceedings of Inter-Noise*, Newport Beach, USA.

ALI, S.A. and TAMURA, A., 2002, Road traffic noise mitigation strategies in Greater Cairo, Egypt. *Applied Acoustics*, 63, 1257–65.

——, 2003, Road traffic noise levels, restrictions and annoyance in Greater Cairo, Egypt. *Applied Acoustics*, 64, 815–23.

AMRAM, M., CHVROJKA, V.J., and DROIN, L., 1987, Phase reversal barriers for better noise control at low frequencies: laboratory versus field measurements. *Noise Control Engineering Journal*, 28, 16–23.

ANDERSON, J., SHIERS, D., and SINCLAIR, M., 2002, *The Green Guide to Specification* (Oxford: Blackwell).

ANDERSON, L.M., MULLIGAN, B.E., GOODMAN, L.S., and REGEN, H.Z., 1983, Effects of sounds on preferences for outdoor settings. *Environment and Behavior*, 15, 539–66.

ANSI, 1977, S3.14, *Rating Noise with Respect to Speech Intelligibility*. American National Standard Institute, New York, USA.

——, 1994, S.1, *Acoustical Terminology*. American National Standard Institute, New York, USA.

——, 1997, S3.5, *Methods for the Calculation of the Articulation Index*. American National Standard Institute, New York, USA.

——, 1999a, S1.26, *Method for the Calculation of the Absorption of Sound by the Atmosphere*. American National Standard Institute, New York, USA.

——, 1999b, S12.2, *Criteria for Evaluating Room Noise* American National Standard Institute, New York, USA.

ANSI, 2003a, S12.8, *Methods for Determining the Insertion Loss of Outdoor Noise Barriers*. American National Standard Institute, New York, USA.

——, 2003b, S12.9, Part 1: *Qualities and Procedures for Description and Measurement of Environmental Sound*. Part 2 (2003): *Measurement of Long-Term, Wide-Area Sound*. Part 3 (2003): *Short-Term Measurements with an Observer Present*. Part 4 (1996): *Noise Assessment and Prediction of Long-Term Community Response*. Part 5 (2003): *Sound Level Descriptors for Determination of Compatible Land Use*. American National Standard Institute, New York, USA.

APFEL, R.E., 1998, *Deaf Architects and Blind Acousticians – A Guide to the Principles of Sound Design* (New Haven: Apple Enterprises Press).

ARANA, M. and GARCÍA, A., 1998, A social survey on the effects of environmental noise on the residents of Pamplona, Spain. *Applied Acoustics*, 53, 245–53.

ASDRUBALI, F. and COTANA, F., 2000, Influence of filtering system on high sound insulation ventilating windows. *Proceedings of Inter-Noise*, Nice, France.

ATKINS, 2003, *The London Road Traffic Noise Map*. Atkins Noise and Vibration. Also see www.londonnoisemap.com

ATTENBOROUGH, K., 1988, Review of ground effects on outdoor sound propagation from continuous broadband sources. *Applied Acoustics*, 24, 289–319.

——, 1992, Ground parameter information for propagation modelling. *Journal of the Acoustical Society of America*, 92, 418–27.

ATTENBOROUGH, K. and LI, K.M., 1997, Ground effect for A-weighted noise in the presence of turbulence and refraction. *Journal of the Acoustical Society of America*, 102, 1013–22.

AXELSSON, Ö., BERGLUND, B., and NILSSON, M.E., 2003, Towards green labelling of soundscapes in residential areas. *Proceedings of Euro-Noise*, Naples, Italy.

AYLOR, D., 1972, Sound transmission through vegetation in relation to leaf area, density, leaf width, and breadth of canopy. *Journal of the Acoustical Society of America*, 51, 411–14.

AYLOR, D.E. and MARKS, L.E., 1976, Perception of noise transmitted through barriers. *Journal of the Acoustical Society of America*, 59, 397–400.

BACON, E.N., 1975, *Design of Cities* (revised edn) (London: Thames and Hudson).

BAI, Y., 2005, *Acoustic Comfort in Renal Dialysis Units*. BArch dissertation, School of Architecture, University of Sheffield, UK.

BALL, L.M., 1942, Air raid siren field tests. *Journal of the Acoustical Society of America*, 13, 10–13.

BALLAS, J.A., 1993, Common factors in the identification of an assortment of brief everyday sounds. *Journal of Experimental Psychology: Human Perception and Performance*, 19, 250–67.

BARRIGÓN, J.M., GÓMEZ, V., MÉNDEZ, J.A., VÍLCHEZ-GÓMEZ, R., and TRUJILLO, J., 2002, An environmental noise study in the city of Cáceres, Spain. *Applied Acoustics*, 63, 1061–70.

BARRIGÓN-MORILLAS, J.M., GÓMEZ-ESCOBAR, V., VAQUERO, J.M., MÉNDEZ-SIERRA, J.A., and VÍLCHEZ-GÓMEZ, R., 2005, Measurement of noise pollution in Badajoz city, Spain. *Acustica united with Acta Acustica*, 91, 797–801.

BARRON, M., 1983, Auditorium acoustic modelling now. *Applied Acoustics*, 16, 279–90.

BATEMAN, I.J., DAY, B., LAKE, I., and LOVETT, A.A., 2001, *The Effect of Road Traffic on Residential Property Values: A Literature Review and Hedonic Pricing Study*. Report commissioned by the Scottish Executive Development Department, Edinburgh.

BEGAULT, D.R., 1990, *3-D Sound for Virtual Reality and Multimedia* (San Diego, CA: Academic Press Professional).

BEIDL, C.V. and STÜCKLSCHWAIGER, W., 1997, Application of the AVL-annoyance index for engine noise quality development. *Acustica united with Acta Acustica*, 83, 789–95.

BERANEK, L.L., 1954, *Acoustics* (New York: McGraw-Hill).

——, 1957, Revised criteria for noise in buildings. *Noise Control*, 3, 19–27.

——, 1989, Balanced noise criterion curves. *Journal of the Acoustical Society of America*, 86, 650–64.

BERANEK, L.L., BLAZIER, W.E., and FIGWER, J.J., 1971, Preferred noise criterion (PNC) curves and their application to rooms. *Journal of the Acoustical Society of America*, 50, 1223–8.

BERGLUND, B., 1998, Community noise in a public health perspective. *Proceedings of Inter-Noise*, Christchurch, New Zealand.

BERGLUND, B. and NILSSON, M., 1998, Total annoyance models for community noises explicated. *Proceedings of Inter-Noise*, Christchurch, New Zealand.

BERGLUND, B., BERGLUND, U., and LINDVALL, T., 1975, Scaling loudness, noisiness, and annoyance of aircraft noise. *Journal of the Acoustical Society of America*, 57, 930–4.

——, 1976, Scaling loudness, noisiness, and annoyance of community noises. *Journal of the Acoustical Society of America*, 60, 1119–25.

BERGLUND, B., BERGLUND, U., GOLDSTEIN, M., and LINDVALL, T., 1981, Loudness (or annoyance) summation of combined community noises. *Journal of the Acoustical Society of America*, 70, 1628–34.

BERGLUND, B., LINDVALL, T., and SCHWELA, D.H., 1999, *Guidelines for Community Noise*. World Health Organization report.

BERGLUND, B., ERIKSEN, C.A., and NILSSON, M.E., 2001, Perceptual characterization of soundscapes in residential areas. *Proceedings of the 17th International Conference on Acoustics (ICA)*, Rome, Italy.

BÉRILLON, J. and KROPP, W., 2000, A theoretical model to consider the influence of absorbing surfaces inside the cavity of balconies. *Acustica united with Acta Acustica*, 86, 485–94.

BERTONI, D., FRANCHINI, A., MAGNONI, M., TARTONI, P., and VALLET, M., 1993, Reaction of people to urban traffic noise in Modena, Italy. *Proceedings of the 6th Congress on Noise as a Public Health Problem, Noise, and Man*, Nice, France.

BIES, D.A. and HANSEN, C.H., 2003, *Engineering Noise Control: Theory and Practice* (3rd edn) (London: E&FN Spon).

BJORKMAN, M. and RYLANDER, R., 1996, Maximum noise levels in road traffic noise. *Proceedings of Inter-Noise*, Liverpool, UK.

BLAUERT, J. and JEKOSCH, U., 1997, Sound-quality evaluation – a multi-layered problem. *Acustica united with Acta Acustica*, 83, 747–53.

BLUMRICH, R. and HEIMANN, D., 2002, A linearized Eulerian sound propagation model for studies of complex meteorological effects. *Journal of the Acoustical Society of America*, 112, 446–55.

BODDEN, M., 1997, Instrumentation for sound quality evaluation. *Acustica united with Acta Acustica*, 83, 775–83.

BODDEN, M. and HEINRICHS, R., 2001, Moderators of sound quality of complex sounds with multiple tonal components. *Proceedings of the 17th International Congress on Acoustics (ICA)*, Rome, Italy.

BÖRGERS, C., GREENGARD, C., and THOMANN, E., 1992, The diffusion limit of free molecular flow in thin plane channels. *The SIAM Journal on Applied Mathematics*, 52, 1057–75.

BOTTELDOOREN, D., 1994, Acoustical finite-difference time-domain simulation in a quasi-Cartesian grid. *Journal of the Acoustical Society of America*, 95, 2313–19.

BOTTELDOOREN, D. and VERKEYN, A., 2002, Fuzzy models for accumulation of reported community noise annoyance from combined sources. *Journal of the Acoustical Society of America*, 112, 1496–508.

BOTTELDOOREN, D., VERKEYN, A., and LERCHER, P., 2001, How can we distinguish exposure and expectation effects in integrated soundscape analyses? *Proceedings of the 17th International Congress on Acoustics (ICA)*, Rome, Italy.

BOUGDAH, H., EKICI, I., and KANG, J., 2006, An investigation into rib-like noise reducing devices. *Journal of the Acoustical Society of America* (provisionally accepted).

BRADLEY, J.S. and JONAH, H., 1979, The effects of site selected variables on human response to traffic noise. Part I: type of housing by traffic noise level. *Journal of Sound and Vibration*, 66, 589–604.

BRADLEY, M., 2005, *The Acoustics of the Octagon Centre*. BArch dissertation, School of Architecture, University of Sheffield, UK.

BRISTOW, A.L. and WARDMAN, M., 2005, Valuing aircraft noise: influential variables. *Proceedings of Inter-Noise*, Rio de Janeiro, Brazil.

BROWN, A.L. and LAM, K.C., 1987, Urban noise surveys. *Applied Acoustics*, 20, 23–39.

BSI, 1992, British Standard BS EN 20140.10: *Measurement of Sound Insulation in Buildings and of Building Elements: Laboratory Measurement of Airborne Sound Insulation of Small Building Elements*. British Standards Institution, London, UK.

——, 1997, British Standard BS4142: *Method for Rating Industrial Noise Affecting Mixed Residential and Industrial Areas*. British Standards Institution, London, UK.

——, 2004, British Standard BS EN 13141–1: *Ventilation for Buildings – Performance Testing of Components/Products for Residential Ventilation*. Part 1: *Externally and Internally Mounted Air Transfer Devices*. British Standards Institution, London, UK.

BULL, M., 2000, *Sounding out the City: Personal Stereos and the Management of Everyday Life* (London: Berg Publishers).

BULLEN, R., 1979, Statistical evaluation of the accuracy of external sound level predictions arising from models. *Journal of Sound and Vibration*, 65, 11–28.

BULLEN, R. and FRICKE, F., 1976, Sound propagation in a street. *Journal of Sound and Vibration*, 46, 33–42.

BUNDESKABINETT, 1998, *TA Lärm – Technische Anleitung zum Schutz gegen Lärm*. Berlin, Germany.

BURNS, S.H., 1979, The absorption of sound by pine trees. *Journal of the Acoustical Society of America*, 65, 658–61.

BUTLER, G.F., 1974, A note on improving the attenuation given by a noise barrier. *Journal of Sound and Vibration*, 32, 367–9.

CARLES, J.L., BERNALDEZ, F.G., and DE LUCIO, J.V., 1992, Audiovisual interactions and soundscape preferences. *Landscape Research*, 17, 52–6.

CARLES, J.L., BARRIO, I.L., and DE LUCIO, J.V., 1999, Sound influence on landscape values. *Landscape and Urban Planning*, 43, 191–200.

CHEN, B., 2002, *Acoustic Comfort in Shopping Centre Atrium Spaces – A Case Study in Sheffield Meadowhall*. MSc dissertation, School of Architecture, University of Sheffield, UK.

CHEN, B. and KANG, J., 2003, Acoustic comfort in shopping centre atrium spaces. *Architectural Science Review*, 47, 107–14.

CHIEN, C.F. and CARROLL, M.M., 1980, Sound source above a rough absorbent plane. *Journal of the Acoustical Society of America*, 67, 827–9.

CHINA EPA (Environmental Protection Agency), 1995, *Conspectus of Environmental Protection in the 21st Century in China* (Beijing: China Environmental Sciences Press).

CHOURMOUZIADOU, K. and KANG, J., 2003, Effects of surface characteristics on the acoustic environment of ancient outdoor performance spaces. Research Symposium: Acoustic Characteristics of Surfaces: Measurement, Prediction and Applications. *Proceedings of the Institute of Acoustics (IOA)*, London, UK.

——, 2006, Simulation of boundary diffraction in open-air theatres. *Proceedings of the Auditorium Acoustics 2006*, Copenhagen, Denmark.

CHRISTIE, D.J. and GLICKMAN, C.D., 1980, The effects of classroom noise on children: evidence for sex difference. *Psychology in the Schools*, 17, 405–8.

CHRISTOPHERS, C., 2003, *Acoustics in Spaces of Worship – A Case Study in Sheffield*. BArch dissertation, School of Architecture, University of Sheffield, UK.

CHUNG, A., 2004, *Acoustics Comfort in Train Stations – Case Studies in Sheffield and Derby Station*. BArch dissertation, School of Architecture, University of Sheffield, UK.

COGGER, N., 2003, Action on environmental noise – review of the 2002 autumn conference. *Acoustics Bulletin*, 28(1), 8–19.

COHEN, M.F. and GREENBERG, D.P., 1985, The hemi-cube: a radiosity solution for complex environments. *Computer Graphics*, 19, 31–40.

COMSOL, A.B., 2004, *FEMLAB User's Guide*, Sweden.

CORBIN, A., 1998, *Village Bells: Sound and Meaning in the 19th Century French Countryside* (New York: Columbia University Press).

COTANA, F., 1999, Experimental data and performances of new high sound insulation ventilating windows. *Proceedings of Inter-Noise*, Florida, USA.

COX, T.J. and D'ANTONIO, P., 2004, *Acoustic Absorbers and Diffusers: Theory, Design and Application* (London: Spon Press).

CROOME, D.J., 1977, *Noise, Building and People* (Oxford: Pergamon Press).

CUMMINGS, A., 1992, The effects of a resonator array on the sound field in a cavity. *Journal of Sound and Vibration*, 154, 25–44.

DAIGLE, G. and STINSON, M., 2005, Attenuation of environmental noise: comparing the predictions of ISO 9613–2 and theoretical models. *Proceedings of Inter-Noise*, Rio de Janeiro, Brazil.

DAL, 1987, *Joint Nordic Method: Environmental Noise from Industrial Plants – General Prediction Method*. Danish Acoustical Laboratory Technical Report 32.

D'ANTONIO, P. and KONNERT, J., 1984, The reflection phase grating diffuser: design theory and application. *Journal of the Audio Engineering Society*, 32, 228–38.

DATAKUSTIK, 2005, *Cadna/A for Windows – User Manual*, Munich.

DAVID, M. and HESSLER, P.E., 2003, *Summary of Noise Modelling Methodology and Results – Report for BP Cherry Point Cogeneration* (Virginia: Hessler Associates Inc.).

DAVIES, H.G., 1978, Multiple-reflection diffuse-scattering model for noise propagation in streets. *Journal of the Acoustical Society of America*, 64, 517–21.

DAVIES, W.J., HEPWORTH, P., MOORHOUSE, A., and OLDFIELD, R., 2005, *Noise from Pubs and Clubs*. Report for the UK Department for Environment, Food and Rural Affairs (DEFRA).

DAVISON, B., 1957, *Neutron Transport Theory* (London: Oxford University Press).

DAY, B.F., FORD, R.D., and LORD, P., 1969, *Building Acoustics* (London: Elsevier).

DE BERG, M., VAN KREVELD, M., OVERMARS, M., and SCHWARZKOPF, O., 1997, *Computational Geometry* (Berlin: Springer).

DE COENSEL, B., BOTTELDOOREN, D., and DE MUER, T., 2003, 1/f noise in rural and urban soundscapes. *Acustica united with Acta Acustica*, 89, 287–95.

DE COENSEL, B., DE MUER, T., YPERMAN, I., and BOTTELDOOREN, D., 2005, The influence of traffic flow dynamics on urban soundscapes. *Applied Acoustics*, 66, 175–94.

DE MUER, T., 2005, *Policy Supporting Tools for Urban Noise Assessment*. PhD dissertation, Department of Information Technology, University of Ghent, Belgium.

DE RUITER, E., 2000, Noise control in the compact city. *Proceedings of the 7th International Congress on Sound and Vibration (ICSV)*, Garmisch-Partenkirchen, Germany.

——, 2004, *Reclaiming Land from Urban Traffic Noise Impact Zones – The Great Canyon*. PhD dissertation, Technical University of Delft, The Netherlands.

DE SALIS, M.H.F., OLDHAM, D.J., and SHARPLES, S., 2002, Noise control strategies for naturally ventilated buildings. *Building and Environment*, 37, 471–84.

DELANY, M.E., RENNIE, A.J., and COLLINS, K.M., 1978, Scale model investigations of traffic noise propagation. *Journal of Sound and Vibration*, 56, 325–40.

DI CARLO, R., 2005, *Acoustics in the Crucible Theatre, Sheffield*. BArch dissertation, School of Architecture, University of Sheffield, UK.

DICKINSON, P., 1993, *Noise Assessment and Management in Indonesia*. Report SEA/EH 461, Regional Office for South-East Asia, World Health Organization, New Delhi, India.

DIETZE, L., 2000, Learning is living: acoustic ecology as pedagogical ground – a report on experience. *Soundscape: The Journal of Acoustic Ecology*, 1, 20–1.

DIN, 1987, *DIN 18005: Berücksichtigung des Schallschutzes im Städtebau*. Deutsches Institut für Normung (German Institute for Standardization), Berlin.

DONAVAN, P.R., 1976, *Sound Propagation in Urban Spaces*. PhD dissertation, Massachusetts Institute of Technology, Cambridge, USA.

DRUMM, I.A. and LAM, Y.M., 2000, The adaptive beam-tracing algorithm. *Journal of the Acoustical Society of America*, 107, 1405–12.

DU, Z., 2002, *Acoustic Comfort in Library Reading Rooms*. MSc dissertation, School of Architecture, University of Sheffield, UK.

DUBOIS, D., 2000, Categories as acts of meaning: the case of categories in olfaction and audition. *Cognitive Science Quarterly*, 1, 35–68.

DUBOIS, D. and DAVID, S., 1999, A cognitive approach of urban soundscapes. *The 137th Meeting of the Acoustical Society of America and Forum Acusticum*, Berlin (abstract published in *Journal of Acoustical Society of America*, 105, 1281).

ECAC, 1997, *Report on Standard Method of Computing Noise Contours around Civil Airports*. European Civil Aviation Conference, ECAC.CEAC Doc. 29.

EGAN, M.D., 1988, *Architectural Acoustics* (New York: McGraw-Hill).

EKICI, I., 2004, *Road Traffic Noise Barrier Design: Measurements and Models Concerning Multiple–Walls and Augmented Earth Mounds*. PhD dissertation, Sheffield Hallam University, UK.

EKICI, I. and BOUGDAH, H., 2004, A review of research on environmental noise barriers. *Building Acoustics*, 10, 289–323.

ELDRED, K.M., 1975, Assessment of community noise. *Noise Control Engineering Journal*, 3, 88–95.

——, 1988, *Noise at the year 2000. Proceedings of the 5th International Congress on Noise as a Public Health Problem*, Stockholm.

ELLERMEIER, W., EIGENSTETTER, M., and ZIMMER, K., 2001, Psychoacoustic correlates of individual noise sensitivity. *Journal of the Acoustical Society of America*, 109, 1464–73.

EMBLETON, T.F.W., 1963, Sound propagation in homogeneous deciduous and evergreen woods. *Journal of the Acoustical Society of America*, 35, 1119–25.

EMBRECHTS, J.J., ARCHAMBEAU, D., and STAN, G., 2001, Determination of the scattering coefficient of random rough diffusing surfaces for room acoustics applications. *Acustica united with Acta Acustica*, 87, 482–94.

EU, 1996, *Future Noise Policy*. European Commission Green Paper, Brussels.

——, 2002, *Directive (2002/49/EC) of the European Parliament and of the Council – Relating to the Assessment and Management of Environmental Noise*.

EVANS, B., 1994, Windows as climate modifiers. *The Architects' Journal*, 31.

FASTL, H., 2001, Neutralizing the meaning of sound for sound quality evaluation. *Proceedings of the 17th International Congress on Acoustics (ICA)*, Rome, Italy.

FASTL, H. and ZWICKER, E., 1987, Lautstärkepegel bei 400 Hz: psychoakustische Messung und Berechnung nach ISO 532B. *Proceedings of DAGA*.

FIDELL, S., 1978, Nationwide urban noise survey. *Journal of the Acoustical Society of America*, 64, 198–206.

FIDELL, S., SILVATI, L., and PEARSONS, K., 1998, Noticeability of a decrease in aircraft noise. *Noise Control Engineering Journal*, 46, 49–56.

FIELD, A., 2000, *Discovering Statistics Using SPSS for Windows* (London: SAGE).

FIELD, C.D. and FRICKE, F.R., 1998, Theory and applications of quarter-wave resonators: a prelude to their use for attenuating noise entering buildings through ventilation openings. *Applied Acoustics*, 53, 117–32.

FIELDS, J.M., 1984, The effect of numbers of noise events on people's reactions to noise: an analysis of existing survey data. *Journal of the Acoustical Society of America*, 75, 447–67.

——, 1993, Effect of personal and situational variables on noise annoyance in residential areas. *Journal of the Acoustical Society of America*, 93, 2753–63.

——, Reactions to environmental noise in an ambient noise context in residential areas. *Journal of the Acoustical Society of America*, 104, 2245–60.

FIELDS, J.M. and WALKER, J.G., 1982, The response to railway noise in residential areas in Great Britain. *Journal of Sound and Vibration*, 85, 177–255.

FIELDS, J.M., DEJONG, R.G., BROWN, A.L., FLINDELL, I.H., GJESTLAND, T., JOB, R.F.S., KURRA, S., LERCHER, P., SCHUEMER-KOHRS, A., VALLET, M., and YANO, T., 1997, Guidelines for reporting core information from community noise reaction surveys. *Journal of Sound and Vibration*, 206, 685–95.

FINEGOLD, L.S., VON GIERKE, H.E., MCKINLEY, R.L., and SCHOMER, P.D., 1998, Addressing the effectiveness of noise control regulations and policies. *Proceedings of Noise Effects '98*, Sydney, Australia.

FLETCHER, H. and MUNSON, W.A., 1933, Loudness, its definition, measurement and calculation. *Journal of the Acoustical Society of America*, 5, 82–108.

FLUENT, 2005, *Fluent 5 Manual*, Lebanon, NH.

FOLEY, J.D., VAN DAM, A., FEINER, S.K., and HUGHES, J.F., 1990, *Computer Graphics*: *Principle and Practice* (2nd edn) (Reading, MA: Addison-Wesley Publishing Company).

FRANSSEN, E.A.M., VAN DONGEN, J.E.F., RUYSBROEK, J.M.H., VOS, H., and STELLATO, R., 2004, *Noise Annoyance and Perceived Environmental Quality – Inventory 2003*. RIVM Report 815120001, The Netherlands.

FRUSTHORFER, B., 1983, Daytime noise stress and subsequent night sleep: interference with sleep patterns, endocrine function and serotoninergic system. *Proceedings of the 4th International Congress on Noise as a Public Health Problem*, Turin, Italy.

FUJIMOTO, K., KATO, T., and UENO, R., 1998, Human response to soundscape in Fukuoka. *Proceedings of Inter-Noise*, Christchurch, New Zealand.

FUJIWARA, K. and FURUTA, N., 1991, Sound shielding efficiency of a barrier with a cylinder at the edge. *Noise Control Engineering Journal*, 37, 5–11.

FUJIWARA, K., OHKUBO, T., and OMOTO, A.A., 1995, Note on the noise shielding efficiency of a barrier with absorbing obstacle at the edge. *Proceedings of Inter-Noise*, Newport Beach, USA.

FUJIWARA, K., HOTHERSALL, D.C., and KIM, C., 1998, Noise barriers with reactive surfaces. *Applied Acoustics*, 53, 255–72.

FUNKHOUSER, T., CARLBOM, I., ELKO, G., PINGALI, G., SONDHI, M., and WEST, J., 1998, A beam tracing approach to acoustic modeling for interactive virtual environments. *Proceedings of the 25th International Conference on Computer Graphics and Interactive Techniques*, Orlando, USA.

GALLOWAY, W., ELDRED, K., and SIMPSON, M., 1974, *Population Distribution of the United States as a Function of Outdoor Noise*. US Environmental Protection Agency Report No. 550/9–74–009, Washington, DC, USA.

GAVER, W., 1993, What in the world do we hear? An ecological approach to auditory event perception. *Ecological Psychology*, 5, 1–29.

GENUIT, K., 2001, The problem of predicting noise annoyance as a function of distance. *Proceedings of the 17th International Congress on Acoustics (ICA)*, Rome, Italy.

GHARABEGIAN, A., 1995, Improving sound wall performance using route silent. *Proceedings of Inter-Noise*, Newport Beach, USA.

GIFFORD, R., 1996, *Environment Psychology* (Boston: Allyn and Bacon).

GILBERT, K. and DI, X., 1993, A fast Green's function method for one-way sound propagation in the atmosphere. *Journal of the Acoustical Society of America*, 94, 2343–52.

GIRDNER, J.H., 1897, To abate the plague of city noises. *North American Review*, 165, 463.

GJESTLAND, T., 1998, Regional differences in noise annoyance assessments. *Proceedings of Inter-Noise*, Christchurch, New Zealand.

GLASER, B. and STRAUSS, A., 1967, *The Discovery of Grounded Theory*: *Strategies For Qualitative Research* (Chicago: Aldine).

GOLDSTEIN, J., 1979, Descriptors of auditory magnitude and methods of rating community noise. In Peppin R.J. and Rodman C.W. (eds) *Community Noise* (Philadelphia: American Society for Testing and Materials), pp. 38–72.

GOMES, M.H.A. and GERGES, S.N.Y., 2001, Modelling of room acoustics parameters using MLS technique and numerical simulation. *Proceedings of 7th International IBPSA Conference*, Rio de Janeiro, Brazil.

GOTTLOB, D., 1995, Regulations for community noise. *Noise News International*, December, 223–36.

GRIFFIN, M.J. and HOWARTH, H.V.C., 1990, Subjective response to combined noise and vibration: summation and interaction effects. *Journal of Sound and Vibration*, 143, 443–54.

GRIFFITHS, I.D., LANGDON, F.J., and SWAN, M.A., 1980, Subjective effects of traffic noise exposure: reliability and seasonal effects. *Journal of Sound and Vibration*, 71, 227–40.

GUASTAVINO, C. and CHEMINÉE, P., 2004, A psycholinguistic approach to the ecological validity of experimental settings: the case of low frequency perception. *Food Quality and Preference*, 15, 884–6.

GUASTAVINO, C. and KATZ, B., 2004, Perceptual evaluation of multi-dimensional spatial audio reproduction. *Journal of Acoustical Society of America*, 116, 1105–15.

GUASTAVINO, C., KATZ, B., POLACK, J.D., LEVITIN, D., and DUBOIS, D., 2005, Ecological validity of soundscape reproduction. *Acustica united with Acta Acustica*, 91, 333–41.

GULIAN, E. and THOMAS, J.R., 1986, The effects of noise, cognitive set and gender on mental arithmetic performance. *British Journal of Psychology*, 77, 503–11.

GUSKI, R., 1997, Psychological methods for evaluating sound quality and assessing acoustic information. *Acustica united with Acta Acustica*, 83, 765–74.

——, 1998, Psychological determinants of train noise annoyance. *Proceedings of Euro-Noise*, Munich, Germany.

HALL, F.H., BIRNIE, S., TAYLOR, S., and PALMER, J., 1981, Direct comparison of community response to road traffic noise and to aircraft noise. *Journal of the Acoustical Society of America*, 70, 1690–8.

HARRIS, C.M., 1966, Absorption of sound in air versus humidity and temperature. *Journal of the Acoustical Society of America*, 40, 148–59.

HARVEY, L., 2000, Australian forum for acoustic ecology. *Soundscape: The Journal of Acoustic Ecology*, 1, 4.

HASHIMOTO, T. and HATANO, S., 2001, Effect of factors other than sound to the perception of sound quality. *Proceedings of the 17th International Congress on Acoustics (ICA)*, Rome, Italy.

HATANO, S., HASHIMOTO, T., KIMURA, Y., and TANAKA, T., 2001, Sound quality evaluation of construction machine. *Proceedings of the 17th International Congress on Acoustics (ICA)*, Rome, Italy.

HAWKINS, R., 1999, *Review of Studies on External Costs of Noise*. Report for the Environmental Protection Economics Division of the (former) UK Department for Environment, Transport and the Regions (DETR).

HECHT-NIELSEN, R., 1990, *Neurocomputers* (Reading, MA: Addison-Wesley Publishing Company).

HECKBERT, P.S. and HANRAHAN, P., 1984, Beam tracing polygonal objects. *Proceedings of the 11th International Conference on Computer Graphics and Interactive Techniques*, Minneapolis, USA.

HEISLER, G.M., MCDANIEL, O.H., HODGDON, K.K., PORTELLI, J.J., and GLEASON, S.B., 1987, Highway noise abatement in two forests. *Proceedings of Noise-Con – National Conference on Noise Control Engineering*, New York, USA.

HELLBRÜCK, J., KATO, T., ZEITLER, A., SCHICK, A., KUWANO, S., and NAMBA, S., 2001, Loudness scaling of traffic noise: perceptual and cognitive factors. *Proceedings of the 17th International Congress on Acoustics (ICA)*, Rome, Italy.

HELLER, O., 1985, Hörfeldaudiometrie mit dem Verfahren der Kategorienunterteilung (KU). *Psychologische Beiträge*, 27, 478–93.

HEUTSCHI, K., 1995, Computermodell zur Berechnung von Bebauungszuschlägen bei Straßenverkehrslärm. *Acustica united with Acta Acustica*, 81, 26–35.

HINTON, J., 2000, *A Report on the Production of Noise Maps of the City of Birmingham.* UK Department for Environment, Food and Rural Affairs (DEFRA), London.

HINTON, J. and BLOOMFIELD, A., 2000, Local noise mapping: the future? *Proceedings of the Institute of Acoustics (IOA),* UK.

HIRAMATSU, K., 1999, Activities and impacts of soundscape association of Japan. *Proceedings of Inter-Noise,* Fort Lauderdale, USA.

HIRAMATSU, K., MATSUI, T., and MINOURA, K., 2001, Environment similarity index concerning sonic environment – toward the evaluation of sonic environment. *Proceedings of the 17th International Congress on Acoustics (ICA),* Rome, Italy.

HO, S.S.T., BUSH-VISHNIAC, I.J., and BLACKSTOCK, D.T., 1997, Noise reduction by a barrier having random edge profile. *Journal of the Acoustical Society of America,* 100, 2669–76.

HODGSON, M.R., 1991, Evidence of diffuse surface reflections in rooms. *Journal of the Acoustical Society of America,* 89, 765–71.

HODGSON, M.R. and ORLOWSKI, R.J., 1987, Acoustic scale modelling of factories. Part I: background, instrumentation, and procedures. *Journal of Sound and Vibration,* 113, 29–46.

HONG KONG EPD, 2003, *Guidelines on Design of Noise Barriers.* Hong Kong Environmental Protection Department, Highways Department, Government of the Hong Kong SAR.

HOROSHENKOV, K.V., HOTHERSALL, D.C., and MERCY, S.E., 1999, Scale modelling of sound propagation in a city street canyon. *Journal of Sound and Vibration,* 223, 795–819.

HOSSAM EL-DIEN, H. and WOLOSZYN, P., 2004, Prediction of the sound field into high-rise building facades due to its balcony ceiling form. *Applied Acoustics,* 65, 431–40.

HOTHERSALL, D.C. and TOMLINSON, S.A., 1997, Effects of high-sided vehicles on the performance of noise barriers. *Journal of the Acoustical Society of America,* 102, 998–1003.

HOTHERSALL, D.C., CHANDLER-WILDE, S.N., and HAJMIRZAE, N.M., 1991a, Efficiency of single noise barriers. *Journal of Sound and Vibration,* 146, 303–22.

HOTHERSALL, D.C., CROMBIE, D.H., and CHANDLER-WILDE, S.N., 1991b, The performance of T-profile and associated noise barriers. *Applied Acoustics,* 32, 269–87.

HOTHERSALL, D.C., HOROSHENKOV, K.V., and MERCY, S.E., 1996, Numerical modelling of the sound field near a tall building with balconies near a road. *Journal of Sound and Vibration,* 198, 507–15.

HOUTGAST, T. and STEENEKEN, H.J.M., 1973, The modulation transfer function in room acoustics as a predictor of speech intelligibility. *Acustica,* 28, 66–73.

——, 1985, A review of the MTF concept in room acoustics and its use for estimating speech intelligibility in auditoria. *Journal of the Acoustical Society of America,* 77, 1069–77.

HOWARD, D.M. and ANGUS, J., 1996, *Acoustics and Psychoacoustics* (Oxford: Focal Press).

HTOURIS, S., 2001, A comparative interpretation of soundscape and noise. *Proceedings of the 17th International Congress on Acoustics (ICA),* Rome, Italy.

HUANG, J., 2003, *Accuracy and Efficiency in Noise-Mapping.* MSc dissertation, School of Architecture, University of Sheffield, UK.

HUANG, Z., 2004, *Environmental Noise Situation of Residential Areas and the Change with City Expansion in Xi'an City, China.* MArch dissertation, School of Architecture, University of Sheffield, UK.

HUFSCHMIDT, M.M., JAMES, D.E., MEISTER, A.S., BROWEN, B.T., and DIXON, J.A., 1983, *Environment, Natural Systems, and Development: An Economic Valuation Guide* (Baltimore: Johns Hopkins University Press).

HUISMAN, W.H.T. and ATTENBOROUGH, K., 1991, Reverberation and attenuation in a pine forest. *Journal of the Acoustical Society of America,* 90, 2664–77.

HUTCHINS, D.A., JONES, H.W., and RUSSELL, L.T., 1984a, Model studies of barrier performance in the presence of ground surfaces. Part I: thin perfectly reflecting barriers. *Journal of the Acoustical Society of America,* 75, 1807–16.

——, 1984b, Model studies of barrier performance in the presence of ground surfaces. Part II: different shapes, *Journal of the Acoustical Society of America,* 75, 1817–26.

IIDA, K., KONDOH, Y., and OKADO, Y., 1984, Research on a device for reducing noise. In *Transportation Research Record 983* (Washington, DC: Transportation Research Board), pp. 51–4.

INGARD, U., 1953, A review of the influence of meteorological conditions on sound propagation. *Journal of the Acoustical Society of America*, 25, 405–11.

IP, G., 2005, *Acoustic Comfort Evaluation in Conventional Office Spaces and Relationship with Health and Safety*. BArch dissertation, School of Architecture, University of Sheffield, UK.

IRWIN, J.D. and GRAF, E.R., 1979, *Industrial Noise and Vibration Control* (Englewood Cliffs, NJ: Prentice-Hall).

ISEI, T., 1980, Absorptive noise barrier on finite impedance ground. *Journal of the Acoustical Society of Japan*, 1, 3–10.

ISEI, T., EMBLETON, T.F.W., and PIERCY, J.E., 1980, Noise reduction by barriers on finite impedance ground. *Journal of the Acoustical Society of America*, 67, 46–58.

ISHIMARU, A., 1997, *Wave Propagation and Scattering in Random Media* (New York: IEEE Press and Oxford University Press).

ISMAIL, M.R. and OLDHAM, D.J., 2005, A scale model investigation of sound reflection from building façades. *Applied Acoustics*, 66, 149–73.

ISO, 1993, ISO 9613: *Attenuation of Sound during Propagation Outdoors*. Part 1 (1993): *Calculation of the Absorption of Sound by the Atmosphere*. Part 2 (1996): *General Method of Calculation*. International Organization for Standardization, Genève, Switzerland.

——, 1994a, ISO 3744: *Determination of Sound Power Levels of Noise Using Sound Pressure – Engineering Method in an Essentially Free Field over a Reflecting Plane*. International Organization for Standardization, Genève, Switzerland.

——, 1994b, ISO 8297: *Determination of Sound Power Levels of Multi Source Industrial Plants for Evaluation of Sound Pressure Levels in the Environment – Engineering Method*. International Organization for Standardization, Genève, Switzerland.

——, 1995, ISO 3746: *Determination of Sound Power Levels of Noise Sources Using an Enveloping Measurement Surface over a Reflecting Plane*. International Organization for Standardization, Genève, Switzerland.

——, 1996, ISO 717–1: *Rating of Sound Insulation in Buildings and of Building Elements*: *Airborne Sound Insulation*. International Organization for Standardization, Genève, Switzerland.

——, 1997, ISO 3382: *Measurement of the Reverberation Time of Rooms with Reference to Other Acoustical Parameters*. International Organization for Standardization, Genève, Switzerland.

——, 1998, ISO 362: *Engineering Method for the Measurement of Noise Emitted by Accelerating Road Vehicles*. International Organization for Standardization, Genève, Switzerland.

——, 1999, ISO 3741: *Determination of Sound Power Levels of Noise Sources Using Sound Pressure – Precision Methods for Reverberation Rooms*. International Organization for Standardization, Genève, Switzerland.

——, 2000, ISO 3747: *Determination of Sound Power Levels of Noise Sources Using Sound Pressure – Comparison Method In Situ*. International Organization for Standardization, Genève, Switzerland.

——, 2003a, ISO 1996: *Acoustics – Description, Measurement and Assessment of Environmental Noise*. Part 1 (2003): *Basic Quantities and Assessment Procedures*. Part 2 (1998): *Acquisition of Data Pertinent to Land Use*. Part 3 (1987): *Application to Noise Limits*. International Organization for Standardization, Genève, Switzerland.

——, 2003b, ISO 226: *Acoustics – Normal Equal-Loudness-Level Contours*. International Organization for Standardization, Genève, Switzerland.

——, 2003c, ISO 3745: *Determination of Sound Power Levels of Noise Sources Using Sound Pressure – Precision Methods for Anechoic and Hemi-Anechoic Rooms*. International Organization for Standardization, Genève, Switzerland.

——, 2004, ISO 17497–1: *Acoustics – Sound-Scattering Properties of Surfaces – Part 1: Measurement of the Random-Incidence Scattering Coefficient in a Reverberation Room*. International Organization for Standardization, Genève, Switzerland.

IU, K.K. and LI, K.M., 2002, The propagation of sound in narrow street canyons. *Journal of the Acoustical Society of America*, 112, 537–50.

IWAMIYA, S. and YANAGIHARA, M., 1998, Features of the soundscape in Fukuoka city, a major city in Japan, recognized by foreign residents. *Proceedings of Inter-Noise*, Christchurch, New Zealand.

JAKOB, A. and MÖSER, M., 2003a, Active control of double-glazed windows – Part I: feedforward control. *Applied Acoustics*, 64, 163–82.

——, 2003b, Active control of double-glazed windows – Part II: feedback control. *Applied Acoustics*, 64, 183–96.

JANCZUR, R., WALERIAN, E., and CZECHOWICZ, M., 2001a, Sound levels forecasting for city-centres. Part III: a road lane structure influence on sound level within urban canyon. *Applied Acoustics*, 62, 493–512.

——, 2001b, Sound levels forecasting for city-centres. Part IV: vehicles steam parameters influence on sound level distribution within a canyon street. *Applied Acoustics*, 62, 645–64.

JÄRVILUOMA, H., 2000, Acoustic environments in change: five village soundscapes revisited. *Soundscape: The Journal of Acoustic Ecology*, 1, 25.

JIN, B.J., KIM, H.S., KANG, H.J., and KIM, J.S., 2001, Sound diffraction by a partially inclined noise barrier. *Applied Acoustics*, 62, 1107–21.

JOB, R.F.S., 1988, Community response to noise: a review of factors influencing the relationship between noise exposure and reaction. *Journal of the Acoustical Society of America*, 83, 991–1001.

JOB, R.F.S., HATFIELD, J., CARTER, N.L., PEPLOE, P., TAYLOR, R., and MORRELL, S., 1999, Reaction to noise: the roles of soundscape, enviroscape and psychscape. *Proceedings of Inter-Noise*, Fort Lauderdale, USA.

JOHNSON, M.E., ELLIOT, S.J., BAEK, K.-H., and GARCIA-BONITO, J., 1998, An equivalent source technique for calculating the sound field inside an enclosure containing scattering objects. *Journal of the Acoustical Society of America*, 104, 1221–31.

JONASSON, H., SANDBERG, U., VAN BLOKLAND, G., EJSMONT, J., WATTS, G., and LUMINARY, M., 2004, *Source Modelling of Road Vehicles*. Deliverable 9 of the Harmonoise Report, European Commission.

JONASSON, H.G. and STOREHEIER, S., 2001, *Nord 2000: New Nordic Prediction Method for Road Traffic Noise*. SP Rapport 2001:11.

JONES, D.L., 1994, Windows of change. *Building Services*, January, 26–27.

JONES, H.W., STREDULINSKY, D.C., and VERMEULEN, P.J., 1980, An experimental and theoretical study of the modelling of road traffic noise and its transmission in the urban environment. *Applied Acoustics*, 13, 251–65.

JONES, R.C., 1946, A fifty horsepower siren. *Journal of the Acoustical Society of America*, 18, 371–87.

JOPSON, I., 2002, The accuracy of noise mapping, *Acoustics Bulletin*, 27(5), 46–7.

JOYNT, J.L.R., 2005, *A Sustainable Approach to Environmental Noise Barrier Design*. PhD dissertation, School of Architecture, University of Sheffield, UK.

JOYNT, J.L.R. and KANG, J., 2002, The integration of public opinion and perception into the design of noise barriers. *Proceedings of the PLEA*, Toulouse, France.

——, 2003, The use of noise mapping, public participation and the full integration of community opinions, in the design of noise barriers to achieve a construction which serves not only objective reductions in noise but also subjective. *Proceedings of Euro-Noise*, Naples, Italy.

——, 2006, A customised lifecycle assessment model for noise barrier design. *Proceedings of the Institute of Acoustics (IOA)*, Southampton, UK.

KAHRS, M., 1998, *Applications of Digital Signal Processing to Audio and Acoustics* (Boston: Kluwer Academic Publishers).

KANG, J., 1988, Experiments on the subjective assessment of noise reduction by absorption treatments. *Chinese Noise and Vibration Control*, (5), 20–8 (in Chinese).

——, 1995, Experimental approach to the effect of diffusers on the sound attenuation in long enclosures. *Building Acoustics*, 2, 391–402.

KANG, J., 1996a, Acoustics in long enclosures with multiple sources. *Journal of the Acoustical Society of America*, 99, 985–9.

——, 1996b, Modelling of train noise in underground stations. *Journal of Sound and Vibration*, 195, 241–55.

——, 1996c, Reverberation in rectangular long enclosures with geometrically reflecting boundaries. *Acustica united with Acta Acustica*, 82, 509–16.

——, 1996d, Sound attenuation in long enclosures. *Building and Environment*, 31, 245–53.

——, 1996e, The unsuitability of the classic room acoustical theory in long enclosures. *Architectural Science Review*, 39, 89–94.

——, 1997a, A method for predicting acoustic indices in long enclosures. *Applied Acoustics*, 51, 169–80.

——, 1997b, Acoustics in long underground spaces. *Tunnelling and Underground Space Technology*, 12, 15–21.

——, 2000a, Modelling the acoustic environment in city streets. *Proceedings of the PLEA*, Cambridge, UK.

——, 2000b, Sound field in urban streets with diffusely reflecting boundaries. *Proceedings of the Institute of Acoustics (IOA)*, Liverpool, UK.

——, 2000c, Sound field resulting from diffusely reflecting boundaries: comparison between various room shapes. *Proceedings of the 7th International Congress on Sound and Vibration*, Garmisch-Partenkirchen, Germany.

——, 2000d, Sound propagation in street canyons: comparison between diffusely and geometrically reflecting boundaries. *Journal of the Acoustical Society of America*, 107, 1394–404.

——, 2001, Sound propagation in interconnected urban streets: a parametric study. *Environment and Planning B*: *Planning and Design*, 28, 281–94.

——, 2002a, *Acoustics of Long Spaces*: *Theory and Design Guide* (London: Thomas Telford Publishing).

——, 2002b, Computer simulation of the sound fields in urban squares: comparison between diffusely and geometrically reflecting boundaries. *Proceedings of the 32nd International Acoustical Conference (IAC) – European Acoustics Association (EAA) Symposium "ACOUSTICS BANSKÁ ŠTIAVNICA 2002"*, Slovakia.

——, 2002c, Effectiveness of architectural changes and urban design options on the noise reduction in urban streets with diffusely reflecting boundaries. *Proceedings of the 9th International Congress on Sound and Vibration*, Orlando, USA.

——, 2002d, Numerical modelling of the sound field in urban streets with diffusely reflecting boundaries. *Journal of Sound and Vibration*, 258, 793–813.

——, 2002e, Numerical modelling of the speech intelligibility in dining spaces. *Applied Acoustics*, 63, 1315–33.

——, 2002f, Prediction and improvement of the conversation intelligibility in dining spaces. *Proceedings of the Institute of Acoustics (IOA)*, Salford, UK.

——, 2002g, Reverberation in rectangular long enclosures with diffusely reflecting boundaries. *Acustica united with Acta Acustica*, 88, 77–87.

——, 2002h, Comparison of sound fields in regularly-shaped, long and flat enclosures with diffusely reflecting boundaries. *International Journal of Acoustics and Vibration*, 7, 165–71.

——, 2003a, Acoustic comfort in 'non-acoustic' buildings: a review of recent work in Sheffield. *Proceedings of the Institute of Acoustics (IOA)*, Oxford, UK.

——, 2003b, Acoustic comfort in urban open public spaces. *Il comfort ambientale nella progettazione degli spazi urbani*, Sesto San Giovanni, Italy.

——, 2004a, *Acoustic Simulation and Comfort in Urban Open Public Spaces*. Sub-final report for European Commission project RUROS – Rediscovering the urban realm and open spaces, School of Architecture, University of Sheffield, UK.

——, 2004b, Application of radiosity method in acoustic simulation. *Proceedings of the 18th International Conference on Acoustics (ICA)*, Kyoto, Japan.

——, 2005a, Numerical modelling of the sound fields in urban squares. *Journal of Acoustical Society of America*, 117, 3695–706.

——, 2005b, Urban acoustics – guest editorial. *Applied Acoustics*, 66, 121–2.

KANG, J. and BROCKLESBY, M.W., 2003, Application of micro-perforated absorbers in developing novel window systems for optimum acoustic, ventilation and daylighting performance. *Proceedings of the Institute of Acoustics (IOA)*, Oxford, UK.

——, 2004a, Design of acoustic windows with micro-perforated absorbers. *Proceedings of the 18th International Conference on Acoustics (ICA)*, Kyoto, Japan.

——, 2004b, Feasibility of applying micro-perforated absorbers in acoustic window systems. *Applied Acoustics*, 66, 669–89.

KANG, J. and DU, Z., 2003, Sound field and acoustic comfort in library reading rooms. *Proceedings of the 10th International Congress on Sound and Vibration*, Stockholm, Sweden.

KANG, J. and FUCHS, H.V., 1999, Predicting the absorption of open weave textiles and micro-perforated membranes backed by an airspace. *Journal of Sound and Vibration*, 220, 905–20.

KANG, J. and HUANG, J., 2002, *Noise-Mapping Case Studies*. Report for BP, School of Architecture, University of Sheffield, UK.

——, 2005, Noise-mapping: accuracy and strategic application. *Proceedings of Inter-Noise*, Rio de Janeiro, Brazil.

KANG, J. and LI, Z., 2006, Numerical simulation of an acoustic window system using finite element method. *Acustica united with Acta Acustica* (under revision).

KANG, J. and NEUBAUER, R., 2001, Predicting reverberation time: comparison between analytic formulae and computer simulation. *Proceedings of the 17th International Congress on Acoustics*, Rome, Italy.

KANG, J. and OLDHAM, D., 2003, Effects of trees and vegetations in street canyons. *En:abIE Meeting*, Hull, UK.

KANG, J. and YANG, W., 2002, Soundscape in urban open public spaces. *World Architecture*, 144, 76–9 (in Chinese).

KANG, J. and ZHANG, M., 2002, Semantic differential analysis on the soundscape of urban open public spaces. *Proceedings of the First Pan-American/Iberian Meeting on Acoustics*, Cancun, Mexico, published by the Iberoamericana De Acústica. (Abstract published in *Journal of the Acoustical Society of America*, 112, 2435.)

KANG, J. and ZHANG, M., 2003, Acoustic simulation and soundscape in urban squares. *Proceedings of the 10th International Congress on Sound and Vibration*, Stockholm, Sweden.

——, 2005, *Semantic Differential Analysis on the Soundscape of Open Urban Public Spaces: A Cross-Cultural Study*. British Academy project final report, School of Architecture, University of Sheffield, UK.

KANG, J., GRASBY, P., DERRICK, M., FRANKS, L., WILLIAMS, P., FLINDELL, I., and HARSHAM, K., 2001a, *Environmental Performance: Noise and Acoustic Management Engineering – Group Guidelines*. BP report, London.

KANG, J., TSOU, J.Y., and LAM, S., 2001b, Sound propagation in urban streets: comparison between the UK and Hong Kong. *Proceedings of the 8th International Congress on Sound and Vibration*, Hong Kong.

KANG, J., MENG, Y., and BROWN, G., 2003a, Sound propagation in micro-scale urban areas: simulation and animation. *Proceedings of Euro-Noise*, Naples, Italy.

KANG, J., YANG, W., and ZHANG, M., 2003b, Soundscape and acoustic comfort in urban open public spaces. *Proceedings of the 19th Annual Meeting of the International Society for Psychophysics*, Cyprus.

——, 2004, Sound environment and acoustic comfort in urban spaces. In Nikolopoulou M. (ed.) *Design Open Spaces in the Urban Environment: A Bioclimatic Approach* (Jointly published by the European Commission and the Centre for Renewable Energy Sources, Athens, Greece), pp. 32–6.

KANG, J., BROCKLESBY, M.W., LI, Z., and OLDHAM, D., 2005, An acoustic window for sustainable buildings. *The 149th Meeting of the Acoustical Society of America*, Vancouver, Cannada. (Abstract published in *Journal of the Acoustical Society of America*, 117, 2379.)

KAPLAN, S., 1987, Aesthetics, affect, and cognition – environmental preference from an evolutionary perspective. *Environment and Behavior*, 19, 3–32.

KARIEL, H.G., 1980, Mountaineers and the general public: a comparison of their evaluation of sounds in a recreational environment. *Leisure Sciences*, 3, 155–67.

KARLSSON, H., 2000, The acoustic environment as a public domain. *Soundscape: The Journal of Acoustic Ecology*, 1, 10–13.

KEELING-ROBERTS, S., 2001, *Creating Acoustic Atmosphere: Case Studies in English Football Stadia*. BArch dissertation, School of Architecture, University of Sheffield, UK.

KEIPER, W., 1997, Sound quality evaluation in the product cycle. *Acustica united with Acta Acustica*, 83, 784–8.

KERBER, G. and MAKAREWICZ, R., 1981, An optical scale model of traffic noise propagation in an urban environment. *Applied Acoustics*, 14, 331–45.

KHAN, M.S., JOHANSSON, O., and SUNDBACK, U., 1996, Evaluation of annoyance response to engine sounds using different rating methods. *Proceedings of Inter-Noise*, Liverpool, UK.

KIHLMAN, T., KROPP, W., ÖHRSTRÖM, E., and BERGLUND, B., 2001, Soundscape support to health – a cross-disciplinary research program. *Proceedings of Inter-Noise*, Hague, The Netherlands.

KLÆBOE, R., AMUNDSEN, A.H., FYHRI, A., and SOLBER, G.S., 2004, Road traffic noise – the relationship between noise exposure and noise annoyance in Norway. *Applied Acoustics*, 65, 893–912.

KLÆBOE, R., KOLBENSTVEDT, R., FYHRI, A., and SOLBER, G.S., 2005, The impact of an adverse neighbourhood soundscape on road traffic noise annoyance. *Acustica united with Acta Acustica*, 91, 1039–50.

KO, N.W.M. and TANG, C.P., 1978, Reverberation time in a high-rise city. *Journal of Sound and Vibration*, 56, 459–61.

KOSTEN, C.W. and VAN OS, G.J., 1962, Community reaction criteria for external noises. *National Physical Laboratory Symposium*, No. 12, London.

KOTZEN, B. and ENGLISH, C., 1999, *Environmental Noise Barriers: A Guide to Their Acoustic and Visual Design* (London: E&FN Spon).

KOYASU, M. and YAMASHITA, M., 1973, Scale model experiments on noise reduction by acoustic barrier of a straight line source. *Applied Acoustics*, 6, 233–42.

KRAUSE, B.L., 1993, The nichs hypothesis: a hidden symphony of animal sounds, the originals of musical expression and the health of habitats. *The Explorers Journal*, Winter, 156–60.

KROKSTAD, A., STRØM, S., and SØRSDAL, S., 1968, Calculating the acoustical room response by use of a ray tracing technique. *Journal of Sound and Vibration*, 8, 118–25.

KRUSKAL, J.B. and WISH, M., 1978, *Multidimensional Scaling* (London: Sage Publications).

KRYTER, K.D., 1970, *The Effects of Noise on Man* (New York: Academic Press).

——, 1982, Community annoyance from aircraft and ground vehicle noise. *Journal of the Acoustical Society of America*, 72, 1222–42.

KULOWSKI, A., 1984, Algorithmic representation of the ray tracing technique. *Applied Acoustics*, 18, 449–69.

KURRA, S., MORIMOTO, M., and MAEKAWA, Z.I., 1999, Transportation noise annoyance. A simulated-environment study for road, railway and aircraft noises, Part 1: overall annoyance. *Journal of Sound and Vibration*, 220, 251–78.

KURZE, U. and BERANEK, L.L., 1971, Sound propagation outdoors. In Beranek, L.L. (ed.) *Noise and Vibration Control* (New York: McGraw-Hill Book Company), pp. 164–93.

KURZE, U.J. and ANDERSON, G.S., 1971, Sound attenuation by barriers. *Applied Acoustics*, 4, 35–53.

KUTTRUFF, H., 1975, Zur Berechnung von Pegelmittelwerten und Schwankungsgrößen bei Straßenlärm. *Acustica*, 32, 57–69.

——, 1982, A mathematical model for noise propagation between buildings. *Journal of Sound and Vibration*, 85, 115–28.

KUWANO, S. and NAMBA, S., 1995, Long-term evaluation of noise. *Proceedings of the 15th International Congress on Acoustics (ICA)*, Trondheim, Norway.

——, 2001, Dimension of sound quality and their measurement. *Proceedings of the 17th International Congress on Acoustics (ICA)*, Rome, Italy.

KUWANO, S., NAMBA, S., FLORENTINE, M., ZHENG, D.R., FASTL, H., and SCHICK, A., 1999, A cross-cultural study of the factors of sound quality of environmental noise. *The 137th Meeting of the Acoustical Society of America and Forum Acusticum*, Berlin, Germany. (Abstract published in *Journal of Acoustical Society of America*, 105, 1081.)

LAMBERT, J., SIMONNET, F., and VALLET, M., 1984, Patterns of behaviour in dwellings exposed to road traffic noise. *Journal of Sound and Vibration*, 92, 159–72.

LAMURE, C., 1975, Noise emitted by road traffic. In Alexandre A., Barde J.-Ph., Lamure, C., and Langdon, F.J. (eds) *Road Traffic Noise* (London: Applied Science Publishers), pp. 85–129.

LANG, J., 1988, Symbolic aesthetics in architecture: toward a research agenda. In Nasar, J.L. (ed.) *Environmental Aesthetics* (Cambridge: Cambridge University Press), pp.11–26.

LAWRENCE, A., 1970, *Architectural Acoustics* (London: Elsevier).

LE POLLÈS, T., PICAUT, J., and BÉRENGIER, M., 2004, Sound field modelling in a street canyon with partially diffusely reflecting boundaries by the transport theory. *Journal of the Acoustical Society of America*, 116, 2969–83.

LE POLLÈS, T., PICAUT, J., COLLE, S., BÉRENGIER, M., and BARDOS, C., 2005, Sound-field modelling in architectural acoustics by a transport theory: application to street canyons. *Physical Review E*, 72, 046609–1–17.

LEE, M., 2004, *Acoustic Comfort in Communal Student Accommodation*. BArch dissertation, School of Architecture, University of Sheffield, UK.

LERCHER, P., 1998, Deviant dose-response curves for traffic noise in 'sensitive areas'. *Proceedings of Inter-Noise*, Christchurch, New Zealand.

LERCHER, P. and WIDMANN, U., 2001, Mental health and a complex sound environment in an Alpine valley: a case study. *Proceedings of the 17th International Congress on Acoustics (ICA)*, Rome, Italy.

L'ESPERANCE, A., 1989, The insertion loss of finite length barriers on the ground, *Journal of the Acoustical Society of America*, 86, 179–83.

LEWERS, T., 1993, A combined beam tracing and radiant exchange computer model of room acoustics. *Applied Acoustics*, 38, 161–78.

LI, B. and TAO, S., 2004, Influence of expanding ring roads on traffic noise in Beijing city. *Applied Acoustics*, 65, 243–9.

LI, Z.M., 2004, *Simulation of Acoustic Windows Using FEMLAB*. MSc dissertation, School of Architecture, University of Sheffield, UK.

LIN, C.H., 2000, *Acoustic Survey in Swimming Spaces*. MSc dissertation, School of Architecture, University of Sheffield, UK.

LMS, 2005a, *Raynoise Manual*, Leuven, Belgium.

——, 2005b, *Sysnoise 5.6 Manual*, Leuven, Belgium.

LYON, R.H., 1974, Role of multiple reflections and reverberation in urban noise propagation. *Journal of the Acoustical Society of America*, 55, 493–503.

LYONS, E., 1983, Demographic correlations of landscape preference. *Environment and Behavior*, 15, 487–511.

MAA, D.Y., 1987, Microperforated-panel wideband absorbers. *Noise Control Engineering Journal*, 29, 77–84.

MAEKAWA, Z., 1968, Noise reduction by screens. *Applied Acoustics*, 1, 157–73.

MAFFIOLO, V., DAVID, S., DUBOIS, D., VOGEL, C., CASTELLENGO, M., and POLACK, J.D., 1997, Sound characterization of urban environment. *Proceedings of Inter-Noise*, Budapest, Hungary.

MAFFIOLO, V., DUBOIS, D., DAVID, S., CASTELLENGO, M., and POLACK, J.D., 1998, Loudness and pleasantness in structuration of urban soundscapes. *Proceedings of Inter-Noise*, Christchurch, New Zealand.

MAFFIOLO, V., CASTELLENGO, M., and DUBOIS, D., 1999, Qualitative judgements of urban soundscapes. *Proceedings of Inter-Noise*, Fort Lauderdale, USA.

MAGRAB, E.B., 1975, *Environmental Noise Control* (London: Wiley Interscience Publications).

MAILLARD, J. and GUIGOU-CARTER, C., 2000, Study of passive/active control on openings for natural ventilation in buildings. *Proceedings of Inter-Noise*, Nice, France.

MANNING, C., 2002, A need for bigger noise mapping budgets. *Acoustics Bulletin* 27(4), 9.

MARQUIS-FAVRE, C., PREMAT, E., AUBRÉE, D., and VALLET, M., 2005a, Noise and its effects – a review on qualitative aspects of sound. Part I: notions and acoustic rating. *Acustica united with Acta Acustica*, 91, 613–25.

MARQUIS-FAVRE, C., PREMAT, E., and AUBRÉE, D., 2005b, Noise and its effects – a review on qualitative aspects of sound. Part II: noise and annoyance. *Acustica united with Acta Acustica*, 91, 626–42.

MARTENS, M.J.M., 1980, Foliage as a low-pass filter: experiments with model forests in an anechoic chamber. *Journal of the Acoustical Society of America*, 67, 66–72.

MARTENS, M.J.M. and MICHELSEN, A., 1981, Absorption of acoustic energy by plant leaves. *Journal of the Acoustical Society of America*, 69, 303–6.

MARTÍNEZ-SALA, R., RUBIO, C., GARCÍA-RAFFI, L.M., SÁNCHEZ-PÉREZ, J.V., SÁNCHEZ-PÉREZ, E.A., and LLINARES, J., 2006, Control of noise by trees arranged like sonic crystals. *Journal of Sound and Vibration*, 291, 100–6.

MAST, A., VAN DEN DOOL, T., VAN DER TOORN, J., and WATTS, G., 2005, Array and subtraction methods for characterising vehicle noise sub sources. *Acoustics Bulletin*, 30(5), 26–32.

MATSUMOTO, T., YAMAMOTO, K., and ISHIKITA, H., 2000, Efficiency of highway noise barrier with horizontal louver – a study by full scale model experiment. *Proceedings of Inter-Noise*, Nice, France.

MAURIN, M. and LAMBERT, J., 1990, Exposure of the French population to transport noise. *Noise Control Engineering Journal*, 35, 5–18.

MAY, D.N., 1979, Freeway noise and high-rise balconies. *Journal of Acoustical Society of America*, 65, 699–704.

MAY, D.N. and OSMAN, M.M., 1980a, The performance of sound absorptive, reflective and T-profile noise barriers in Toronto. *Journal of Sound and Vibration*, 71, 65–71.

——, 1980b, Highway noise barriers: new shapes. *Journal of Sound and Vibration*, 71, 73–101.

MCGRATH, D., 1995, Huron – a digital audio convolution workstation. Preprint 4023, *AES 5th Australia Regional Convention*, Australia.

MEHRABIAN, A., 1976, *Public Places and Private Spaces – The Psychology of Work, Play, and Living Environments* (New York: Basic Books Inc. Publisher).

MENG, Y. and KANG, J., 2004, Fast simulation of sound fields for urban square animation. *Proceedings of Inter-Noise*, Prague, Czech Republic.

MENG, Y., KANG, J., HUERTA, R., and ZHANG, M., 2005, Soundscape animation of urban open spaces in an interactive virtual environment. *Proceedings of the 12th International Congress on Sound and Vibration*, Lisbon, Portugal.

MENGE, C.W., 1978, Sloped barriers for highway noise control. *Proceedings of Inter-Noise*, San Francisco, USA.

MENOUNOU, P. and BUSCH-VISHNIAC, I.J., 2000, Jagged edge noise barriers. *Building Acoustics*, 7, 179–200.

MEUNIER, S., BOULLET, I., and RABAU, G., 2001, Loudness of impulsive environmental sounds. *Proceedings of the 17th International Congress on Acoustics (ICA)*, Rome, Italy.

MIEDEMA, H.M.E. and VOS, H., 1998, Exposure-response relationships for transportation noise. *Journal of the Acoustical Society of America*, 104, 3432–45.

——, 1999, Demographic and attitudinal factors that modify annoyance from transportation noise. *Journal of the Acoustical Society of America*, 105, 3336–44.

MIZUNO, K., SEKIGUCHI, H., and IIDA, K., 1984, Research on a noise control device, 1st report. *Bulletin of Japanese Society of Mechanical Engineers*, 27, 1499–1505.

——, 1985, Research on a noise control device, 2nd report. *Bulletin of Japanese Society of Mechanical Engineers*, 28, 2737–43.

MOHAJERI, R., 1998, *Investigation of an Intelligent Control System for Attenuation of Transport Noise*. PhD dissertation, Department of Mechanical and Mechatronic Engineering, University of Sydney, Australia.

MOHSEN, E.A. and OLDHAM, D.J., 1977, Traffic noise reduction due to the screening effect of balconies on a building façade. *Applied Acoustics*, 10, 243–57.

MOMMERTZ, E. and VORLÄNDER, M., 1995, Measurement of scattering coefficients of surfaces in the reverberation chamber and in the free field. *Proceedings of the 15th International Conference on Acoustics (ICA)*, Trondheim, Norway.

MOORE, G.R., 1984, *An Approach to the Analysis of Sound in Auditoria*. PhD dissertation, University of Cambridge, UK.

MOREIRA, N. and BRYAN, M., 1972, Noise annoyance susceptibility. *Journal of Sound and Vibration*, 21, 449–62.

MÖSER, M., 2004, *Engineering Acoustics: An Introduction to Noise Control* (Berlin: Springer).

MOUGHTIN, C., 2003, *Urban Design: Street and Square* (3rd edn) (London: Architectural Press).

MUDRI, L. and LENARD, J.D., 2000, Comfortable and/or pleasant ambience: conflicting issues? *Proceeding of PLEA*, Cambridge, UK.

MULHOLLAND, K.A., 1979, The prediction of traffic noise using a scale model. *Applied Acoustics*, 12, 459–78.

MULLIGAN, B.E., LEWIS, S.E., FAUPEL, M.L., GOODMAN, L.S., and ANDERSON, L.M., 1987, Enhancement and masking of loudness by environmental factors vegetation and noise. *Environment and Behavior*, 19, 411–43.

MULTIGEN-PARADIGM, 2005, www.multigen.com

MURADALI, A. and FYFE, K.R., 1998, A Study of 2D and 3D barrier insertion loss using improved diffraction-based methods. *Applied Acoustics*, 53, 49–75.

NAGEL, K. and SCHRECKENBERG, M., 1992, A cellular automaton model for freeway traffic. *Journal de Physique I*, 2, 2221–9.

NAMBA, S., KUWANO, S., and KATO, T., 1996, Trade-off effect between the number of events and L. in the evaluation of train noise. *Proceedings of Inter-Noise*, Liverpool, UK.

NATHANAIL, C. and GUYOT, F., 2001, Parameters influencing perception of highway traffic noise. *Proceedings of the 17th International Congress on Acoustics (ICA)*, Rome, Italy.

NAVRUD, S., 2002, *The State-of-the-Art on Economic Valuation of Noise*. Final Report to European Commission Directorate-General (DG) Environment.

NELSON, J.P., 1980, Airport and property values: a survey of recent evidence. *Journal of Transport Economics and Policy*, 14, 37–52.

——, 1982, Highway noise and property values: a survey of recent evidence. *Journal of Transport Economics and Policy*, 16, 117–38.

——, 1987, *Transportation Noise Reference Book* (London: Butterworths).

NICHOLAS, J. and DAIGLE, G.A., 1986, Experimental study of a slow-wave guide barrier on finite impedance ground. *Journal of the Acoustical Society of America*, 80, 869–76.

NIKOLOPOULOU, M., KANG, J., KATZSCHNER, L., and SCUDO, G., 2003, The thermal and acoustic environment in open urban spaces: approaches and findings. *The 1st EU Greencluster Conference*, Leipzig, Germany.

OCHIAI, H., TOKITA, Y., and YAMADA, S., 1999, Study on evaluation method of infra and low frequency noise. *Proceedings of Inter-Noise*, Fort Lauderdale, USA.

OCHMANN, M., 1995, The source simulation technique for acoustic radiation problems. *Acustica united with Acta Acustica*, 81, 512–27.

ODPM, 1994, *Planning Policy Guidance (PPG) 24: Planning and Noise*. Office of the Deputy Prime Minister, UK.

OECD, 1991, *Fighting Noise in the 1990s*. Organisation for Economic Co-operation and Development, Paris, France.

——, 1995, *Urban Travel and Sustainable Development*. Organisation for Economic Co-operation and Development – European Conference of Ministers of Transport, Paris, France.

ÖGREN, M. and FORSSEN, J., 2004, Modeling a city canyon problem in a turbulent atmosphere using an equivalent sources approach. *Applied Acoustics*, 65, 629–42.

ÖGREN, M. and KROPP, W., 2004, Road traffic noise propagation between two dimensional city canyons using an equivalent sources approach. *Acustica united with Acta Acustica*, 90, 293–300.

OKUBO, T. and FUJIWARA, K., 1998, Efficiency of a noise barrier on the ground with an acoustically soft cylindrical edge. *Journal of Sound and Vibration*, 216, 771–90.

OLDHAM, D.J., KANG, J., and BROCKLESBY, M.W., 2005a, Modelling the acoustical and airflow performance of simple lined ventilation apertures. *Proceedings of the 12th International Congress on Sound and Vibration*, Lisbon, Portugal.

——, 2005b, Numerical modelling of the acoustical performance of a simple lined aperture in a thick wall. *Proceedings of Inter-Noise*, Rio de Janeiro, Brazil.

——, 2005c, Modelling the acoustical and airflow performance of natural ventilation inlet and outlet units. *The 149th Meeting of the Acoustical Society of America*, Vancouver, Cannada. (Abstract published in *Journal of the Acoustical Society of America*, 117, 2379.)

——, 2005d, Modeling the acoustical and airflow performance of simple lined apertures. *Building Acoustics*, 12, 275–90.

OLLERHEAD, J.B., 1978, Predicting public reaction to noise from mixed sources. *Proceedings of Inter-Noise*, San Francisco, USA.

OSGOOD, C.E., 1952, The nature and measurement of meaning. *Psychological Bulletin*, 49, 197–237.

OSGOOD, C.E., SUCI, G.J., and TANNENBAUM, P.H., 1957, *The Measurement of Meaning* (Urbana: University of Illinois Press).

OSTASHEV, V., WILSON, D., LIU, L., ALDRIDGE, D., SYMONS, N., and MARLIN, D., 2005, Equations for finite-difference, time-domain simulation of sound propagation in moving inhomogeneous media and numerical implementation. *Journal of the Acoustical Society of America*, 117, 503–17.

PAGE, R.A., 1997, Noise and helping behavior. *Environment and Behavior*, 9, 311–34.

PAPOULIS, A., 1991, *Probability, Random Variables and Stochastic Process* (3rd edn) (New York: McGraw-Hill).

PARKIN, P.H. and SCHOLES, W.E., 1965, The horizontal propagation of sound from a jet engine close to the ground at Hatfield. *Journal of Sound and Vibration*, 2, 353–74.

PARRY, G.A., PYKE, J.R., and ROBINSON, C., 1993. The excess attenuation of environmental noise sources through densely planted forest. *Proceedings of the Institute of Acoustics (IOA)*, Bath, UK.

PATSOURAS, C.H., FASTLE, H., PATSOURAS, D., and PFAFFELHUBER, K., 2001, Psychoacoustic sensation magnitudes and sound quality rating of upper middle class cars' idling noise. *Proceedings of the 17th International Congress on Acoustics (ICA)*, Rome, Italy.

PATTERSON, D.W., 1996, *Artificial Neural Networks: Theory and Application* (London: Prentice-Hall).

PEASE, J. (ED.), 2002, Pressure groups round on strategy vacuum. *Noise Management*, 23, 1.

PENN, C.N., 1979, *Noise Control* (London: Shaw & Sons).

PICAUT, J. and SIMON, L., 2001, A scale model experiment for the study of sound propagation in urban areas. *Applied Acoustics*, 62, 327–40.

PICAUT, J., SIMON, L., and HARDY, J., 1999, Sound field modelling in streets with a diffusion equation. *Journal of the Acoustical Society of America*, 106, 2638–45.

PICAUT, J., LE POLLÈS, T., LÕHERMITE, P., and GARY, V., 2005, Experimental study of sound propagation in a street. *Applied Acoustics*, 66, 149–73.

PICCOLO, A., PLUTINO, D., and CANNISTRARO, G., 2005, Evaluation and analysis of the environmental noise of Messina, Italy. *Applied Acoustics*, 66, 447–65.

PIRINCHIEVA, R.K., 1991, The influence of barrier size on its sound diffraction. *Journal of Sound and Vibration*, 148, 183–92.

PLOVSING, B. and KRAGH, J., 2000, *Nord: Comprehensive Outdoor Sound Propagation Model – Propagation in Atmosphere without Significant Refraction*. Technical Report AV 1849/00, Danish Electronics, Light and Acoustics (DELTA).

PORTEOUS, J.D., 1996, *Environmental Aesthetics – Ideas, Politics and Planning* (London: Routledge Press).

PORTEOUS, J.D. and MASTIN, J.F., 1985, Soundscape. *Journal of Architectural and Planning Research*, 2, 169–86.

PORTER, N.D., FLINDELL, I.H., and BERRY, B.F., 1998, *Health Effect-Based Noise Assessment Methods: A Review and Feasibility Study*. The UK National Physical Laboratory Report CMAM 16.

POWELL, C.A., 1979, *A Summation and Inhibition Model of Annoyance Response to Multiple Community Noise Sources*. NASA Technical Paper 1479.

PRALL, D., 1929, *Aesthetic Judgement* (New York: Crowell).

PREIS, A., 1996, Measurement of annoyance components. *Proceedings of Inter-Noise*, Liverpool, UK.

PRICE, M.A., ATTENBOROUGH, K., and NICHOLAS, W., 1988, Sound attenuation through trees: measurements and models. *Journal of the Acoustical Society of America*, 84, 1836–44.

PUSLAR-CUDINA, M. and CUDINA, M., 1999, Noise and colors. *Proceedings of the 6th International Congress of Sound and Vibration*, Copenhagen, Denmark.

QUADSTONE, 2005, www.paramics-online.com

RAHIM, M.A., 2001, *Acoustic Ray-Tracing for a Single Urban Street Canyon with Strategic Design Options*. MSc dissertation, Department of Computer Science, University of Sheffield, UK.

RAIMBAULT, M., BÉRENGIER, M., and DUBOIS, D., 2001, Common factors in the identification of urban soundscapes pilot study in two French cities: Lyon and Nantes. *Proceedings of the 17th International Congress on Acoustics*, Rome, Italy.

——, 2003, Ambient sound assessment of urban environments: field studies in two French cities. *Applied Acoustics*, 64, 1241–56.

RATHE, E.J., 1969, Note on two common problems of sound attenuation. *Journal of Sound and Vibration*, 10, 472–9.

RAWLINS, A.D., 1976, Diffraction of sound by a rigid screen with a soft or perfectly absorbing edge. *Journal of Sound and Vibration*, 45, 53–67.

RECUERO, M., BLANCO-MARTIN, E., and GRUNDMAN, J., 1996, Study of the acoustical environment of a city. *Proceedings of Inter-Noise*, Liverpool, UK.

ROBINSON, D.W., 1971, Towards a unified system of noise assessment. *Journal of Sound and Vibration*, 14, 279–98.

ROBINSON, D.W. and DADSON, R.S., 1956, A redetermination of the equal loudness relation for pure tones. *British Journal of the Applied Physics*, 7, 166–81.

ROSEN, S., 1974, Hedonic prices and implicate markets: product differentiation in pure competition. *Journal of Political Economy*, 82, 34–55.

ROSSING, T.D., 1990, *The Science of Sound* (Massachusetts: Addison-Wesley).

RUSPA, G., 2001, Sound effects within a wood. *Proceedings of the 17th International Congress on Acoustics (ICA)*, Rome, Italy.

RYLANDER, R., SORENSEN, S., and KAJLAND, A., 1972, Annoyance reaction from aircraft noise exposure. *Journal of Sound and Vibration*, 24, 419–44.

SABINE, W.C., 1993, *Collected Papers on Acoustics* (Los Altos, USA: Peninsula Publishing).

SAC, 1993, China National Standards GB 3096–93: *Standard of Environmental Noise of Urban Area*. Standardization Administration of China (SAC), Beijing, China.

SALOMONS, E., 1998, Improved Green's function parabolic equation method for atmospheric sound propagation. *Journal of the Acoustical Society of America*, 104, 100–11.

SANDBERG, U. and EJSMONT, J.A., 2002, *Tyre/Road Noise Reference Book* (Harg, Sweden: Informex).

SANDERS, M., GUSTANSKI, J., and LAWTON, M., 1974, Effect of ambient illumination on noise level of groups. *Journal of Applied Psychology*, 59, 527–8.

SANOFF, H., 2000, *Community Participation Methods in Design and Planning* (New York: Wiley).

SATO, T., 1993, A path analysis of the effect of vibration on road traffic noise annoyance. *Proceedings of the 6th International Congress on Noise as a Public Health Problem*, Nice, France.

SATO, T., YANO, T., YAMASHITA, T., KAWAI, K., RYLANDER, R., BJÖRKMAN, M., and ÖHRSTRÖM, E., 1998, Cross-cultural comparison of community responses to road traffic noise in Gothenburg, Sweden, and Kumamoto, Japan, Part II: casual modelling by path analysis. *Proceedings of the 7th International Congress on Noise as a Public Health Problem*, Sydney, Australia.

SATTLER, M.A. and ROTT, J.A.A., 1996, Social survey on traffic noise for the city of Porto Alegre, Brazil. *Proceedings of Inter-Noise*, Liverpool, UK.

SCHAFER, R.M., 1977a, *Five Village Soundscapes* (Vancouver: A.R.C Publications).

——, 1977b, *The Tuning of the World* (New York: Knopf). (Reprinted as *Our Sonic Environment and the Soundscape: The Tuning of the World*. Destiny Books, 1994.)

SCHARF, B., 1978, Loudness. In Carterette, E.C. and Friedman, M.P. (eds) *Handbook of Perception* (New York: Academic Press), pp. 187–242.

SCHAUDINISCHKY, L.H., 1976, *Sound, Man and Building* (London: Applied Science Publishers Ltd).

SCHRÖDER, E., 1973, Nachhall in geschlossenen bebauten Straßen. *Lärmbekämpfung*, 17, 11–18.

SCHROEDER, M.R., 1975, Diffuse sound reflection by maximum length sequences. *Journal of the Acoustical Society of America*, 57, 149–51.

SCHULTE-FORTKAMP, B., 1996, Combined methods to investigate effects of noise exposure and subjective noise assessment. *Proceedings of Inter-Noise*, Liverpool, UK.

——, 2001, The quality of acoustic environments and the meaning of soundscapes. *Proceedings of the 17th International Congress on Acoustics (ICA)*, Rome, Italy.

SCHULTE-FORTKAMP, B. and NITSCH, W., 1999, On soundscapes and their meaning regarding noise annoyance measurements. *Proceedings of Inter-Noise*, Fort Lauderdale, FL, USA.

SCHULTE-FORTKAMP, B., MUCKEL, P., CHOUARD, N., and ENSEL, L., 1999, Subjective evaluation methods: the meaning of the context in evaluation of sound quality and an appropriate test procedure. *Proceedings of the 6th International Congress of Sound and Vibration*, Copenhagen, Denmark.

SCHULTZ, T.J., 1978, Synthesis of social surveys on noise annoyance. *Journal of the Acoustical Society of America*, 64, 377–405.

——, 1982, *Community Noise Rating* (London: Applied Science Publishers).

SCHWARTZ, H., 1995, Noise and silence: the soundscape and spirituality. *Proceedings of Realizing the Ideal: The Responsibility of the World's Religions – Section IV: Religion and the Ideal Environment*, Seoul, Korea.

SERGEEV, M.V., 1979, Scattered sound and reverberation on city streets and in tunnels. *Soviet Physics – Acoustics*, 25, 248–52.

SEZNEC, R., 1980, Diffraction of sound around barriers: use of the boundary elements technique. *Journal of Sound and Vibration*, 73, 195–209.

SHAO, W., LEE, H.P., and LIM, S.P., 2001, Performance of noise barriers with random edge profiles. *Applied Acoustics*, 62, 1157–70.

SHAW, E.A.G. and OLSON, N., 1972, Theory of steady-state urban noise for an ideal homogeneous city. *Journal of the Acoustical Society of America*, 51, 1781–93.

SHIELD, B., 2002, How can we be sure noise maps are accurate? *Acoustics Bulletin*, 27(3), 39.

SHIMA, H., WATANABE, T., YOKOI, T., MIZUNO, K., MATSUMOTO, K., YAMAMOTO, M., and MIYAMA, T., 1998, Branched noise barriers. *Proceedings of Inter-Noise*, Christchurch, New Zealand.

SIEGEL, R. and HOWELL, J., 1981, *Thermal Radiation Heat Transfer* (2nd edn) (Washington, DC: Hemisphere).

SILLION, F.X. and PUECH, C., 1994, *Radiosity and Global Illumination* (San Francisco: Morgan Kaufmann Publishers).

SKINNER, C.J. and GRIMWOOD, C.J., 2005, The UK noise climate 1990–2001: population exposure and attitudes to environmental noise. *Applied Acoustics*, 66, 231–43.

SLUTSKY, S. and BERTONI, H.L., 1988, Analysis and programs for assessment of absorptive and tilted parallel barriers. *Transportation Research Record*, 1176, 13–22.

SMITH, B.R., 2000, *The Acoustic World of Early Modern England: Attending to the O-factor* (Chicago: University of Chicago Press).

SOMMERFELD, A., 1971, *Thermodynamics and Statistical Mechanics* (London: Academic Press).

SOUTHWORTH, M., 1969, The sonic environment of cities. *Environment and Behavior*, 1, 49–70.

STANNERS, D. and BORDEAU, P. (eds), 1995, *Europe's Environment the Dobris Assessment. Noise and Radiation* (Copenhagen, Denmark: European Environment Agency), pp. 359–74.

STEEMERS, K., NIKOLOPOULOU, M., KANG, J., COMPAGNON, R., and KATZSCHNER, L., 2003, Mapping urban comfort: a synopsis of the RUROS project tools. *The 2nd EU Greencluster Conference*, Athens, Greece.

STEENACKERS, P., MYNCKE, H., and COPS, A., 1978, Reverberation in town streets. *Acustica*, 40, 115–19.

STEENEKEN, H.J.M. and HOUTGAST, T., 1980, A physical method for measuring speech-transmission quality. *Journal of the Acoustical Society of America*, 67, 318–26.

STEIGLITZ, K., 1996, *A Digital Signal Processing Primer with Applications to Digital Audio and Computer Music* (New York: Addison-Wesley).

STEPAN, C., 2003, *Church Acoustics – A Case Study of Two Churches*. BArch dissertation, School of Architecture, University of Sheffield, UK.

STEPAN, C., CHRISTOPHERS, C., and KANG, J., 2003, Acoustic measurements and subjective surveys of five churches in Sheffield. *Proceedings of the Institute of Acoustics (IOA)*, UK.

STOCKER, J. and CARRUTHERS, D., 2003, How accurate is a noise map created using air quality source data? *Acoustics Bulletin*, 28(2), 14–17.

STRAUSS, A.L. and CORBIN, J., 1990, *Basics of Qualitative Research: Grounded Theory Procedures and Techniques* (California: Sage).

SUSINI, P., MCADAMS, S., and WINSBERG, S., 1999, A multidimensional technique for sound quality assessment. *Acustica united with Acta Acustica*, 85, 650–6.

SUTHERLAND, L.C., BRADEN, M.H., and COLMAN, R., 1973, *A Programme for the Measurement of Environmental Noise in the Community and its Associated Human Response*. Report No. DOT–TST–74–5, Department of Transportation, Office of Noise Abatement, Washington, DC, USA.

SUZUKI, Y. and TAKESHIMA, H., 2004, Equal-loudness-level contours for pure tones. *Journal of the Acoustical Society of America*, 116, 918–31.

TAMURA, A., 1998, An environmental index based on inhabitants' recognition of sounds. *Proceedings of the 7th International Congress on Noise as a Public Health Problem*, Sydney, Australia.

TAYLOR, S.M., 1984, A path model of aircraft noise annoyance. *Journal of Sound and Vibration*, 96, 243–60.

THE SCOTTISH OFFICE, 1999, *Planning and Noise*. Planning Advice Note 56, Edinburgh.

THOMAS, L., 2000, *Sound Propagation in Interconnected Urban Streets: A Ray Tracing Model*. MSc dissertation, Department of Computer Science, University of Sheffield, UK.

THORSSON, P., 2006, Application of linear transport to sound propagation in cities. *Acustica united with Acta Acustica* (to be published).

THORSSON, P. and ÖGREN, M., 2005, Macroscopic modelling of urban traffic noise – influence of absorption and vehicle flow distribution. *Applied Acoustics*, 66, 195–209.

THORSSON, P., ÖGREN, M., and KROPP, W., 2004, Noise levels on the shielded side in cities using a flat city model. *Applied Acoustics*, 65, 313–23.

TOMPSETT, R., 2002, Noise mapping – accuracy is our priority. *Acoustics Bulletin*, 27(4), 9.

TONIN, R., 1996, A method of strategic traffic noise impact analysis. *Proceedings of Inter-Noise*, Liverpool, UK.

TRANCIK, R., 1986, *Finding Lost Space: Theory of Urban Design* (New York: Van Nostrand Reinhold Company).

TRUAX, B. (ed.), 1999, *Handbook for Acoustic Ecology* (Cambridge Street Publishing), CD-ROM version.

TRUAX, B., 2001, *Acoustic Communication* (2nd edn) (Westport, CT: Greenwood Press).

TSAI, K.T. and LAI, P.R., 2001, The research of the interactions between the environmental sound and sight. *Proceedings of the 17th International Congress on Acoustics (ICA)*, Rome, Italy.

TURNER, R.K., PEARCE, D.W., and BATEMAN, I.J., 1994, *Environmental Economics: An Elementary Introduction* (Hemel Hempstead: Harvester Wheatsheaf).

TURNER, S. and HINTON, J., 2002, Noise mapping is a strategic exercise. *Acoustics Bulletin*, 27(5), 46.

TWERSKY, V., 1983, Reflection and scattering of sound by correlated rough surfaces. *Journal of the Acoustical Society of America*, 73, 85–94.

UK DEFRA, 1999, *Noise Climate Assessment: A Review of National and European Practices*. UK Department for Environment, Food and Rural Affairs.

——, 2001, *Towards a National Ambient Noise Strategy*. UK Department for Environment, Food and Rural Affairs.

——, 2003, *Traffic Noise Mapping in London*. UK Department for Environment, Food and Rural Affairs. (Also see www.noisemapping.org)

UK DfES (Department for Education and Skills), 2003, *Acoustic Design of Schools: A Design Guide*. Building Bulletin 93, The Stationery Office, London.

UK DfT, 1988, *Calculation of Road Traffic Noise (CRTN)*. UK Department for Transport.

——, 1995, *Calculation of Rail Noise (CRN)*. UK Department for Transport.

UK HIGHWAYS AGENCY, 1995, *Design Manual for Roads and Bridges, Volume 10, Environmental Design, Section 5, Environmental Barriers, Part 1, Design Guide for Environmental Barriers*. HMSO, London.

UMNOVA, O., ATTENBOROUGH, K., and LINTON, C.M., 2006, Effects of porous covering on sound attenuation by periodic arrays of cylinders. *Journal of the Acoustical Society of America*, 119, 278–84.

US EPA, 1971, *Fundamentals of Noise, Measurement, Rating Schemes, and Standards*. Environmental Protection Agency (EPA), Washington, DC, USA.

——, 1974, *Information on Levels of Environmental Noise Requisite to Protect Public Health and Welfare with an Adequate Margin of Safety*. EPA/ONAC Report 550/9–74–004, Environmental Protection Agency (EPA), Washington, DC, USA.

——, 1982, *Guidelines for Noise Impact Analysis*. Environmental Protection Agency (EPA), Washington, DC, USA.

US FICUN, 1980, *Guidelines for Considering Noise in Land Use Planning and Control*. US Government Printing Office Report #1981–337–066/8071 – Federal Interagency Committee on Urban Noise (FICUN), Washington, DC, USA.

US NRC, 1977, *Guidelines for Preparing Environmental Impact Statements on Noise*. Report of Working Group 69, National Research Council, Washington, DC, USA.

VALLET, M., 1996, Annoyance after changes in airport noise environment. *Proceedings of Inter-Noise*, Liverpool, UK.

VALLET, M., VERNET, I., CHAMPELOVIER, P., and MAURIN, M., 1996, A road traffic noise index for the night time. *Proceedings of Inter-Noise*, Liverpool, UK.

VAN BEEK, A., BEUVING, M., DITTRICH, M., BEIER, M., ZHANG, X., JONASSON, H., LETOURNEAUX, F., TALOTTE, C., and RINGHEIM, M., 2002, *Rail Sources: State of the Art*. Harmonoise Report HAR12TR–020118-SNCF10, European Commission.

VAN DER HEIJDEN, L.A.M. and MARTENS, M.J.M., 1982, Traffic noise reduction by means of surface wave exclusion above parallel grooves in the roadside. *Applied Acoustics*, 15, 329–39.

VAN RENTERGHEM, T. and BOTTELDOOREN, D., 2002, Effect of a row of trees behind noise barriers in wind. *Acustica united with Acta Acustica*, 88, 869–78.

——, 2003, Numerical simulation of the effect of trees on downwind noise barrier performance. *Acustica united with Acta Acustica*, 89, 764–78.

VAN RENTERGHEM, T., BOTTELDOOREN, D., CORNELIS, W.M., and GABRIELS, D., 2002, Reducing screen-induced refraction of noise barriers in wind by vegetative screens. *Acustica united with Acta Acustica*, 88, 231–8.

VAN RENTERGHEM, T., SALOMONS, E., and BOTTELDOOREN, D., 2005, Efficient FDTD-PE model for sound propagation in situations with complex obstacles and wind profiles. *Acustica united with Acta Acustica*, 91, 671–9.

——, 2006, Parameter study of sound propagation between city canyons with coupled FDTD-PE model. *Applied Acoustics*, 67, 487–510.

VERZINI, A., FRASSONI, C., and ORTIZ, A.H., 1999, A field study about effects of low frequency noises on man. *The 137th Meeting of the Acoustical Society of America and Forum Acusticum*, Berlin, Germany. (Abstract published in *Journal of Acoustical Society of America*, 105, 942.)

VESTA SERVICES, 2004, *Qnet User's Manual*, Winnetka, USA.

VIOLLON, S., 2003, Two examples of audio-visual interactions in an urban context. *Proceedings of Euro-Noise*, Naples, Italy.

VIOLLON, S. LAVANDIER, C., and DRAKE, C., 2002, Influence of visual setting on sound ratings in an urban environment. *Applied Acoustics*, 63, 493–511.

VOLKMANN, J.E. and GRAHAM, M.L., 1942, A survey on air raid alarm signals. *Journal of the Acoustical Society of America*, 14, 1–9.

VORLÄNDER, M. and MOMMERTZ, E., 2000, Definition and measurement of random-incidence scattering coefficients. *Applied Acoustics*, 60, 187–99.

VOS, J., 1992, Annoyance caused by simultaneous impulse, road-traffic, and aircraft sounds: a quantitative model. *Journal of the Acoustical Society of America*, 91, 3330–45.

VOSS, R. and CLARKE, J., 1975, '1/f noise' in music and speech. *Nature*, 258, 317–8.

——, 1978, '1/f noise' in music: music from 1/f noise. *Journal of the Acoustical Society of America*, 63, 258–63.

WALERIAN, E., JANCZUR, R., and CZECHOWICZ, M., 2001a, Sound levels forecasting for city-centers. Part I: sound level due to a road within urban canyon. *Applied Acoustics*, 62, 359–80.

——, 2001b, Sound levels forecasting for city-centers. Part II: effect of source model parameters on sound level in built-up area. *Applied Acoustics*, 62, 461–92.

WALTERS, A.A., 1975, *Noise and Prices* (London: Oxford University Press).

WARDMAN, M. and BRISTOW, A.L., 2004, Noise and air quality valuations: evidence from stated preference residential choice models. *Transportation Research D*, 9, 1–27.

WARREN, D.H., MCCARTHY, T.J., and WELCH, R.B., 1983, Discrepancy and nondiscrepancy methods of assessing visual–auditory interaction. *Perception and Psychophysics*, 33, 413–19.

WASSILIEFF, C., 1988, Improving the noise reduction of picket barriers, *Journal of the Acoustical Society of America*, 84, 645–50.

WATANABE, T. and YAMADA, S., 1996, Sound attenuation through absorption by vegetation. *Journal of the Acoustical Society of Japan*, 17, 175–82.

WATTS, G., 1996a, Acoustic performance of a multiple edge noise barrier profile at motorway sites. *Applied Acoustics*, 47, 47–66.

WATTS, G., 1996b, Acoustic performance of parallel traffic noise barriers. *Applied Acoustics*, 47, 95–119.

——, 2005, Harmonoise models for predicting road traffic noise. *Acoustics Bulletin*, 30(5), 19–25.

WATTS, G. and GODFREY, N.S., 1999, Effects on roadside noise levels of sound absorptive materials in noise barriers. *Applied Acoustics*, 58, 385–402.

WATTS, G. and MORGAN, P.A., 1996, Acoustic performance of an interference-type noise-barrier profile. *Applied Acoustics*, 49, 1–16.

WATTS, G., CROMBIE, D.H., and HOTHERSALL, D.C., 1994, Acoustic performance of new designs of traffic noise barriers: full scale tests. *Journal of Sound and Vibration*, 177, 289–305.

WATTS, G., CHINN, L., and GODFREY, N., 1999, The effects of vegetation on the perception of traffic noise. *Applied Acoustics*, 56, 39–56.

WATTS, G., HOTHERSALL, D.C., and HOROSHENKOV, K.V., 2001, Measured and predicted acoustic performance of vertically louvered noise barriers. *Applied Acoustics*, 62, 1287–311.

WEINSTEIN, N.D., 1978, Individual differences in reactions to noise: a longitudinal study in a college dormitory. *Journal of Applied Psychology*, 63, 458–66.

WEISS, M.A., 1998, *Data Structures and Algorithm Analysis in C++* (Massachusetts: Addison-Wesley).

WENZEL, E., 1992, Launching sounds into space. In Jacobson, L. (ed.) *Cyberarts: Exploring Art and Technology* (San Francisco: Miller Freeman Inc.).

WESTERKAMP, H., 2000, Sound excursion: Plano Pilato, Brasilia. *Soundscape: The Journal of Acoustic Ecology*, 1, 20–1.

WG-HSEA, 2002, *Position Paper on Dose-Response Relationships between Transportation Noise and Annoyance*. European Commission, Office for Official Publications of the European Communities, Luxemburg.

WHITE, F.A., 1975, *Our Acoustic Environment* (New York: Wiley).

WHO, 1995, *Concern for Europe's Tomorrow – Health and the Environment in the WHO European Region*. World Health Organization (WHO) European Centre for Environment and Health, Wissenschaftliche Verlagsgesellschaft, Stuttgart, Germany.

WIENER, F.M., MALME, C.I., and GOGOS, C.M., 1965, Sound propagation in urban areas. *Journal of the Acoustical Society of America*, 37, 738–47.

WILHELMSSON, M., 2000, The impact of traffic noise on the values of single-family houses. *Journal of Environmental Planning and Management*, 43, 799–815.

WILLIAMS, M.M.R., 1971, *Mathematical Methods in Particle Transport Theory* (London: Butterworths).

WILSON, C.E., 1989, *Noise Control* (New York: Happer and Row).

WIRT, L.S., 1979, The control of diffracted sound by means of thnadners (shaped noise barriers). *Acustica*, 42, 73–88.

WRIGHT, J.R., 1995, An exact model of acoustic radiation in enclosed spaces. *Journal of the Audio Engineering Society*, 43, 813–19.

WRIGHTSON, K., 2000, An introduction to acoustic ecology. *Soundscape: The Journal of Acoustic Ecology*, 1, 10–13.

WU, S. and KITTINGER, E., 1995, On the relevance of sound scattering to the prediction of traffic noise in urban streets. *Acustica united with Acta Acustica*, 81, 36–42.

XING, Z. and KANG, J., 2006, Acoustic comfort in residential areas – a cross-cultural study. *Proceedings of the Institute of Acoustics (IOA)*, Southampton, UK.

YAMAMOTO, K., SHONO, Y., OCHIAI, H., and HIRAO, Y., 1995, Measurements of noise reduction by absorptive devices mounted at the top of highway noise barriers. *Proceedings of Inter-Noise*, Newport Beach, USA.

YAMASHITA, M. and YAMAMOTO, K., 1990, Scale model experiments for the prediction of road traffic noise and the design of noise control facilities. *Applied Acoustics*, 31, 185–96.

YAMASHITA, T., YANO, T., and IZUMI, K., 1991, Comparison of community response to road traffic noise in warmer and colder areas in Japan. *Proceedings of Inter-Noise*, Sydney, Australia.

YANG, W., 2000, *Soundscape in Urban Open Public Spaces: Comparison between Beijing and Sheffield*. MSc dissertation, School of Architecture, University of Sheffield, UK.

——, 2005, *An Aesthetic Approach to the Soundscape of Urban Public Open Spaces*. PhD dissertation, School of Architecture, University of Sheffield, UK.

YANG, W. and KANG, J., 2001, Acoustic comfort and psychological adaptation as a guide for soundscape design in urban open public spaces. *Proceedings of the 17th International Congress on Acoustics (ICA)*, Rome, Italy.

——, 2003, A cross-cultural study of the soundscape in urban open public spaces. *Proceedings of the 10th International Congress on Sound and Vibration*, Stockholm, Sweden.

——, 2005a, Acoustic comfort evaluation in urban open public spaces. *Applied Acoustics*, 66, 211–29.

——, 2005b, Soundscape and sound preferences in urban squares. *Journal of Urban Design*, 10, 69–88.

YANO, T., YAMASHITA, T., and IZUMI, K., 1996, Social survey on community response to railway noise-comparison of responses obtained with different annoyance scales. *Proceedings of Inter-Noise*, Liverpool, UK.

YEOW, K.W., 1976, External reverberation times observed in built-up areas. *Journal of Sound and Vibration*, 48, 438–40.

——, 1977, Decay of sound levels with distance from a steady source observed in a built-up area. *Journal of Sound and Vibration*, 52, 151–4.

YEOW, K.W., POPPLEWELL, N., and MACKAY, J.F.W., 1977, Method of predicting L_{eq} created by urban traffic. *Journal of Sound and Vibration*, 53, 103–9.

YOSHIDA, T., OSADA, Y., KAWAGUCHI, T., HOSHIYAMA, Y., YOSHIDA, K., and YAMAMOTO, K., 1997, Effects of road traffic noise on inhabitants of Tokyo. *Journal of Sound and Vibration*, 205, 517–22.

YU, C. and KANG, J., 2005, Acoustics and sustainability in the built environment: an overview and two case studies. *Proceedings of Inter-Noise*, Rio de Janeiro, Brazil.

YU, L., 2003, *Predicting Sound Field and Acoustic Comfort in Urban Open Spaces Using Neural Network*. MSc dissertation, School of Architecture, University of Sheffield, UK.

YU, L. and KANG, J., 2005a, Neural network analysis of soundscape in urban open spaces. *The 149th Meeting of the Acoustical Society of America*, Vancouver, Cannada. (Abstract published in *Journal of the Acoustical Society of America*, 117, 2591.)

——, 2005b, Soundscape evaluation in city open spaces using artificial neural network. *UIA 2005 – XXII World Congress of Architecture*, Istanbul, Turkey.

YUKI, M.R., 2000, *Towards a Literary Theory of Acoustic Ecology: Soundscapes in Contemporary Environmental Literature*. PhD dissertation, University of Nevada, USA.

ZANNIN, P.H.T., DINIZ, F.B., and BARBOSA, W.A., 2002, Environmental noise pollution in the city of Curitiba, Brazil. *Applied Acoustics*, 63, 351–8.

ZEITLER, A. and HELLBRÜCK, J., 1999, Sound quality assessment of everyday-noises by means of psychophysical scaling. *Proceedings of Inter-Noise*, Fort Lauderdale, USA.

——, 2001, Semantic attributes of environmental sounds and their correlations with psychoacoustic magnitudes. *Proceedings of the 17th International Congress on Acoustics (ICA)*, Rome, Italy.

ZHANG, M. and KANG, J., 2004, Evaluation of urban soundscape by future architects. *The 147th Meeting of the Acoustical Society of America*, New York, USA. (Abstract published in *Journal of Acoustical Society of America*, 115, 2497.)

——, 2006a, Towards the evaluation, description and creation of soundscape in urban open spaces. *Environment and Planning B: Planning and Design* (accepted).

——, 2006b, Semantic differential analysis of the soundscape in urban open public spaces. *Building and Environment* (submitted).

ZHEN, C.J. (ed.), 2000, *Handbook of Environmental Engineering: Environmental Noise Control* (Beijing: Higher Education Press) (in Chinese).

ZIMMER, K. and ELLERMEIER, W., 1999, Psychometric properties of four measures of noise sensitivity:

a comparison. *Journal of Environmental Psychology*, 19, 295–302.

ZUBE, E.H., PITT, D.G., and EVANS, G.W., 1983, A life span development study of landscape assessment. *Journal of Environmental Psychology*, 3, 115–28.

ZWICKER, E. and FASTL, H., 1999, *Psychoacoustics – Facts and Models* (Berlin: Springer).

ZWICKER, E. and FELDKELLER, R., 1967, *Das Ohr als Nachrichtenempfänger* (Stuttgart: S. Hirzel Verlag).

Index

2.5d propagation 173
3d visual environment 145

absolute magnitude estimation 25
absorber 8–10, 11, 15, 131, 178, 180, 182–3, 189, 192, 213, 215–16, 236–7, 239
absorber: membrane 8, 10; panel 8, 10; perforated panel 8, 9–10; porous 8, 9, 180
absorption coefficient 8, 9, 11, 18, 109, 111, 121, 123, 130, 133, 134, 135, 142, 144, 156, 157, 199, 213, 214, 215, 216, 219, 225, 227, 230, 236–7: angle-independent 117; normal-incidence 8; oblique-incidence 8; random-incidence 8; Sabine 8, 18; statistical 8
absorptive louver 189
acceleration 148, 173
accommodation coefficient 127, 130
acoustic: animation 134–5, 139; atmosphere 103–4; comfort 7, 43, 57–71, 102, 105, 106; enclosure 178–9; event 22, 94; screening 13; window 182–5
active: control 180; sound 82, 98–9
admittance 113
aerial photograph 153, 168
aerodynamic noise 8
age 3, 23, 57, 60, 71, 73–6, 78, 81, 85–6, 90–1, 94
air: absorption 142, 143, 144, 145; attenuation 13; exchange 182; movement 182–3; quality 91, 164; turbulence 8
air raid siren 145
airborne sound 12–13
aircraft 4, 22, 27, 30, 31, 32, 39, 41, 147–8
airflow 180, 181, 182, 183
ambient noise 22, 31, 83, 196
amplitude 5, 17–18, 22, 112, 189
analogous scale 25
anechoic chamber 142, 143, 144, 168, 177
angle of incidence 10, 12, 112, 151, 199
annoyance 4, 6, 21–6, 30–3, 35, 37, 47, 48, 102, 191
area source 6, 153, 168
articulation index 31
artificial intelligence 95
artificial neural network 26, 44, 94–7

artificial sound 72, 74–6, 78, 102
aspect ratio 180, 205, 233–4
atmospheric: absorption 149, 171; attenuation 125, 174
attitude 23, 35, 38, 39, 46, 72, 78
auralisation 134, 141
aural–visual interaction 23, 47–9, 100
axial mode 20

background noise 19, 26, 30, 31, 35, 36, 49, 55, 83, 102, 140–1, 142, 229
back-scattering 172
backward masking 5
balanced noise criterion curve 31
balcony 131, 176
barrier 12, 13, 15–17, 38, 100, 150–1, 153, 154, 175, 176, 180, 185–98, 200, 229: absorbent 192; arrow-profile 187, 188; bio- 194; branched 187; cantilevered 192, 193; dispersive 191; fir tree profile 187, 188; galleried 192, 193; helium filled 187; interfering type 187; jagged edged 189; lightweight 187; multiple-edged 187, 188; phase interference 187, 188; phase reversal 187, 188; picket 189, 190; reactive 187, 188; sloped 191; suspended panel 193; T profile 187, 188, 189; thick 17, 188; U section 187; vertically louvered 189, 190; Y profile 187, 188
beam tracing 114, 135, 173, 174, 199
behaviour 24, 46, 96: sound 142
binaural sound 140
bird 45, 48, 52, 58, 72, 73, 75, 84, 98, 99, 134, 140
boundary element method 130, 131, 132
building arrangement 158, 161, 175, 213–17, 236–44
building envelope 169, 175, 178–85

category partitioning scale method 36
cavity resonator 179
centre frequency 3
children's shouting 49, 72, 73, 76, 77, 79, 105
coherent 112, 113, 144, 189, 246,
coincidence effect 12

combined wall 178
communication 23, 31, 44, 88, 89, 105: quality
   102
community noise 6, 24, 25, 32–3, 38, 43
community noise equivalent level 30
compensation 24: numerical 144
complaint 35, 37
composite noise rating 31
construction 21, 24, 37, 39, 41, 44, 45, 48–9, 50,
   72, 73, 78, 80, 84, 112, 134, 148, 187, 194
corrected noise level 30
coupling 131, 132
courtyard 131, 176
cradle-to-gate 195
critical frequency 12
crossings 72, 73, 78, 84, 148, 174
cross street 122, 141, 200, 203–4, 210–12, 216
cultural 24, 32, 38, 43, 44, 45, 47, 49, 51, 52, 60,
   63, 70, 72, 76, 78, 81, 94, 98
cylindrical sound propagation 14

day–evening–night level 21, 29–30
daylighting 182
day–night average sound level 21, 37
day–night equivalent sound level 28–9
decay 17, 18–19, 30, 58, 134, 229, 236: curve
   18–19, 100, 110, 123, 127, 145, 200, 202,
   207–8, 209, 210, 213, 215, 219, 221, 227, 230,
   233, 236
deceleration 148
decibel 2, 4–5, 28
demographic factor 23, 43, 46, 71, 73, 85, 96,
   102
design process 96, 141, 194–5, 198
dieselness 83
diffraction 12, 15, 16, 17, 132, 142, 150–1, 154,
   166, 171, 173, 174, 185, 187, 189, 192, 200,
   246: edge 262
diffuse sound field 8, 18, 19, 173
diffusely reflecting boundary 10, 107, 114, 115,
   123, 133–4, 141, 143, 177, 199, 200–1, 202–3,
   204, 205–17, 219–21, 223, 225–31, 233–9
diffusion 11, 114–15, 127, 128, 134, 144, 172,
   177, 191, 224: coefficient 11, 12, 135, 137–9,
   141, 177, 199, 208, 222, 223, 225, 227–9;
   equation 128, 129, 172
digital sound rendering 134
direct sound 17, 19, 109–10, 113, 121, 137, 140,
   144, 150, 155, 160, 162, 166, 175, 193, 200,
   204, 208, 210–12, 216, 224, 227, 232, 236, 238
directivity: correction 149; factor 6, 19, 134;
   index 6–7, 151
discrimination scale 25
dominant source model 26
double diffraction 151, 185, 192
double-leaf boundary 12–13
dry signal 139, 142
dummy head 140
dynamic range 99, 103

early decay time (EDT) 18–19, 100, 102, 110,
   121, 133, 135, 137, 139, 199, 202, 204, 205,

207, 208–11, 217, 221, 223, 225–37, 239,
   243–6
early reflection 11, 17, 208, 217, 227, 233,
   235–6
earth mound 192
echo 94, 134
echogram 18, 94, 113
economic 6, 21, 23–5, 32, 33, 47, 195
EDT see early decay time
education level 24, 38, 57, 87
effective free area 180–1
effective height 16, 190–1
effective perceived noise level 30
eigenfrequency 19
element normalised level difference 180
embodied energy 195
emitter 119, 163
empirical formula 133–4
enclosure 18, 19, 123, 178, 179, 192, 201, 204,
   225, 244–5
energy exchange 115, 120, 121, 122
energy summation model 26
engine noise 8
environment assessment 153
environment similarity index 94
environmental load 24
equal-loudness-level contour 3–4
equal-sensation matching method 25
equivalent continuous sound level 21, 27–8
equivalent sources method 130, 131–2
exchange coefficient 129, 130
expansion chamber 179
experience 5, 24, 396, 47, 60, 72, 76, 78, 81,
   88, 94, 105, 154
Eyring formula 18, 133, 229, 233, 236

far field 134, 135, 172, 199, 204, 208, 229, 230,
   232–5
fast Fourier transform 133, 174
finite difference time domain method 130,
   132–3, 246
flat city model 169–71
fluctuation noise exposure time 27, 30
fluctuation strength 5
flutter echo 227, 233
focus 45, 48, 94, 175
fog 15
foliage 98, 152, 176, 177
football stadium 103–4
foreground sound 45, 58, 60, 64, 67, 72, 89, 96
forest 102, 177
forward masking 5
fountain 51–2, 54, 55, 56, 58, 64, 67, 68, 73,
   84, 89, 98, 102, 134, 140, 141, 164, 246
fractional octave band 3
fractionation method 25
free field 3, 4, 8, 11, 13, 14, 107, 149, 196
frequency 1, 3, 4–13, 31, 45, 58, 93, 113, 142,
   144, 148, 171, 173, 177, 180, 187, 246: band
   3, 246; weighting 4, 5

gate-to-grave 196

gender 24, 46, 76–7, 85, 94, 195
geographical information system 153, 164, 168, 173
geometrical divergence 149
geometrically reflecting boundary 107, 110, 114–15, 141, 177, 199, 200, 202–9, 213, 214, 217, 219, 220, 221, 224–7, 229, 231–9
glare 94
grassland 176
grating diffuser 10
green space 48, 84, 90, 99
grid: factor 153; interpolation 164, 168
ground: attenuation 15, 150; effect 15, 16, 149, 150, 171, 174, 177; factor 150; reflection 16, 111, 150–1
grounded theory 194

habit 24
health 6, 23, 24, 27, 32–3, 38, 41, 44, 182
hearing impairment 6, 33
hedonic method 24
Helmholtz equation 112
hemispherical sound propagation 14
horn 42
house price 24–5
humidity 13, 14, 15, 18, 49, 56, 70, 94, 96, 143, 145, 217, 222

image source method 107–13, 114, 155–7, 199
immission 32, 36
impedance 12, 15, 131, 132, 142, 144, 145: tube 8
impulse response 17, 19, 114, 139, 142, 144, 145
impulsiveness 6, 36
income 24, 38
independent effects model 26
industrial 6, 32, 33, 34, 35, 36, 38, 42, 147, 148, 152, 164, 168–9
insertion loss 16, 17, 150, 187
intelligent window 182
interactive window 182
interference 6, 15, 17, 31, 37, 107, 112–13, 143, 146, 187–8, 189, 199
interview 38, 39, 49, 50, 54, 55, 56, 57, 58, 59, 81, 83, 105, 194
intrusiveness 6, 26
inverse square law 14, 175
isopsophic index 31
isotropic scattering 172

junction 145, 148, 173, 199, 204, 216

keynote 45, 58, 60, 81

Lambert 114, 128–9, 199
landscape 23, 44, 48, 76, 82, 94, 96, 98, 100, 169, 192–4
leaf 177
legislation 4, 32, 35, 36, 37, 39
lifecycle assessment 194, 195–6
light 3, 23, 48, 115, 143, 183

line source 6, 14, 17, 158, 168, 175, 191, 200, 219, 224, 225, 246
linear transport model 171–3
lined band 179
lined duct 179
listening test 4, 49, 139
liveliness 104
long enclosure 201, 204
longitudinal: profile 189; wave 1
loss factor 131
loudness 3–6, 26, 27, 31, 45, 46, 58, 94, 102, 104, 196
loudspeaker 7, 98, 102, 140, 143, 146, 182
louver 182, 185, 189, 190
lower limiting frequency 3
luminous 143

macropreference 78
macroscale 107, 147–74
magnitude estimation method 25, 196
marital status 24
masking 5, 31, 67, 70, 99
mass law 12
maximum: error margin 163, 164; length sequence 10, 143; search radius 164
mean: beam length 117; free path length 172
mesoscale 169
meteorological 17, 29, 145, 147, 149, 151–2, 153, 155, 174
micromodel 173
microperforated 10, 182, 192
microphone array method 8
micropreference 81
microscale 107–46, 153, 155, 157, 171
microsimulation 173
middle region 150
modal approach 131
model reverberation chamber 142
molecular flow 125
molecular relaxation 14
moving-medium 132
multiple: reflections 107, 112, 115, 121, 123, 132, 152, 154, 155, 157, 171, 173, 174, 177, 178, 190, 191, 192, 208, 217, 229, 235, 238, 239; scattering theory 172, 177; sources 26–7, 95, 134, 142, 200, 204
music 41, 45, 48, 56, 58, 67, 69, 70, 72, 73, 74, 84, 98, 100, 140–1: from passenger cars 72, 73; from stores 56, 72, 73, 75, 98; live 98; played on street 47, 49, 56, 72, 73, 76, 77, 79, 134

N curve 32
natural: frequency 19, 185; sound 44, 46, 48, 58, 72, 74, 75, 76, 84, 85, 90, 98, 102, 140; ventilation 179, 182
near field 99, 134, 135, 137, 156, 199, 200, 204, 208, 210, 211, 217, 225, 227, 229, 232, 233, 235, 236
neighbourhood 23, 32, 37, 46
neuron 95
node 95, 96, 115, 172, 173, 174

noise: and number index 31; climate 21, 38–42; control 37, 38, 43, 44, 169, 200; criterion 31; descriptor 30, 37; exposure category 34–5; exposure forecast 31; map 34, 147, 153, 154, 162, 163, 164, 167, 168; pollution level 30; rating number 32; sensitivity depreciation index 24
noise-mapping 30, 147–74: algorithm 147, 154, 155, 157; application 147, 166–9; software 56, 134, 147, 152–3, 157, 168–9
noisiness 3–4, 27: index 31
non-refracting atmosphere 132
normal frequency 19
noy 4
nuisance 35, 37

oblique mode 20
octave band 3, 31, 135, 144, 149, 150, 154
one-third octave band 31, 145, 146, 148, 149, 153, 155
ordered category scale 25
origin–destination matrix 173

PA system 98, 104, 105
paired comparison 25, 140
parabolic equation method 130, 132–3
parametric study 110, 135–9, 225, 246
parapet 246
partial enclosure 192
partially diffusely reflecting 125
particle: dynamics 171; transport 125;
particle-tracing 113, 114
passive: control 180; sound 82, 98
patch 110, 115–23, 131, 132, 199, 213, 216, 229, 233
perceived noise level 4, 27, 30, 31, 197
percentile loudness 6
perception 1, 3, 6, 23, 25, 26, 35, 44, 47, 48, 49, 67, 94, 95, 96, 100, 102, 135, 140, 194, 195, 196–7, 246
perceptual dimension 25
period 1, 10, 11, 17, 22, 27
phase 1, 2, 6, 10, 15, 112, 113, 125, 187–9
phon 3, 4
phonon 125
physical scale modelling 107, 142–5
pink noise 3, 25
pitch strength 5, 6
pitched roof 157, 158, 160
plane: source 6, 14; wave 6
planning 34, 35, 44, 155, 175–8, 195, 215
planning permission 34
pleasantness 5, 45, 197
pneumatic noise source 145
podium 175
point source 6, 13, 14, 16, 17, 108, 112, 117, 118, 122, 123, 130, 134, 135, 143, 144, 148, 149, 155, 168, 171, 173, 175, 189, 191, 199, 200, 202–5, 210–11, 216, 219, 223, 224, 225, 229, 231, 233
polarity profile method 25
portable anechoic chamber 168

postmasking 5
preferred noise criterion curve 31
preferred speech interference level 31
privacy 102
profession 57, 60, 78
profiled façade 246
propagation model 147, 174
property value 6, 24
propulsion noise 148
psychological 21, 23–5, 32, 43, 46, 81, 82, 83, 93, 94, 96, 196, 197
public participation 134, 194–5

quarter-wavelength: resonator 180; resonance 187
questionnaire survey 49, 56–8, 81, 105, 194

radiant interchange 115
radiation: balance 115; exchange 115
radiosity model 107, 114–24, 133, 134, 158
railway 8, 21, 23, 33, 49, 53, 147, 154, 167, 168
rain 15, 102, 104
rapid speech transmission index 31
ratio: production method 25; scale 25
ray number 114, 135, 136, 137, 139
rayleigh integral 131
ray tracing 107, 113–14, 120, 142
receiver region 150
reciprocity 173
recyclability 196
reference: intensity 2; pressure 2
reflection 10, 15, 16, 18, 107, 108, 111, 112, 114, 115, 121, 123, 125, 127–9, 144, 151, 174, 190–2, 204–8, 213, 216, 227–9: order 114, 135, 137, 138, 139, 141, 153, 155, 156, 162, 163, 164, 166, 167, 168; path 212; pattern 10, 94, 199, 204–8, 213, 227–9
reflection coefficient 8, 128, 151: complex 112; spherical wave 112
refraction 15, 132, 152, 189, 199
regional difference 24
regulation 21, 32–8, 47
relaxation 14, 50–2, 88, 89
resonance 10, 12, 131, 177, 183, 187
resonant frequency 9, 10, 19, 187
resonator 8, 9, 179, 180
reverberant sound 18, 19
reverberation: chamber 11, 142, 144, 180, 182, 183; radius 19; time (RT) 18, 31, 123, 130, 202, 205, 209, 219, 221, 230–2, 234–5, 237–8
rib-like structure 187
road: elevated 192–3; in cuttings 192–3; network 166, 171, 173; surface texture 148, 173, 175
rolling noise 8, 148
roof 6, 153, 157–8, 160, 164, 174, 178, 217, 218–19, 246
room: acoustics 1, 11, 17–20, 134, 139, 143; mode 17, 19–20
room-noise criterion 31
roughness 5, 11, 58, 115

roundabout 84, 174
RT *see* reverberation time
RT/EDT ratio 236
RT30 18–19, 199

Sabine formula 18
scatterer 172, 177
scattering 11, 115, 125, 152, 172, 173, 177, 191,
   192, 233, 246: coefficient 11, 115; strength
   172
Schroeder diffuser 10–11
self-protection building 176
semantic differential method/technique 24, 25,
   43, 83–91
semi-anechoic chamber 142, 182, 183
semi-structured interview 57, 194
shadow zone 189, 192
sharpness 5, 6, 26, 58
shopping mall 100–2
shrub 100, 177
side street 200, 204
signal-to-noise ratio 31, 140, 141, 143, 144
silencer 179–81, 182
single particle distribution function 125
situational variable 22
skateboarding 56, 99
sleep 6, 24, 33, 39, 41
snow 15, 37
social 6, 21, 24, 32, 43, 44, 47, 70, 76, 82, 83, 84,
   85, 88, 90, 92–6
social survey 39
solar 183, 185
solid angle 10, 117
sone 4
sonic crystal effect 177
sound: content 81; exposure level 30;
   identifiability 5; intensity 1, 2, 6, 8, 13, 14, 30,
   143; intensity level 2; intensity technique 8;
   level 2–5, 12, 17, 18, 21, 22, 26, 27–30, 32, 40,
   43, 54–5, 58, 60–4, 65, 68, 70, 71, 78, 82, 93,
   94, 98, 99, 100, 102, 105, 134, 142, 146, 155,
   171, 174, 177, 208, 217; level meter 5, 27, 146;
   level rise time 22; particle 125, 127, 128, 129,
   172, 173; pleasantness 5, 44, 45, 67, 70, 72, 73,
   82, 83, 84, 85, 88, 90, 102; power 1–2, 6, 8, 13,
   14, 117, 123, 134, 148, 149, 151, 155, 164,
   168, 171, 219, 225; power level 2, 6, 8, 14,
   123, 134, 148, 149, 151, 155, 164, 168, 219,
   225; preference 43, 46, 47, 57, 71–81; pressure
   2–3, 4, 5, 27, 189; pressure level 1, 92, 93,
   102, 104, 107, 123–4, 160, 161, 169, 170, 171,
   199, 203, 213, 214, 226, 233; quality 5–6, 24,
   44, 46, 47, 48, 83, 84, 89; sensitivity 46, 81;
   sequence 25, 94; source 1, 2, 5, 6–8, 17, 18,
   22, 23, 28, 36, 43, 44, 45, 47, 49, 50, 56, 64,
   67, 70, 83, 84, 89, 92, 93, 94, 96, 98, 102, 104,
   118, 121, 125, 128, 130, 134, 139, 140, 141,
   143, 144, 145, 153, 163, 164, 165, 171, 196,
   199, 200, 246; velocity 1, 125; wave 1–2, 8, 9,
   10, 12, 13, 14, 15, 17, 107, 112, 143, 177, 187,
   191, 217
soundmark 45, 58, 60, 72, 84, 98, 99

soundscape 43–106, 134, 141, 244, 246:
   description 43, 91–4; design 43, 88, 96–100;
   evaluation 43, 45–9, 84, 85, 88–9, 94–6, 97;
   walk 83–4
source: directivity 6, 151, 169; model 26, 134,
   147, 148, 155–7; region 132, 150
spark 123, 143, 144
spatiality 89
spectral masking 5
spectrum 3, 21, 44, 67, 93, 96, 98, 99, 144, 174,
   187, 222
speech: interference 6, 31; interference level
   31; transmission index 31
spherical source 6
splitter panel 189
square 47, 49, 53, 54–8, 60, 61, 63, 64, 69, 71,
   72–3, 78, 79, 81–2, 84, 90–1, 94, 98, 99,
   100, 107, 110–11, 114, 115, 122, 133, 135,
   137, 139, 140–2, 153, 155, 156, 157, 164,
   175, 177, 199–246
standard 11, 21, 30, 32–8, 40, 41, 147, 153, 168
standing wave 19
statistical sound level 27, 174
steady-state SPL 93, 110, 121, 123, 130, 173,
   199
stimulus–response compatibility 5
stochastic particle bounce method 177
street canyon 107, 108–9, 112, 113, 115, 118,
   122, 128, 129, 134, 143, 144, 145, 147, 155:
   parallel 131–2
street/square furniture 94, 100, 229
street/square height 107
structure-borne sound 12, 185
subjective: evaluation 21, 25, 56, 60, 61, 63, 64,
   65, 66, 67, 68, 69, 70, 71, 94, 95, 105; test 4,
   139–41, 143
subjectively corrected model 26
summation and inhibition model 26
surface wave 187
surrounding speech 49, 50–3, 58, 72, 73, 75,
   78, 79
sustainability 196, 198
swimming space 104–5

tangential mode 20
temperature: 13, 14–15, 18, 23, 49, 56, 70, 91,
   94, 143, 145, 152, 155, 199, 222; gradient
   15, 152
temporal: distribution 67, 128, 129; masking 5
threshold 2, 4, 5, 25, 46, 64
tolerance 64, 82, 102
tonal component 22, 36, 45, 93, 169
tonality 6, 36, 45
total noise load 31
total room constant 19
traffic: flow 28, 29, 145, 171; noise index 30
transmission: coefficient 8, 178; factor 173;
   loss 12, 13, 17, 169, 171, 178
transport equation 125, 128, 129, 172
transport theory 107, 125–30, 172
tree 23, 100, 172, 176, 177, 194, 229
turbulent: atmosphere 131, 132; scattering 246

tyres and road interaction 81

upper limiting frequency 31
urban: open public space 43, 49–58, 60, 64, 67, 70, 81–2, 83–8, 89, 90, 91–4, 96–100; texture 175, 217, 218
user 46–7, 174

validation 122–3, 145, 154, 155, 164, 165
vector summation model 26
vegetation 15, 100, 150, 154, 176–8, 191, 192, 194, 196, 197, 199, 217, 229
vehicle: category 148; flow density 131
ventilation 32, 102, 103, 182, 183: opening 178, 179–81
viaduct 192
vibration 1, 8, 10, 12, 23, 48, 177, 185
view 23, 56, 70, 91, 100, 149, 174
virtual reality 124, 140, 197
virtual source 131

viscosity effect 14
visual: perception 48; screening 196; shielding 196

water 45, 48, 49, 50–3, 54, 58, 67, 70, 72, 73, 76, 78, 79, 98, 99, 102, 143, 150, 195: flow 48; model 143
wave equation 130, 131, 132, 172
wavelength 1, 6, 10, 12, 17, 19, 107, 131, 142, 143, 150, 173, 180, 187, 190, 199
weather condition 15, 152
weighted equivalent continuous perceived noise level 31
white noise 3, 146
wind 14–15, 23, 45, 48, 56, 58, 70, 94, 96, 102, 104, 132, 151, 155, 179, 189: effect 125; gradient 15, 199; speed 13, 14, 145, 151, 152, 194, 246

zone category 36

Printed and bound by CPI Group (UK) Ltd, Croydon, CR0 4YY

01/11/2024

01782604-0002